Enzyme Structure
and Mechanism

Alan Fersht

MRC Laboratory of Molecular Biology, Cambridge

Enzyme Structure and Mechanism

W. H. Freeman and Company
Reading and San Francisco

Library of Congress Cataloging in Publication Data

Fersht, Alan, 1943–
  Enzyme structure and mechanism.

  Bibliography: p.
  Includes index.
  1. Enzymes. I. Title.
QP601.F42   547'.758   77-6441
ISBN 0-7167-0189-8
ISBN 0-7167-0188-X pbk.

Set in Linotron Times 10/12 pt
by the Universities Press, Belfast, U.K.
Printed in the United States of America

To P. J. FERSHT *in memoriam*
W. P. JENCKS
M. F. PERUTZ

Preface

During the past two decades the advances in X-ray crystallography, transient kinetics, and the study of chemical catalysis have revolutionized our ideas on enzyme catalysis and mechanism. It is the intention of this text to provide a brief account of these developments for senior undergraduate students and postgraduates who have attended courses in chemistry and biochemistry. The philosophical and theoretical aspects of this book centre upon how the interactions of an enzyme with its substrates lead to enzyme catalysis and specificity, and upon the relationship between structure and mechanism. The experimental approaches emphasized are those involving the direct study of enzymes as molecules. As such, there is a strong emphasis on pre-steady-state kinetics where enzymes are handled in substrate quantities and enzyme-bound intermediates observed directly. The steady-state kinetics of multisubstrate enzymes and the detailed chemistry of coenzymes and cofactors are discussed in only a cursory manner.

There have been two guiding rules in the preparation of this book. The first is to discuss general principles and ideas using specific enzymes as examples. (Although to avoid overloading the more theoretical chapters on kinetics, most of the illustrative examples are presented in a separate chapter). The second is to stick closely to examples where hard evidence is available and to avoid speculation and woolly evidence. In consequence, the discussion of detailed chemical mechanisms is generally restricted to enzymes whose tertiary structures have been solved by X-ray crystallography. Similarly, the discussion of the theoretical aspects of allosteric proteins is very much restricted to haemoglobin because it is the only example where good (or any) evidence is available on the nature of the interactions of positive cooperativity.

The references cited tend to be those of the most recent reviews or papers where more extensive bibliographies are given, and also those of the original papers in order to maintain a historical perspective. Illustrative examples have been taken where possible from the files of the MRC

Laboratory of Molecular Biology because of their ready availability and uniform quality of presentation. In this context I must thank Annette Lenton for both the illustrations she has prepared especially for this book and also for those prepared for other members of the laboratory whose files I have shamelessly raided.

I am particularly indebted to W. P. Jencks, H. B. F. Dixon, R. H. Gutfreund, K. F. Tipton, and R. S. Mulvey for their critical comments on the entire manuscript, and also to M. F. Perutz and D. M. Blow for their comments on individual chapters. I wish to thank The Royal Society, the American Chemical Society, Cornell University Press, Academic Press, John Wiley and Alan R. Liss for permission to reproduce illustrations.

A.F.

Contents

Chapter 1

The three-dimensional structure
of enzymes

In 1930 Haldane wrote a book on enzymes that is still worth reading today.[1] The most striking feature of this book to me is that so much was then known about the properties and action of enzymes yet so little was known about the enzymes themselves: the question of whether or not enzymes are proteins was still the subject of raging controversy. The knowledge was so one-sided because there were no means of studying enzymes directly. All the information had been deduced indirectly from the effects of enzymes on their substrates. Nevertheless, the foundations of modern steady-state kinetics had been laid in a little over thirty years, the first cell-free enzyme extract having been prepared by Büchner in 1897.

In order to proceed further, it was necessary to isolate purified enzymes in *substrate* quantities and examine them directly. This was accomplished in 1926 when Sumner crystallized urease from jack-bean extracts. Soon afterwards (1930–6), Northrop and Kunitz crystallized pepsin, trypsin, and chymotrypsin. This provided the material to prove finally that enzymes are proteins and to allow the development of the techniques of modern protein chemistry, the sequencing of proteins pioneered by Sanger, the solution of the three-dimensional structure of proteins pioneered by Perutz and Kendrew, and the use of rapid-reaction kinetics which had been initiated by Roughton in 1923.

The major part of the present book deals with the direct study of enzymes in substrate quantities, taking up the story from where Haldane was forced to stop. In this first chapter we shall discuss the general features of the most significant advance in our knowledge of enzymes, their three-dimensional structure. Also, since lysozyme and the serine proteases have been the testing grounds of so much of the experiment and theory that is discussed in later chapters, some aspects of the structure of these enzymes are described briefly. The structures and mechanisms of individual enzymes are discussed in Chapter 12.

TABLE 1.1. *The common amino acids*

Amino acid (Three- and one-letter symbols, molecular weight)	Side chain, R ($R\mathrm{CH(NH_3^+)CO_2^-}$)	pK_as
Glycine (Gly, G, 75)	$\mathrm{H-}$	2·35, 9·78
Alanine (Ala, A, 89)	$\mathrm{CH_3-}$	2·35, 9·87
Valine (Val, V, 117)	$\begin{array}{l}\mathrm{CH_3}\\ \quad\mathrm{CH-}\\ \mathrm{CH_3}\end{array}$	2·29, 9·74
Leucine (Leu, L, 131)	$\begin{array}{l}\mathrm{CH_3}\\ \quad\mathrm{CHCH_2-}\\ \mathrm{CH_3}\end{array}$	2·33, 9·74
Isoleucine (Ile, I, 131)	$\begin{array}{l}\mathrm{CH_3CH_2}\\ \quad\quad\mathrm{CH-}\\ \quad\mathrm{CH_3}\end{array}$	2·32, 9·76
Phenylalanine (Phe, F, 165)	$\langle\bigcirc\rangle\!-\!\mathrm{CH_2-}$	2·16, 9·18
Tyrosine (Tyr, Y, 181)	$\mathrm{HO}\!-\!\langle\bigcirc\rangle\!-\!\mathrm{CH_2-}$	2·20, 9·11, 10·13
Tryptophan (Trp, W, 204)	indole$-\mathrm{CH_2-}$	2·43, 9·44
Serine (Ser, S, 105)	$\mathrm{HOCH_2-}$	2·19, 9·21
Threonine (Thr, T, 119)	$\begin{array}{l}\mathrm{HO}\\ \quad\mathrm{CH-}\\ \mathrm{CH_3}\end{array}$	2·09, 9·11
Cysteine (Cys, C, 121)	$\mathrm{HSCH_2-}$	1·92, 8·35, 10·46
Methionine (Met, M, 149)	$\mathrm{CH_3SCH_2CH_2-}$	2·13, 9·28
Asparagine (Asn, N, 132)	$\mathrm{H_2NC(\!=\!O)CH_2-}$	2·1, 8·84
Glutamine (Gln, Q, 146)	$\mathrm{H_2NC(\!=\!O)CH_2CH_2-}$	2·17, 9·13
Aspartic acid (Asp, D, 133)	$^-\mathrm{O_2CCH_2-}$	1·99, 3·90, 9·90
Glutamic acid (Glu, E, 147)	$^-\mathrm{O_2CCH_2CH_2-}$	2·10, 4·07, 9·47
Lysine (Lys, K, 146)	$\mathrm{H_3N^+(CH_2)_4-}$	2·16, 9·18, 10·79

TABLE 1.1. *The common amino acids* (*Contd.*)

Amino acid (Three- and one-letter symbols, molecular weight)	Side chain, R ($R\mathrm{CH(NH_3^+)CO_2^-}$)	pK_as
Arginine (Arg, R, 174)	$\mathrm{H_2N}\overset{+}{=}\mathrm{C-NH(CH_2)_3-}$ with $\mathrm{H_2N}$	1·82, 8·99, 12·48
Histidine (His, H, 155)	imidazole $\mathrm{CH_2-}$	1·80, 6·04, 9·33
Proline (Pro, P, 115)	pyrrolidine ring with $\mathrm{CO_2^-}$, C, H, $\overset{+}{\mathrm{N}}\mathrm{H_2}$	1·95, 10·64

pK_as from *Data for biochemical research*, R. M. C. Dawson, D. C. Elliott, W. H. Elliott, and K. M. Jones, Oxford University Press (1969)

A. The primary structure of proteins

The major constituent of proteins is an unbranched polypeptide chain consisting of L-α-amino acids linked together by amide bonds between the α-carboxyl of one residue and the α-amino group of the next. Usually only the 20 amino acids listed in Table 1.1 are involved.

$$\mathrm{H_2NCHC}\overset{O}{-}\mathrm{NHCHC}\overset{O}{-}\mathrm{NHCHC}\overset{O}{-}\mathrm{NHCHC}\overset{O}{-}\mathrm{NH-}$$
$$\underset{R_1}{} \quad \underset{R_2}{} \quad \underset{R_3}{} \quad \underset{R_4}{} \tag{1.1}$$

The primary structure is defined by the sequence in which the amino acids form the polymer. By convention, the sequence is written as in (1.1), beginning with the N-terminus written on the left.

Although the primary structures of almost all intracellular proteins consist of linear polypeptide chains, many extracellular proteins contain covalent —S—S— crossbridges from having two cysteine residues linked by their thiols. This either creates loops in the main polypeptide chain due to intrachain bridges or links different chains together (Fig. 1.1). In these latter cases, the multiple chains are derived from a single chain precursor that has become covalently modified by proteolysis, examples being insulin from proinsulin and chymotrypsin from chymotrypsinogen. These bridges may be cleaved by reduction with thiols. It is of interest that single polypeptide chains may often be reduced and denatured, and then reversibly oxidized and renatured. But reduction and denaturation

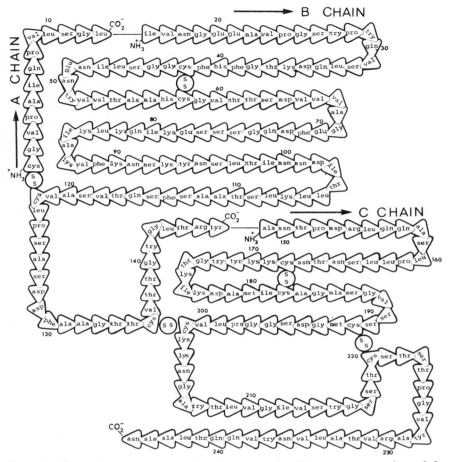

FIG. 1.1. The primary structure of α-chymotrypsin. The enzyme consists of 3 chains linked together by S—S bridges. However, the chains are derived from the single polypeptide chain of chymotrypsinogen by the excision of residues Ser-13, Arg-14, Thr-147, and Asn-148. (Courtesy of D. M. Blow)

of proteins derived from the covalent modifications of precursors is generally irreversible.

The spontaneous refolding of single polypeptide chains that have not been covalently modified leads to an important principle: the genetically encoded sequence of a protein determines its three-dimensional structure.

B. The three-dimensional structure of enzymes

1. X-ray diffraction methods

The importance of X-ray diffraction methods in enzymology cannot be overstated. Not only have they provided the experimental basis of our present knowledge of the structure of proteins, but they have proved to be the single most important factor in the investigation of enzyme mechanisms. In this section we shall briefly discuss the type of information that can be derived from X-ray diffraction experiments.

X-rays are scattered on striking the electrons of atoms in a similar manner to the way that light waves are scattered by the engraved lines of a diffraction grating. The regular lattice of a crystal acts as a three-dimensional diffraction grating towards a monochromatic beam of X-rays, giving a pattern of diffracted rays in directions where the scattered rays reinforce and do not destructively interfere. The structure of the crystal, or more precisely the distribution of its electron density, may be calculated from the diffraction pattern by Fourier transformation. This requires knowledge of the intensities and directions of the diffracted rays, which are easily measured on photographic film as a pattern of spots or by a diffractometer, and also of their *phases*. Determination of the phase of each diffracted ray (an essential requirement for the Fourier transformation) is the most difficult problem. Ways of circumventing this problem which were used for simple crystal structures could not be applied to proteins. The solution of the *phase problem* was the stumbling block that held up protein crystallography until Perutz and his co-workers applied the method of *isomorphous replacement* in 1954.[2] In this method, a heavy metal atom is bound at specific sites in the crystal without disturbing its structure or packing. The metal scatters X-rays more strongly than the atoms of the protein and adds its scattering power to every diffracted ray. Information about the phases of the diffracted rays from the protein can then be obtained from the changes in intensity, depending on whether they are reinforced or diminished by the scattering from the heavy atom. Several different isomorphous substitutions are needed to give an accurate determination of the phases.

Once the phase and amplitude of every diffracted ray are known, the electron density of the protein may be calculated. The structure of the protein is obtained by first fitting into this density wire models of the amino acid residues of the primary structure, and then by computerized refinement.

Accuracy and resolution
The degree of accuracy that is attained depends on both the quality of the data and the *resolution*. The term resolution is best illustrated for our purposes by the data in Table 1.2. At low resolution, 4–6 Å, the electron density map reveals little more in most cases than the overall shape of the molecule. At 3·5 Å, it is often possible to follow the course of the polypeptide backbone, but there may be ambiguities. At 3·0 Å, it is possible in favourable cases to begin to resolve the amino-acid side chains, and, with some uncertainties, to fit the sequence to the electron density. At 2·5 Å, the position of atoms may often be fitted with an accuracy of ±0·4 Å. In order to locate atoms to 0·2 Å, a resolution of

TABLE 1.2. *Resolution and structural information*

Resolution (Å) (1 Å = 0·1 nm)	Structural features observable in a good map
5·5	Overall shape of molecule. Helices as rods of strong intensity.
3·5	The main chain (usually with some ambiguities).
3·0	Start to resolve the side chains.
2·5	Side chains well resolved. The plane of the peptide bond resolved. Atoms located to about ±0·4 Å.
1·5	Atoms located to about ±0·1 Å. The present limit of protein crystallography.
0·77	Bond lengths in small crystals measured to 0·005 Å.

about 1·9 Å and very well-ordered crystals are necessary. What this means in practical terms is illustrated in Fig. 1.2. The reflections required for high-resolution analysis are those that have been diffracted through the greater angles. But it is seen that the intensities rapidly decrease at higher resolution. Some crystals diffract better than that shown in Fig. 1.2, the majority worse. A further point is that the number of reflections to be analysed increases as the third power of the resolution; an increase from 3 to 1·5 Å increases the amount of data to be collected by a factor

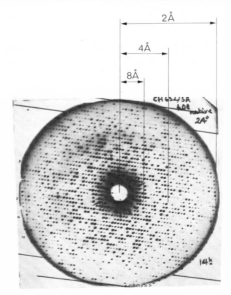

FIG. 1.2. X-ray diffraction pattern from a crystal of α-chymotrypsin showing the data required for various resolutions

of eight and the total effort required by an even larger factor due to the poorer quality of the data.

The isomorphous replacement method becomes ineffective for protein crystals beyond a resolution of 2 to 2·5 Å since the addition of the heavy atom may cause some alteration in structure and because of the problem of observing small changes in intensities which are already very weak. However, computer methods have now been developed to refine a structure using the measured intensities at higher resolution and the model structure which has been determined at 2 to 2·5 Å resolution.

2. The structural building blocks

Even before the structure of any crystalline protein had been solved by X-ray crystallography, Pauling and Corey had worked out the structures of the units that have subsequently been found to be the basic building blocks of the architecture of proteins.[3] They first solved the structures of crystals of small peptides to find the dimensions and geometry of the peptide bond. Then, by building precisely constructed models, they found structures that could fit the X-ray diffraction patterns of fibrous proteins. The diffraction patterns of fibres do not consist of the lattice of points found from crystals, but of a series of lines corresponding to the distances between constantly recurring elements of structure. This method of building models was later used by Watson and Crick to solve the double helical nature of DNA.

a. The peptide bond[3]

The X-ray diffraction studies on the crystals of small peptides showed that the peptide bond is planar and *trans* (Fig. 1.3). This structure has been found for all peptide bonds in proteins apart from a few rare exceptions. This planarity is due to a considerable delocalization of the lone pair of electrons of the nitrogen on to the carbonyl oxygen. The C—N bond is consequently shortened and has double-bond character. Twisting of the bond breaks this and loses the 75 to 88 kJ/mol (18 to 21 kcal/mol) of delocalization energy.[3,4]

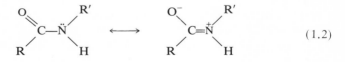

$$(1.2)$$

b. Secondary structure: α-helix and β-sheet[3]

The polypeptide chains of fibrous proteins are found to be organized into hydrogen bonded structures, known as secondary structure. In these ordered regions, any buried carbonyl oxygen forms a hydrogen bond with an amino NH group. This is done in two major ways; by forming α-helices (found from the fibre diffraction studies on α-keratin) or

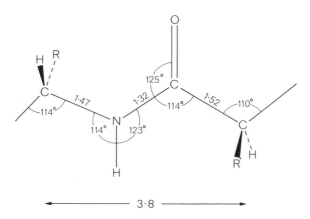

FIG. 1.3. The peptide bond. Distances are in Å units. (Dimensions from ref. 3)

β-sheets (found in β-keratin). The polypeptide chain of a globular protein also folds upon itself to form local regions of secondary structure.

The α-helix is illustrated in Fig. 1.4. It is a stable structure, each amide group making a hydrogen bond with the third amide group away from it in either direction. The C=O groups are parallel to the axis of the helix and point almost straight at the NH groups to which they are hydrogen bonded. The side chains of the amino acids point away from the axis. There are 3·6 amino acids in each turn of the helix. The rise of the helix per turn, the pitch, is about 5·4 Å.

An extended polypeptide chain (Fig. 1.5) can make complementary hydrogen bonds with a parallel extended chain. This in turn can match up with another extended chain to build up a sheet. There are two stable arrangements; the parallel β-pleated sheet in which all the chains are aligned in the same direction, and the antiparallel β-pleated sheet in which the chains alternate in direction (see Fig. 1.6). The repeating unit of a planar peptide bond linked to a tetrahedral carbon produces a pleated structure.

Correlations have been made from statistical analyses and experiments with synthetic polymers to show that certain amino acids have a propensity for forming helices, and others for destabilizing them.[5] Proline cannot be incorporated into a helix without seriously distorting it.

c. Bends in the main chain[6-9]

Another frequently observed structural unit occurs when the main chain sharply changes direction using a 'β-bend' composed of four successive residues. In these, the C=O group of residue i is hydrogen bonded to the NH of residue $i+3$ instead of $i+4$ in the α-helix. A 180° change of

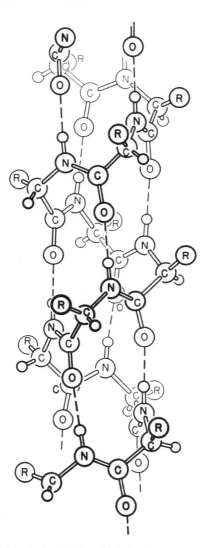

FIG. 1.4. The right-handed α-helix which is found in proteins. (Reprinted from Linus Pauling: *The nature of the chemical bond.* © 1939, 1940, third edition © 1960 by Cornell University. Used by permission of Cornell University Press.)

direction, which can link two antiparallel strands of pleated sheet, can be achieved in two ways (Types I and II, Fig. 1.7). Type II has the restriction that glycine must be the third residue in the sequence. A distortion of Type I gives a bend, Type III, which can be repeated indefinitely to form a helix known as the $3\cdot0_{10}$ helix (a tighter, less stable, helix than the α, with $3\cdot0$ residues per turn, forming hydrogen bonded loops of 10 atoms).

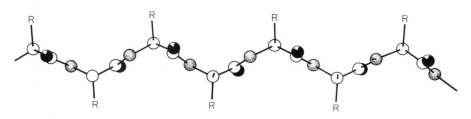

FIG. 1.5. An extended polypeptide chain. The hydrogen bonds are made perpendicular to the plane of the paper so that the sheet made from successive parallel chains is pleated

3. The assembly of proteins from the building blocks

There was great excitement when the structures of the first two proteins were solved. The single polypeptide chain of myoglobin was found to consist of a series of α-helical rods connected together without any β-sheet. Then haemoglobin was found to be just like four myoglobin chains (see Fig. 1.8). But, as more protein structures were solved, the story rapidly complicated. Lysozyme and carboxypeptidase, the first enzymes whose structures were known, consist of α-helices and a β-sheet. Soon afterwards, α-chymotrypsin was found to consist almost entirely of β-sheet with only two regions of α-helix. It became apparent that general rules of protein folding would be hard to find. But, now that nearly one hundred structures have been solved, the first steps are being taken in this direction. Certain recurring features have been noted. Methods of classifying structures are being developed.

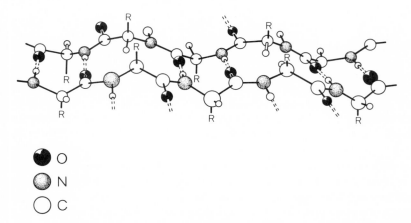

● O

● N

○ C

FIG. 1.6. Two strands of β-sheet. Other strands may be added in a similar manner

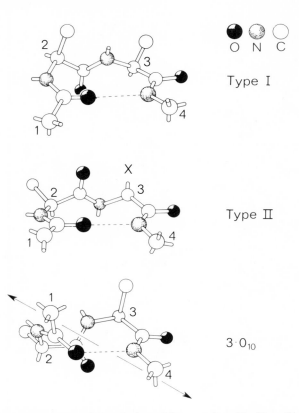

Type I

Type II

$3 \cdot 0_{10}$

FIG. 1.7. The three types of bends found in proteins. Residue X in Type II is always glycine. (J. J. Birktoft and D. M. Blow, *J. molec. Biol.* **68,** 187 (1972))

It has been suggested that most proteins can be considered crudely as a layered sandwich structure, with each layer consisting of either α-helices or β-sheet. Levitt and Chothia have very recently proposed a simple and instructive way of representing structures in this manner (Fig. 1.9).[10] The helices and sheets pack by stacking their amino-acid side chains. The internal packing is good, with the side chains of one piece of secondary structure fitting between those of another to give a hydrophobic region of similar density to a hydrocarbon liquid or wax.[11] There are few hydrogen bonds in these regions: in carboxypeptidase A there are only seventeen hydrogen bonds connecting the eight helices to each other and the sheet, less than two per helix. The buried hydrogen-bond donors and acceptors are invariably paired. The strands that connect the helices and sheets are usually exposed to solvent and often short. Globular proteins thus have hydrophobic cores with their charged groups on the surface.

It has been pointed out by Crick[12] that two α-helices (whose chains run

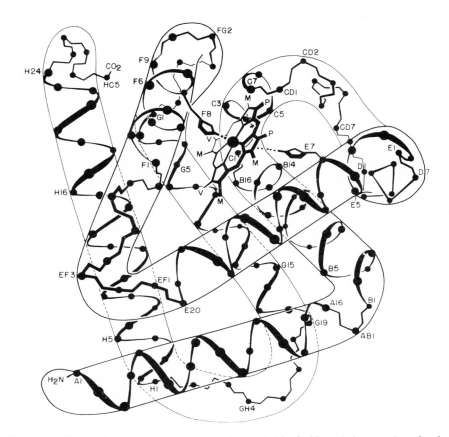

FIG. 1.8. Comparison of the structures of myoglobin (left) and the β sub-unit of haemoglobin (right). The overall folds of the two are essentially identical. The main differences are in the N and C terminals and in the contact interfaces of haemoglobin. (Redrawn and modified from R. E. Dickerson *The Proteins* (ed. H. Neurath), Academic Press (1964))

in the same direction) can be packed together if they incline at 20° to each other so that the side chains of one may pack between those of the other. This is calculated for 3·60 amino acids per turn of helix. For a tighter helix of 3·55 residues per turn, this angle narrows to about 10°. It has been noted by Chothia[13] that the β-sheet is invariably twisted and that this is energetically more favourable than being planar (Fig. 1.10).

Proteins are not loose and floppy structures, but their residues are packed as tightly as crystalline amino acids.[14-16] The packing density (the fraction of the total space occupied by the atoms) is about 0·75 for proteins compared with values of 0·7 to 0·78 found for crystals. Close-packed spheres have a packing density of 0·74 and infinite cylinders 0·91.

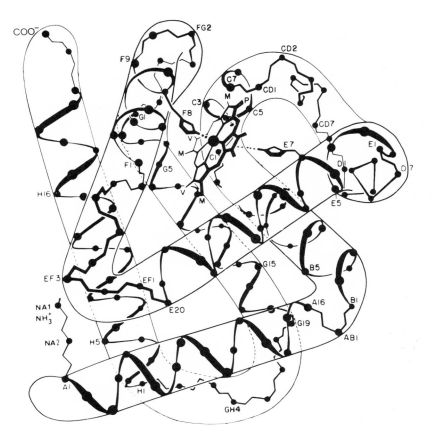

FIG. 1.8 (right).

4. Quaternary and tertiary structure

The three-dimensional structure of a protein composed of a single polypeptide chain or covalently linked chains is known as its *tertiary structure.* Many proteins are composed of sub-units that are not covalently linked together. The overall organization of these sub-units is known as the *quaternary structure.* The three-dimensional structure of each sub-unit is still referred to as its tertiary structure. A change in quaternary structure means that the sub-units move relative to each other.

The sub-units may be identical polypeptide chains or chemically different. As a general rule, the contact interfaces between the sub-units are as closely packed as the interiors of proteins and hydrogen bonding groups, and ions are paired on these surfaces.[17] The change of quaternary structure in haemoglobin involves a shift from one set of close-packed interfaces to another.

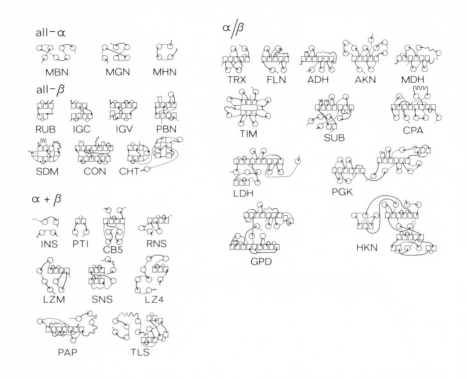

Fig. 1.9. Classification of protein structure according to Levitt and Chothia.[10] The protein is viewed from a direction where most segments of secondary structure are seen end on. Each strand of a β-sheet is represented by a square. The front end of each α-helix is represented by a circle. The segments which are close in space are close together in the diagram. The segments are connected by bold or thin arrows (from N to C terminal) to indicate whether the connection is at the near or far end. The approximate scale is: diameter of α-helix $= 5$ Å, β-strand $= 5 \times 4$ Å, separation of helices $= 10$ Å, separation of hydrogen bonded β-strands 5 Å, separation of non-hydrogen bonded strands $= 10$ Å.

Abbreviations: MHN, myohaemerythrin; MGN, myogen; MBN, myoglobin; RUB, rubredoxin; IGC, immunoglobulin constant region; IGV, immunoglobulin variable region; PBN, prealbumin; SDM, superoxide dismutase; CON, conconavalin A; CHT, chymotrypsin; INS, insulin; PTI, pancreatic trypsin inhibitor; CB5, cytochrome b_5; RNS, ribonuclease; LZM, lysozyme; SNS, staphyloccocal nuclease; LZ4, T4 lysozyme; PAP, papain; TLS, thermolysin; TRX, thioredoxin; FLN, flavodoxin; ADH, alcohol dehydrogenase coenzyme domain; AKN, adenyl kinase; TIM, triosephosphate isomerase; SUB, subtilisin; CPA carboxypeptidase; LDH, lactate dehydrogenase; PGK, phosphoglycerate kinase; GPD, glyceraldehyde-3-phosphate dehydrogenase; HKN, hexokinase; MDH, malate dehydrogenase

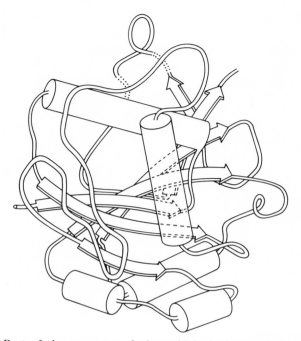

FIG. 1.10. Part of the structure of glyceraldehyde-3-phosphate dehydrogenase from *B. stearothermophilus* showing the assembly from the helices (cylinders) and parallel β-sheet (arrows). Note the twist in the β-sheet (courtesy of G. Biesecker and A. Wonacott)

5. Are the crystal and solution structures of an enzyme identical?

This is a very pertinent question since the structural details we have discussed and the mechanisms we shall deal with later are based on crystal structures. This question was once difficult to answer but the evidence is now available to say that the answer is generally yes. The reason why the 'yes' is qualified is that there is some controversy about the active site of carboxypeptidase (see Chapter 12), but this appears to concern only the position of a mobile side chain of a tyrosine residue.

The identity of solution and crystal structure is supported by the following evidence: (a) Some enzymes have been crystallized from different solvents and in different forms, but their structures remain essentially identical (e.g. subtilisin[18,19] and lysozyme[20]). (b) Classes of similar enzymes have similar structures (e.g. the serine proteases—see later). (c) In α-chymotrypsin, areas which are in contact between the dimer in solution are also in contact in the crystalline dimer.[21,22] (d) In some cases the crystals retain enzymic activity (e.g. ribonuclease[23,24] and carboxypeptidase A[25]). It is difficult to compare activities in solution and in the crystalline state because of the problems of diffusing the substrates into the crystals. But, in one ingenious example, this problem was overcome by crystallizing an enzyme-bound intermediate (indolylacryloyl-chymotrypsin—see below, Section D2) at a pH where it is stable.[26] On changing the pH to increase the reactivity, the intermediate is found to hydrolyse with the same first-order rate constant as in solution.

However, where there is an equilibrium between more than one conformation of the enzyme in solution, crystallization will select out only one of the conformations. α-Chymotrypsin has a substantial fraction of an inactive conformation present under the conditions of crystallization,[27] but only the active form of the enzyme crystallizes.

C. Families of enzymes

1. The serine proteases[28–34]

The initial excitement of discovering that the two oxygen-binding proteins haemoglobin and myoglobin have a common tertiary structure as well as a common function was rekindled when the same was found for the mammalian serine proteases. The enzymes are so called because they

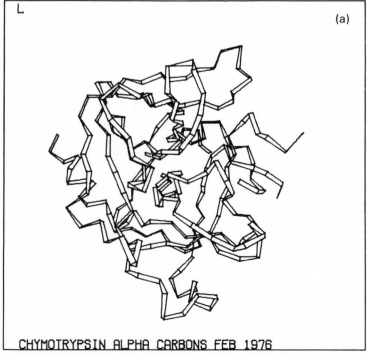

L

(a)

CHYMOTRYPSIN ALPHA CARBONS FEB 1976

FIG. 1.11. Ribbon diagrams of the polypeptide chains of (a) chymotrypsin, (b) elastase, and (c) trypsin. The α-carbons are at the pleats in the ribbon. The small differences that there are occur in the external loops (courtesy of J. Smith)

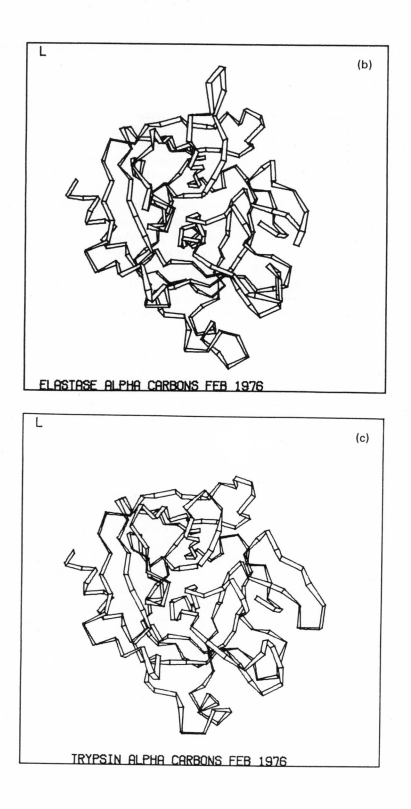

ELASTASE ALPHA CARBONS FEB 1976

TRYPSIN ALPHA CARBONS FEB 1976

have a uniquely reactive serine residue that reacts irreversibly with organophosphates such as diisopropyl fluorophosphate. The major pancreatic enzymes, trypsin, chymotrypsin, and elastase, are kinetically very similar, hydrolysing peptides and synthetic ester substrates. Their activities peak at around pH 7·8 and fall off at low pH with a pK_a of about 6·8. In all three cases an 'acylenzyme' is formed during reaction in which the carboxyl portion of the substrate esterifies the hydroxyl of the reactive serine.

The major difference between them is specificity. Trypsin is specific for peptides and esters of the amino acids lysine and arginine; chymotrypsin for the large hydrophobic side chains of phenylalanine, tyrosine, and tryptophan; elastase for the small hydrophobics such as alanine. When the structures of the crystalline enzymes were solved it was found that the polypeptide backbones of all three are essentially superimposable (Fig. 1.11) apart from some small additions and deletions in the chain. The difference in their specifities is due to just a few changes in a pocket that binds the amino-acid side chain. There is a well-defined binding pocket in chymotrypsin for the large hydrophobic side chains.[35] In trypsin, the residue at the bottom of the pocket is an aspartate instead of the Ser-189 of chymotrypsin.[36] The negatively charged carboxylate of Asp-189 forms a salt linkage with the positively charged ammonium or guanidinium at the end of the side chain of lysine or arginine. In elastase, the two glycines at the mouth of the pocket in chymotrypsin are replaced by a bulky valine (Val-216) and threonine (Thr-226).[37] This prevents the entry of large side chains into the pocket and provides a way of binding the small side chain of alanine (see Fig. 1.12).

The remarkable similarity of all three tertiary structures could not have been guessed in advance from a comparison of their sequences. There is extensive homology between their primary structures, but only about 50% of the sequences of elastase and chymotrypsin are composed of chemically identical or similar amino acids to those in trypsin.[34] The crystallographic results now tell us, of course, that this level is highly significant. By this token, it is expected that the plasma serine proteases that occur in the blood clotting cascade system also have very similar tertiary structures (Table 1.3). Closer examination of the sequence homologies shows that 60% of the amino acids in the interiors are conserved, but only 10% of the surface residues. The major differences occur in exposed areas and external loops.

A totally unexpected feature was found in the crystal structure of chymotrypsin. It was known from solution studies that the imidazole ring of His-57 increases the reactivity of Ser-195. But there was no hint that the histidine is hydrogen bonded to the carboxylate of Asp-102 to form a

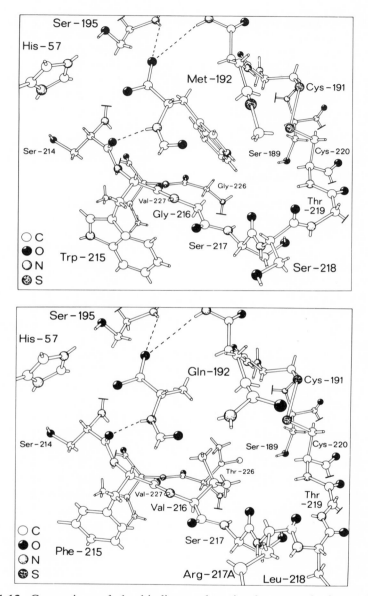

FIG. 1.12. Comparison of the binding pockets in chymotrypsin (top, with N-formyl-L-tryptophan bound) and elastase (bottom, with N-formyl-L-alanine bound). The binding pocket in trypsin is very similar to that in chymotrypsin except that residue 189 is an aspartate to bind positively charged side chains. Note the hydrogen bonds between the substrate and backbone of the enzyme

TABLE 1.3. *Sequence homologies in mammalian serine proteases*[a]

Enzyme	% Homology
Pancreas	
Trypsin	100
Chymotrypsin-A	53
Chymotrypsin-B	49
Elastase	48
Plasma	
Thrombin	38
Factor Xa	50

[a] % Chemical similarity of residues in sequence compared with bovine trypsin. Bovine enzymes apart from porcine elastase. (From B. S. Hartley, *Sym. Soc. Gen. Microbiol.* **24,** 152 (1974))

catalytic triad now called the 'charge relay system'.[38] This has subsequently been found in all serine proteases.

$$-\overset{O^-}{\underset{O}{\overset{|}{C}}} \cdots HN\!\!\diagdown\!\!\diagup N \cdots HO \qquad (1.3)$$

The mammalian serine proteases appear to represent a classic case of *divergent evolution*. All are presumably derived from a common ancestral serine protease some time back in evolution.

2. Serine proteases from microorganisms: convergent evolution

The first crystal structure of a bacterial serine protease to be solved, subtilisin from *Bacillus amyloliquifaciens*, revealed an enzyme of apparently totally different construction from the mammalian serine proteases.[39] This was not unexpected as there is no sequence homology between them. But closer examination shows that they are functionally identical as far as substrate binding and catalysis is concerned. Subtilisin has the charge relay system, the same system of hydrogen bonds for binding the carbonyl oxygen and the acetamido NH of the substrate, and the same series of subsites for binding the acyl portion of the substrate as have the mammalian enzymes (Fig. 1.12 and see later). This appears to be a case of *convergent evolution*. Different organisms, starting from different tertiary structures, have evolved a common mechanism.

More recently, some non-mammalian serine proteases have been

TABLE 1.4. *Species differences in serine proteases*

Enzyme	Species	% Homology
Trypsin	Cow	100
	Dogfish	69
	S. griseus	43
Elastase	Pig	48
	M. sorangium	26
	S. griseus	~20
Subtilisin	B. subtilis	0
	B. amyloliquifaciens	0

Data from Table 1.3 and L. T. J. Delbaere, W. L. B. Hutcheon, M. N. G. James, and W. E. Thiessen, *Nature, Lond.* **257,** 758 (1975)

shown to be 20–50% homologous with their mammalian counterparts (Table 1.4). We now know that this suggests very similar tertiary structure. The crystal structure of the elastase-like protease from *S. griseus* has been solved and, despite its having only 186 amino acids in its sequence compared with 245 in α-chymotrypsin, it is found to have two thirds of the residues in a similar conformation to those in the mammalian enzymes.[40] Possibly, these bacterial enzymes and the pancreatic ones have evolved from a common precursor. But the evolutionary relationships are not clear.

3. The carboxypeptidases[41,42]

Carboxypeptidase A hydrolyses the C-termini of proteins. It is specific for the hydrophobic side chains of phenylalanine, tyrosine, and tryptophan. Carboxypeptidase B is specific for the positively charged side chains of lysine and arginine. They are related in the same way as chymotrypsin and trypsin. 49% of their sequences are identical. Differences in their tertiary structures are confined to the external regions. The major difference is that Ile-255 in the binding pocket of the A form becomes Asp-255 in the B form to bind the positively charged side chains of the basic substrates.

4. The dehydrogenases and domains[43]

One would expect that the NAD^+-dependent dehydrogenases, a class of enzymes with the same chemical function and binding the same cofactor, would form a structurally related family. They appear to do this, but not in the same clear-cut way as do the serine proteases. The tertiary structures of the first two crystal structures to be solved, dogfish lactate dehydrogenase and soluble porcine malate dehydrogenase, were found to be almost superimposable, apart from the first 20 residues of lactate dehydrogenase. It seems likely that these have evolved from a common precursor dehydrogenase. But the subsequent solution of the structures of horse liver alcohol dehydrogenase and lobster glyceraldehyde-3-phosphate dehydrogenase complicated the picture, being extensively different. However, it was noticed that the structure of each of the four enzymes consists of two domains, one of which was similar in all four. This domain binds the NAD$^+$ (Fig. 1.13).

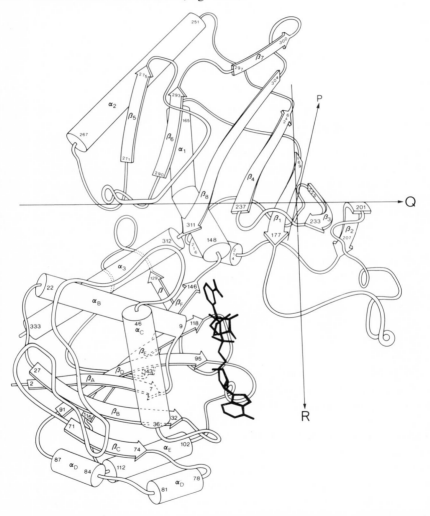

F$_{IG}$. 1.13. The domains in glyceraldehyde-3-phosphate dehydrogenase from *B. stearothermophilus* (G. Biesecker, J. I. Harris, J. C. Thierry, J. E. Walker and A. Wonacott, *Nature, Lond.* **266,** 328 (1977)).

These structural features have led to the hypothesis that the dehydrogenases have evolved from the fusion of a gene coding for the dinucleotide binding domain with a series of genes, each coding for a separate 'catalytic domain'. If this is correct, this gives a very simple means of generating a family of enzymes of diverse specificity.

The nucleotide binding fold is a complicated structure which differs in detail from one dehydrogenase to another. In its idealized form it consists of six strands of parallel β-sheet with four parallel helices running antiparallel to the sheet. This structure occurs in other nucleotide binding proteins such as phosphoglycerate kinase. But it is also found in proteins, such as flavodoxin, that are not involved in nucleotide binding. It is not known whether these similarities are evidence for a common evolutionary precursor protein or are caused by there being only a limited number of ways of folding a polypeptide chain.

D. The structure of enzyme–substrate complexes

The outstanding characteristic of enzyme catalysis is that the enzyme specifically binds its substrates and the reactions take place in the confines of the enzyme–substrate complex. Thus to understand how an enzyme works, we do not only need to know the structure of the native enzyme but also the structures of the complexes of the enzyme with its substrates, intermediates, and products. Once these have been determined we can see how the substrate is bound, what catalytic groups are close to the substrate and what structural changes occur in the substrate and enzyme on binding. There is one obvious difficulty: enzyme–substrate complexes react to give products in fractions of a second whilst the acquisition of X-ray data usually takes several hours. For this reason, it is usual to determine the structures of the complexes of enzymes with the reaction products, inhibitors, or substrate analogues.

1. Methods for examining enzyme–substrate complexes

a. The difference Fourier method
Protein crystals generally contain about 50% solvent and never less than 30%. Often there are channels of solvent leading from the exterior of the crystal to the active site, so that substrates may be diffused into the crystals and bound to the enzyme. Alternatively, it is often possible to cocrystallise the enzyme and substrates. Provided there are no drastic changes in the structure or packing of the enzyme on binding the ligand, it is possible to solve the structure of the complex by the difference Fourier technique. This involves measuring the *differences* between the diffraction patterns of the native crystals and those soaked in a solution of the ligand. The electron density of the bound ligand and any minor

changes in the structure of the enzyme may be obtained without the necessity of solving the whole structure from scratch.

b. Production of stable complexes

The first attempts to determine the structure of a productively bound enzyme–substrate complex used extrapolation from the structures of stable enzyme–inhibitor complexes. The classic example of this, lysozyme, is discussed at the end of the chapter. This may be done in several ways. For example, a portion of the substrate may be bound to the enzyme and the structure of the remainder determined by model building. Alternatively, an unreactive analogue of the substrate may be used in which the reactive bond is modified. A good example of this is the binding of a phosphonate, rather than a phosphate, to ribonuclease (see Chapter 12C).

Methods are also available in some cases for the direct study of productively bound enzyme–substrate complexes under conditions where they do not react. These range from the use of weakly reactive substrates at a pH were the enzyme is largely inactive due to a residue being in the wrong ionic state (see below—indolylacryloyl-chymotrypsin) or at very low temperatures (an elastase acylenzyme—see below), or by using a chemically modified enzyme. An example of the latter, which has yet to be used, is lysozyme in which the $-CO_2H$ of the catalytically important residue Asp-52 is converted to a $-CH_2OH$ (see below and Chapter 12E).

It is also possible on occasion to examine directly a productively bound enzyme–substrate complex under conditions where it is rapidly reacting. This happens when an equilibrium may be set up between substrates and products where the equilibrium position favours the substrates. Adding the enzyme to the equilibrium mixture cannot alter the position of the solution equilibrium so it should be possible to obtain a stable enzyme–substrate complex. However, there is no guarantee that the equilibrium position is the same for the enzyme-bound reagents as it is in solution (Chapter 3). One example of such an equilibrium is that for the hydrolysis of a dipeptide, which *can* be forced to favour synthesis by adding excess amine (1.4).[44,45] This has also been used for NMR experiments.[46]

$$RCONHR' \rightleftharpoons RCO_2^- + H^+ + R'NH_2 \qquad (1.4)$$

Another example is the binding of dihydroxyacetone phosphate to triosephosphate isomerase (Chapter 12G1). This enzyme catalyses a simple equilibrium between two reagents which favours one form. In this context it should be remembered that an enzyme–product complex is also an enzyme–substrate complex—that for the reverse reaction.

Examples of these methods are discussed in Chapter 12. The serine proteases and lysozyme are discussed below because of their historical importance in the development of the ideas and techniques involved, as well as their being extensively used in this text.

2. Example 1: the serine proteases

Peptide and synthetic ester substrates are hydrolysed by the serine proteases by the acylenzyme mechanism (1.5).[47] A non-covalent intermediate is formed first followed by the attack of the hydroxyl of Ser-195

$$E\text{—}OH + RCONHR' \rightleftharpoons E\text{—}OH.RCONHR' \rightleftharpoons$$

$$
\underset{\underset{\displaystyle NHR'}{|}}{\overset{\overset{\displaystyle O^-}{|}}{E\text{—}O\text{—}C\text{—}R}} \rightleftharpoons E\text{—}\underset{+}{OCOR} \rightleftharpoons E\text{—}OH.RCO_2H \quad (1.5)
$$

$$NH_2R'$$

$$E\text{—}OH + RCO_2H$$

on the substrate to give the *tetrahedral intermediate*. This then collapses to give the *acylenzyme*, releasing the amine or alcohol. The acylenzyme then hydrolyses to form the enzyme–product complex. Crystallographic studies have been performed to give the structures of most of the complexes. The rest can be obtained by model building. Working backwards, the structure of the enzyme–product complex, N-formyl-L-tryptophan chymotrypsin (Fig. 1.12), has been solved by diffusing the product into the crystal and then using the difference Fourier technique.[48] The structure of a non-specific acylenzyme, indolylacryloylchymotrypsin, was solved at pH 4 where it is deacylated only very slowly.[49] The substrate, an activated derivative of the acid, indolylacryloylimidazole, was diffused into the crystal where it acylated Ser-195. The structure of carbobenzoxyalanyl-elastase has been solved at −55°.[50,51]

Nature has provided a rare opportunity for solving the structures of the enzyme–substrate complexes of trypsin and chymotrypsin with polypeptides. There are many naturally occurring polypeptide inhibitors that bind to trypsin and chymotrypsin very tightly because they are locked into the conformation that a normal flexible substrate takes up on binding.[52] They do not hydrolyse under normal physiological conditions because the amino group that is released on the cleavage of the peptide is constrained and cannot diffuse away from the active site of the enzyme. On removing the constraints in the pancreatic trypsin inhibitor by reducing an S—S bridge in its polypeptide chain, the peptide bond between Lys-15 and Ala-16 is readily cleaved by trypsin.[53] The structures of trypsin, its complex with the basic pancreatic trypsin inhibitor complex, and the free inhibitor have been solved at resolutions of 1·4, 1·9, and 1·7 Å respectively.[54] These are among the most accurate determinations yet done so that critical atomic positions are known to about 0·1 to 0·2 Å. These and

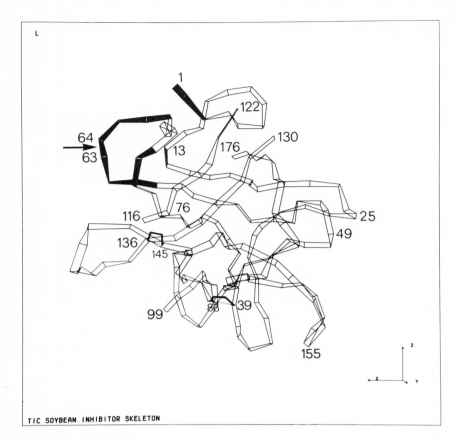

L

TIC SOYBEAN INHIBITOR SKELETON

Fig. 1.14. Ribbon diagram of the polypeptide chain of the soya bean trypsin inhibitor. The shaded regions are those that make contacts with the enzyme. (D. M. Blow, J. Janin, and C. Chothia, *Nature, Lond.* **249,** 54 (1974))

other studies have given the following information about the binding of substrates.[55–65]

a. The binding site

The binding site for a polypeptide substrate consists of a series of subsites across the surface of the enzyme. By convention they are labelled as in (1.6). The substrate residues are called P (for peptide), and the subsites S. Apart from the primary binding site S_1 for the side chains of the aromatic substrates of chymotrypsin or the basic amino acid substrated of trypsin, there is no obvious well-defined cleft or groove for substrate binding. The subsites run along the surface of the protein.

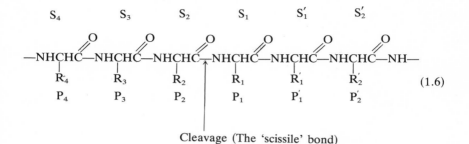

$$\begin{array}{cccccc} S_4 & S_3 & S_2 & S_1 & S_1' & S_2' \end{array}$$

(1.6)

Cleavage (The 'scissile' bond)

b. The primary binding site (S_1)

The binding pocket for the aromatic side chains of the specific substrates of chymotrypsin is a well-defined slit in the enzyme 10–12 Å deep and 3·5–4 Å by 5·5–6·5 Å in cross section.[48] This gives a very snug fit since an aromatic ring is about 6 Å wide and 3·5 Å thick. A methylene group is about 4 Å in diameter (Chapter 9) so that the side chain of lysine or arginine is bound nicely by the same shaped slit in trypsin. Of course, as mentioned in section C1, there is a carboxylate at the bottom of the pocket in trypsin to bind the positive charge on the end of the side chain, and also the mouth of the pocket is blocked somewhat in elastase.

The binding pocket in chymotrypsin may be described as a 'hydrophobic pocket' since it is lined with the non-polar side chains of amino acids. It provides a suitable environment for binding the non-polar or hydrophobic side chains of the substrates. The physical causes of hydrophobic bonding and its strength are discussed in Chapter 9.

The hydrophobic binding site in subtilisin is not a well-defined slit as in the pancreatic enzymes, but more like a shallow groove. However, the following hydrogen bonds are found for all the enzymes (Fig. 1.12).

The carbonyl oxygen of the reactive bond has a binding site between the backbone NH groups of Ser-195 and Gly-193. The hydrogen bonds that are made are very important because the oxygen becomes negatively charged during the reaction. There is also a hydrogen bond between the NH part of the N-acylamino group of the substrate and the C=O of Ser-214.

c. Sites S_1—S_2—S_3

The hydrogen bond between the N-acylamino NH and the carbonyl of Ser-214 initiates a short region of antiparallel β-sheet between the residues Ser-214, Trp-215, and Gly-216 of the enzyme and the amino acids P_1, P_2, and P_3 of the substrate.

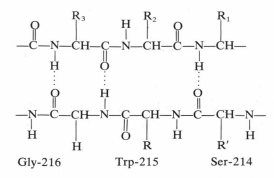

Gly-216   Trp-215   Ser-214

d. Site S'₁—the leaving group site

There is a leaving group site which is constructed to fit L-amino acids.[65]
The contacts with the enzyme are predominantly hydrophobic, accounting
for the lack of exonuclease activity with the enzyme since this would
require binding a —CO_2^- in a non-polar region (Fig. 1.15). ('Leaving
group' is chemists' jargon for a group displaced from a molecule, in this
case from an acyl group.)

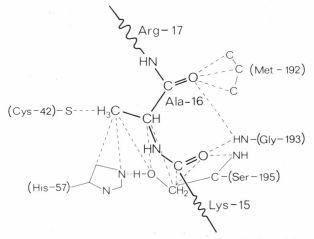

FIG. 1.15. Sketch of the leaving group site in chymotrypsin found from fitting the
model of the pancreatic trypsin inhibitor to the enzyme (ref. 65).

3. Example 2: lysozyme

Lysozyme catalyses the hydrolysis of a polysaccharide that is the major
constituent of the cell wall of certain bacteria. The polymer is formed
from β(1–4) linked alternating units of N-acetylglucosamine (NAG) and
N-acetylmuramic acid (NAM) (Fig. 1.16). The solution of the structure of
lysozyme is one of the triumphs of X-ray crystallography.[66–68] Whereas
solving the crystal structures of the serine proteases represented the

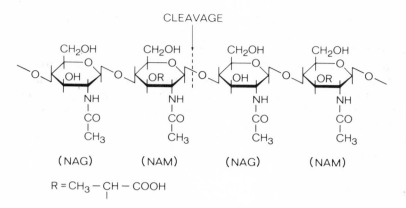

FIG. 1.16. The polysaccharide substrate of lysozyme found in bacterial cell walls.

culmination of a long series of studies stretching back through the history of enzymology, the structure of the little-known enzyme lysozyme stimulated the solution studies. Also, the mechanism of the enzyme, which was previously unknown, was guessed from examining the crystal structure of the native enzyme and the complexes with inhibitors.

Unlike chymotrypsin, lysozyme has a well-defined deep cleft running down one side of the ellipsoidal molecule for binding the substrate. This is partly lined with non-polar side chains if amino acids for binding the non-polar regions of the substrate, and also has hydrogen bonding sites for the acylamino and hydroxyl groups. The cleft is divided into six sites, ABCDEF. NAM residues can bind only in B, D, and F, whilst NAG residues of synthetic substrates may bind in all sites. The bond which is cleaved lies between sites D and E. There are no means at present of solving the structure of a productively bound enzyme–substrate complex in the same way as for chymotrypsin and trypsin since there are no similar natural inhibitors. The method of determining the structure of the complex is based on model building and is typical of the general approach that is used. The structure of the enzyme and the inhibitor (NAG)₃ has been solved.[67] This is a very poor substrate of the enzyme since the structural studies show that it is bound non-productively in the ABC sites, avoiding the DE sites where cleavage takes place. The structure of a productively-bound complex was obtained by building a wire model of the complex of (NAG)₃ with the enzyme and extending the polysaccharide chain by adding further NAG units using chemical intuition about the contacts made with the enzyme. Nowadays, this procedure is done using a computer programme to optimize the fit between the enzyme and substrate. It was found that the bond which is cleaved is placed between the carboxyl groups of Glu-35 and Asp-52 (later proven to be in the un-ionized and ionized forms respectively). How these contribute to the mechanism will be discussed in Chapter 12.

To summarize, the binding sites of lysozyme and the serine proteases are approximately complementary in structure to the structures of the substrates: the non-polar parts of the substrate match up with non-polar side chains of the amino acids; the hydrogen bonding sites on the substrate bind to the backbone NH groups of the protein or, in the case of lysozyme, to the polar side chains of amino acids also. The reactive part of the substrate is firmly held by this binding next to acidic, basic, or nucleophilic groups on the enzyme.

References

1   J. B. S. Haldane, *Enzymes*, Longmans, Green and Co. (1930). MIT Press (1965).
2   D. W. Green, V. Ingram, and M. F. Perutz, *Proc. R. Soc.* **A225,** 287 (1954).
3   L. Pauling, *Nature of the chemical bond*, Cornell University Press (1960).
4   A. R. Fersht, *J. Am. chem. Soc.* **93,** 3504 (1971).
5   P. Y. Chou and G. D. Fasman, *Biochemistry* **13,** 211, 222 (1974).
6   C. M. Venkatachalam, *Biopolymers* **6,** 1425 (1968).
7   P. N. Lewis, F. A. Momany, and H. A. Scheraga, *Proc. natn. Acad. Sci. U.S.A.* **68,** 2293 (1971).
8   I. D. Kuntz, *J. Am. chem. Soc.* **94,** 4009 (1972).
9   J. L. Crawford, W. N. Lipscomb, and C. G. Schellman, *Proc. natn. Acad. Sci. U.S.A.* **70,** 538 (1973).
10   M. Levitt and C. Chothia, *Nature, Lond.* **261,** 552 (1976).
11   I. D. Kuntz, *J. Am. chem. Soc.* **94,** 8568 (1972).
12   F. H. C. Crick, *Acta crystallogr.* **6,** 689 (1953).
13   C. Chothia, *J. molec. Biol.* **75,** 295 (1973).
14   M. H. Klapper, *Biochim. biophys. Acta* **229,** 557 (1971).
15   F. M. Richards, *J. molec. Biol.* **82,** 1 (1974).
16   C. Chothia, *Nature, Lond.* **254,** 304 (1975).
17   C. Chothia and J. Janin, *Nature, Lond.* **256,** 705 (1975).
18   J. D. Robertus, R. A. Alden, and J. Kraut, *Biochem. biophys. Res. Commun.* **42,** 334 (1971).
19   J. Drenth, W. G. T. Hol, J. N. Jansonius, and R. Koekoek, *Cold Spring Harb. Symp. quant. Biol.* **36,** 107 (1971).
20   J. Moult, A. Yonath, W. Traub, A. Smilansky, A. Podjarny, D. Rabinovich, and A. Saya, *J. molec. Biol.* **100,** 179 (1976).
21   C. S. Hexter and F. H. Westheimer, *J. biol. Chem.* **246,** 3928 (1971).
22   A. R. Fersht and J. Sperling, *J. molec. Biol.* **74,** 137 (1973).
23   M. S. Doscher and F. M. Richards, *J. biol. Chem.* **238,** 2399 (1963).
24   J. Bello and E. F. Nowoswiat, *Biochim. biophys. Acta* **105,** 325 (1965).
25   C. A. Spilburg, J. L. Bethune, and B. L. Vallee, *Proc. natn. Acad. Sci. U.S.A.* **71,** 3922 (1974).
26   G. L. Rossi and S. A. Bernhard, *J. molec. Biol.* **49,** 85 (1970).
27   A. R. Fersht and Y. Requena, *J. molec. Biol.* **60,** 279 (1971).

28   G. P. Hess, *The Enzymes* **3,** 213 (1971).
29   B. Keil, *The Enzymes* **3,** 249 (1971).
30   S. Magnusson, *The Enzymes* **3,** 277 (1971).
31   B. S. Hartley and S. Magnusson, *The Enzymes* **3,** 323 (1971).
32   D. M. Blow, *The Enzymes* **3,** 185 (1971).
33   J. Kraut, *The Enzymes* **3,** 547 (1971).
34   B. S. Hartley, *Symposia Soc. gen. Microbiol.* **24,** 151 (1974).
35   B. W. Matthews, P. B. Sigler, R. Henderson, and D. M. Blow, *Nature, Lond.* **214,** 652 (1967).
36   R. M. Stroud, L. M. Kay, and R. E. Dickerson, *J. molec. Biol.* **83,** 185 (1974).
37   D. M. Shotton and H. C. Watson, *Nature, Lond.* **225,** 811 (1970).
38   D. M. Blow, J. J. Birktoft, and B. S. Hartley, *Nature, Lond.* **221,** 337 (1969).
39   C. S. Wright, R. A. Alden, and J. Kraut, *Nature, Lond.* **221,** 235 (1969).
40   L. T. J. Delbaere, W. L. B. Hutcheon, M. N. G. James, and W. E. Thiessen, *Nature, Lond.* **257,** 758 (1975).
41   F. A. Quiocho and W. N. Lipscomb, *Adv. Protein Chem.* **25,** 1 (1971).
42   M. F. Schmid and J. R. Herriott, *J. molec. Biol.* **103,** 175 (1976).
43   M. G. Rossman, A. Liljas, C.-I. Brandén, and L. J. Banaszak, *The Enzymes*, **11,** 61 (1975).
44   A. R. Fersht and Y. Requena, *J. Am. chem. Soc.* **93,** 3499 (1971).
45   A. R. Fersht and M. Renard, *Biochemistry* **13,** 1416 (1974).
46   G. Robillard, E. Shaw, and R. G. Shulman, *Proc. natn. Acad. Sci. U.S.A.* **71,** 2623 (1974).
47   J. Fastrez and A. R. Fersht, *Biochemistry* **12,** 2025 (1973).
48   T. A. Steitz, R. Henderson, and D. M. Blow, *J. molec. Biol.* **46,** 337 (1969).
49   R. Henderson, *J. molec. Biol.* **54,** 341 (1970).
50   A. L. Fink and A. I. Ahmed, *Nature, Lond.* **263,** 294 (1976).
51   T. Alber, G. A. Petsko, and D. Tsernoglou, *Nature, Lond.* **263,** 297 (1976).
52   M. Laskowski, Jr. and R. W. Sealock, *The Enzymes* **3,** 375 (1971).
53   K. A. Wilson and M. Laskowski, Sr., *J. biol. Chem.* **246,** 3555 (1971).
54   M. Rigbi, *Proceedings of the International Conference on Proteinase Inhibitors* (ed. H. Fritz and H. Tschesche), Walter de Gruyter, Berlin, p. 117 (1971).
55   W. Bode, P. Schwager, and R. Huber, *Proceedings of the Tenth FEBS Meeting* **40,** 3 (1975).
56   R. M. Sweet, H. T. Wright, J. Janin, C. M. Chothia, and D. M. Blow, *Biochemistry* **13,** 4212 (1974).
57   J. D. Robertus, J. Kraut, R. A. Alden, and J. J. Birktoft, *Biochemistry* **11,** 4293 (1972).
58   J. Kraut, J. D. Robertus, J. J. Birktoft, R. A. Alden, P. E. Wilcox, and J. C. Powers, *Cold Spring Harb. Symp. quant. Biol.* **36,** 117 (1971).
59   D. M. Shotton, N. J. White, and H. C. Watson, *Cold Spring Harb. Symp. quant. Biol.* **36,** 91 (1971).
60   D. M. Segal, G. H. Cohen, D. R. Davies, J. C. Powers, and P. E. Wilcox, *Cold Spring Harb. Symp. quant. Biol.* **36,** 85 (1971).
61   D. Atlas, S. Levit, I. Schechter, and A. Berger, *FEBS Lett.* **11,** 281 (1970).
62   A. Gertler and T. Hofmann, *Can. J. Biochem.* **48,** 384 (1970).
63   R. C. Thompson, *Biochemistry* **13,** 5495 (1974).

64   W. K. Baumann, S. A. Bizzozero, and H. Dutler, *Eur. J. Biochem.* **39,** 381 (1973).
65   A. R. Fersht, D. M. Blow, and J. Fastrez, *Biochemistry* **12,** 2035 (1973).
66   C. C. F. Blake, D. F. Koenig, G. A. Mair, A. C. T. North, D. C. Phillips, and V. R. Sarma, *Nature, Lond.* **206,** 757 (1965).
67   D. C. Phillips, *Proc. natn. Acad. Sci. U.S.A.* **57,** 484 (1967).
68   C. C. F. Blake, L. N. Johnson, G. A. Mair, A. C. T. North, D. C. Phillips, and V. R. Sarma, *Proc. R. Soc.* **B167,** 378 (1967).

Chapter 2

Chemical catalysis

We know from structural and kinetic studies that enzymes have well-defined binding sites for their substrates, sometimes form covalent intermediates, and generally involve acidic, basic, and nucleophilic groups. In this chapter we shall see why these features are necessary for catalysis and how they are used. The importance of the enzyme–substrate binding energy is dealt with later in Chapter 10.

Many of the concepts in catalysis are based on *transition state theory*. As an elementary knowledge of the theory greatly simplifies the understanding of some ideas and is essential for others, we begin with a discussion of its principles and applications. This is followed by an introduction to the basic principles of chemical catalysis and the factors responsible for the magnitude of enzyme catalysis. Subsequent sections deal with progressively more advanced topics in kinetics and solution catalysis, starting with the factors that determine chemical reactivity, such as what is a good nucleophile and what is a good leaving group in a particular reaction, and the analysis of structure–reactivity relationships. Finally, some topics in kinetics and stereochemistry are dealt with here for convenience.

A. Transition state theory[1-4]

There are several theories to account for chemical kinetics. The simplest is the collision theory, which will be used later in Chapter 4 to calculate the rate constants for the collision of molecules in solution. A more sophisticated theory, which is particularly useful for analysing structure–reactivity relationships, is the transition state theory. The processes by which the reagents collide are ignored: the only physical entities considered are the reagents, or ground state, and the most unstable species on the reaction pathway, the *transition state*. The transition state occurs at the peak in the reaction coordinate diagram (Fig. 2.1), in which the energy of the reagents is plotted as the reaction proceeds. In the transition state chemical bonds are in the process of being made and broken. In

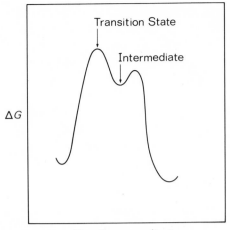

Reaction coordinate

Fɪɢ. 2.1. Transition states occur at the peaks and intermediates in the troughs of the energy profile of a reaction

contrast, *intermediates*, whose bonds are fully formed, occupy the troughs in the diagram. A simple way of deriving the rate of the reaction is to consider that the transition state and the ground state are in thermodynamic equilibrium so that the concentration of the transition state may be calculated from the difference in their energies. The overall reaction rate is then obtained by multiplying the concentration of the transition state by the rate constant for its decomposition. This process is simpler than it sounds because the energy difference between the ground state and the transition state is used only qualitatively, and it may be shown that all transition states decompose at the same frequency for a given temperature.

The analysis for a unimolecular reaction is as follows. Suppose the difference in Gibbs energy between the transition state $X^{\ddagger}$, and the ground state, X, is $\Delta G^{\ddagger}$. Then using a well-known relationship from equilibrium thermodynamics,†

$$[X^{\ddagger}] = [X]\exp(-\Delta G^{\ddagger}/RT). \tag{2.1}$$

† The two thermodynamic equations that are most useful for simple kinetic and binding experiments are: (a) the relationship between the Gibbs-energy change and the equilibrium constant of a reaction,

$$\Delta G = -RT \ln K;$$

(b) the relationship between the Gibbs-energy change and the changes in the enthalpy and entropy,

$$\Delta G = \Delta H - T\Delta S.$$

The frequency at which the transition state decomposes is the same as the vibrational frequency ν of the bond which is breaking. This frequency is obtained from the equivalence of the energies of an excited oscillator calculated from quantum theory ($E = h\nu$), and classical physics ($E = kT$), that is

$$\nu = kT/h \tag{2.2}$$

where k is the Boltzmann constant, and h the Planck constant. At 25°, $\nu = 6{\cdot}212 \times 10^{12} \, \text{s}^{-1}$.

The rate of decomposition of X is thus given by:

$$-d[X]/dt = \nu[X^{\ddagger}] \tag{2.3}$$

$$= [X](kT/h)\exp(-\Delta G^{\ddagger}/RT). \tag{2.4}$$

The first-order rate constant for the decomposition of X is given by

$$k_1 = (kT/h)\exp(-\Delta G^{\ddagger}/RT). \tag{2.5}$$

The Gibbs energy of activation $\Delta G^{\ddagger}$ may be separated into enthalpic and entropic terms, if required, by using another relationship from equilibrium thermodynamics,

$$\Delta G^{\ddagger} = \Delta H^{\ddagger} - T\Delta S^{\ddagger} \tag{2.6}$$

(where $\Delta H^{\ddagger}$ is the enthalpy, and $\Delta S^{\ddagger}$ the entropy, of activation).

$$k_1 = (kT/h)\exp(\Delta S^{\ddagger}/R)\exp(-\Delta H^{\ddagger}/RT). \tag{2.7}$$

(A more rigorous approach includes a factor known as the transmission coefficient, but this is generally close to unity and may be ignored.)

1. Significance and application of transition state theory

The importance of transition state theory is that it relates the rate of a reaction to the difference in Gibbs energy between the transition state and the ground state. This is especially important when considering the relative reactivities of pairs of substrates, or comparing the rate of a given reaction under different sets of conditions. Under some circumstances the ratio of rates may be calculated, or, more generally, the trends in reactivity estimated qualitatively. For example, the alkaline hydrolysis of an ester, such as phenyl acetate, involves the attack of the negatively charged hydroxide ion on the neutral ground state. This means that in the transition state of the reaction there must be some negative charge transferred to the ester. We can predict that p-nitrophenyl acetate is more reactive than phenyl acetate since the nitro group is electron withdrawing and will stabilize the negatively charged transition state with respect to the neutral ground state. Consider also the spontaneous decomposition of tertiary-butyl bromide into the tertiary-butyl carbonium ion and the bromide ion. The transition state of this reaction must be dipolar. Therefore a polar solvent, such as water, will stabilize the transition state, and a non-polar solvent, such as diethyl ether, will destabilize the transition state with respect to the ground state.

In Chapters 10 and 11 it will be seen how the transition state theory may be used quantitatively in enzymic reactions to analyse structure reactivity and specificity relationships involving discrete changes in the structure of the substrate.

2. The Hammond postulate[5]

A useful guide in the application of transition state theory or the analysis of structure–reactivity data is the Hammond postulate. This states that if there is an unstable intermediate on the reaction pathway, the transition state for the reaction will resemble the structure of this intermediate. The reasoning behind this is that the unstable intermediate will be in a small dip at the top of the reaction coordinate diagram. This is a useful way of guessing the structure of the transition state for predicting the types of stabilization it requires. For example, a carboxonium ion is an inter-mediate in the reaction catalysed by lysozyme (Fig. 2.2). Since these are known to be unstable high-energy compounds, the transition state is assumed to resemble the structure of the carboxonium ion. One cannot really apply the Hammond postulate to bimolecular reactions. These involve two molecules condensing to form one transition state and a large part of the Gibbs energy change is caused by the loss in entropy (see Section B4). The Hammond postulate applies mainly to energy differ-ences and works best with unimolecular reactions. The postulate is sometimes extended to cases where there are large energy differences between the reagents and products. If the products are very unstable the

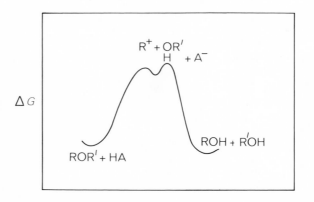

FIG. 2.2. The transition state for the cleavage of an acetal resembles the car-bonium ion intermediate which occupies the small 'dip' at the top of the energy diagram

transition state is presumed to resemble them. The same applies if the reagents are very unstable. The postulate is less reliable in these situations.

B. Principles of catalysis

1. Where, why, and how catalysis is required

To illustrate the importance of catalysis, let us examine the uncatalysed reactions which may be catalysed by chymotrypsin and lysozyme.

The uncatalysed attack of water on an ester leads to a transition state in which a positive charge develops on the attacking water molecule and a negative charge on the carbonyl oxygen:

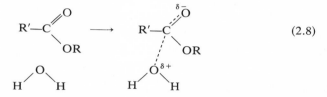

$$(2.8)$$

The uncatalysed hydrolysis of an acetal involves a transition state which is close in structure to a carbonium ion and an alkoxide ion.

$$(2.9)$$

In both reactions the transition state is very unfavourable because of the unstable positive and negative charges which are developed. Stabilization of these charges catalyses the reaction by lowering the energy of the transition state. This can be done in the case of the positive charge developing on the attacking water molecule by transferring one of the protons to a base during the reaction. This is known as *general base catalysis.*

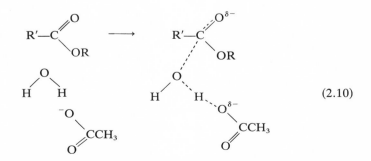

$$(2.10)$$

General base catalysis by the acetate ion

Similarly, the negative charge developed on the alcohol expelled from the acetal can be stabilized by proton transfer from an acid. This is known as *general acid catalysis*.

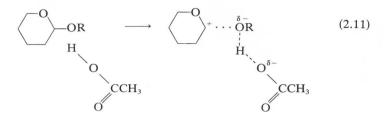

$$(2.11)$$

General acid catalysis by acetic acid

This acid–base catalysis is termed 'general' to distinguish it from *specific acid* or *base catalysis* in which the catalyst is the proton or hydroxide ion.

Positive and negative charges may also be stabilized by *electrostatic catalysis*. The positively charged carbonium ion cannot be stabilized by general base catalysis as it does not ionize. But it can be stabilized by the electric field from a negatively charged carboxylate ion. The negative charge on an oxyanion may also be stabilized by a positively charged metal ion, such as Zn^{2+} or Mg^{2+}. The stabilization of a negative charge, i.e. an electron, is known as *electrophilic catalysis*.

The above types of catalysis function by stabilizing the transition state of the reaction without changing the mechanism. Catalysis may also involve a different reaction pathway. A typical example is *nucleophilic catalysis* in an acyl transfer or hydrolytic reaction. The hydrolysis of acetic anhydride is greatly enhanced by pyridine because of the rapid formation of the highly reactive acetylpyridinium ion.

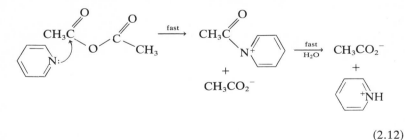

$$(2.12)$$

For nucleophilic catalysis to be efficient, the nucleophile must be more nucleophilic than the one it replaces, and the intermediate must be more reactive than the parent compound.

Nucleophilic catalysis is a specific example of *covalent catalysis*: the substrate is transiently modified by forming a covalent bond with the catalyst to give a reactive intermediate. There are also many examples of electrophilic catalysis by covalent modification. In the reactions of pyridoxal phosphate, Schiff base formation, and thiamine pyrophosphate, it will be seen later that electrons are stabilized by delocalization.

2. General acid–base catalysis

a. Detection and measurement

The general species catalysis of the hydrolysis of an ester is measured from the increase in the hydrolytic rate constant with increasing concentration of the acid or base. This is usually done at constant pH by maintaining a constant ratio of the acidic and basic forms of the catalyst. It is important to keep the ionic strength of the reaction medium constant since many reactions are sensitive to changes in salt concentration. In order to tell whether the catalysis is due to the acidic or basic form of the catalyst, it is necessary to repeat the measurements at a different buffer ratio. For ester hydrolysis, it is found that the increase in rate is generally proportional to the concentration of the basic form, i.e. general base catalysis. The slope of the plot of the rate constant against the concentration of base gives the second-order rate constant k_2 for the general base catalysis (see Fig. 2.3).

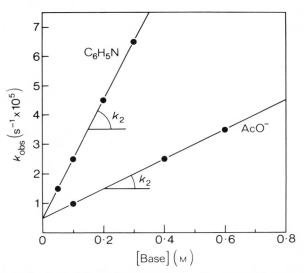

FIG. 2.3. Determination of the rate constants for the general base catalysis of the hydrolysis of ethyl dichloroacetate (W. P. Jencks and J. Carriuolo, *J. Amer. chem. Soc.* **83,** 1743 (1961)). The first-order rate constants for the hydrolysis at various concentrations of the base are plotted against its concentration. The slope of the linear plot is the second-order rate constant for the general base catalysis (k_2). The intercept at zero buffer concentration is the 'spontaneous' hydrolysis rate constant for the particular pH. A plot of the spontaneous rate constant against pH gives the rate constants for the H^+ and OH^- catalysis. It is seen that pyridine is a more effective catalyst than the weaker base, acetate ion

b. The efficiency of acid–base catalysis: the Brönsted equation[6]
It is found experimentally that the general base catalysis of the hydrolysis
of an ester is proportional to the basic strength of the catalyst.[7] The
second-order rate constant k_2 for the dependence of the rate of hydrolysis
on the concentration of the base, is given by the equation

$$\log k_2 = A + \beta \mathrm{p} K_a. \qquad (2.13)$$

Eqn (2.13) is an example of the Brönsted equation. β is known as the
Brönsted β-value. It measures the sensitivity of the reaction to the $\mathrm{p}K_a$ of
the conjugate acid of the base. A is a constant for the particular reaction.

Brönsted equations are also common in general acid catalysis, as for
example in the hydrolysis of certain acetals.[8–10] In acid catalysis, α is
used, where

$$\log k_2 = A - \alpha \mathrm{p} K_a, \qquad (2.14)$$

rather than the β for base catalysis. α and β are always between 0 and 1
for acid–base catalysis (apart from some peculiar carbon acids)[11] because
complete transfer of a proton gives a value of 1 and no transfer a
value of 0. The usual values for ester hydrolysis are 0·3 to 0·5, and for
acetal hydrolysis, about 0·6.

Some values of the rate enhancement by general base and general acid
catalysis are listed in Tables 2.1 and 2.2. These values are calculated from

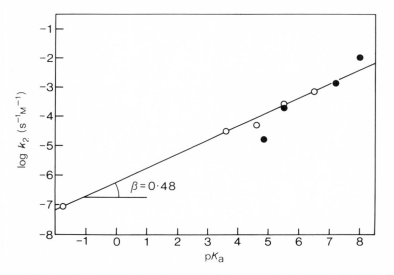

Fɪɢ. 2.4. The Brönsted plot for the general base catalysis of the hydrolysis of
ethyl dichloroacetate. The logarithms of the second-order rate constants obtained
from the plot of Fig. 2.3 are plotted against the $\mathrm{p}K_a$ of the conjugate acid of the
catalytic base. The slope is the β-value. Note that the points for amine bases (●)
fall on the same line as those for oxyanion bases (○) showing that the catalysis
depends primarily on the basic strength of the base and not its chemical nature

TABLE 2.1. *Influence of β on general base catalysis*

β	$\dfrac{\text{Rate in 1 M solution of base}}{\text{Rate in water}}$	
	$pK_a = 5$	$pK_a = 7$
0	1	1
0·3	2·9	8·6
0·5	44	427
0·7	951	$2·4 \times 10^4$
0·85	$9·7 \times 10^3$	$4·9 \times 10^5$
1	10^5	10^7

the Brönsted equation using pK_as of 15·74 and $-1·74$ for ionization of H_2O and H_3O^+ (i.e. from $[H^+][OH^-] = 10^{-14}$ and the concentration of water $= 55$ M). The magnitude of the catalysis depends strongly on α and β and the pK_a of the catalyst. A crucial factor in whether or not the catalysis is effective is the ionization state of the catalyst under the reaction conditions. An acid of pK_a 5 is a much better general acid catalyst than one of pK_a 7, but at pH 7 only 1% of an acid of pK_a 5 is in the active acid form whilst the remaining 99% is ionized. An acid of pK_a 7 is only 50% ionized at pH 7 and still 50% active. It is seen from Table 2.2 that for $\alpha = 0·85$ or less, an acid of pK_a 7 is a better catalyst at pH 7 than is an acid of pK_a 5. Similarly, a base of pK_a 7 is a more effective catalyst than one of pK_a 9 at pH 7 (at $\beta = 0·85$ or less) due to the inherently more reactive base being mainly protonated at the pH below its pK_a. The most effective acid–base catalysts at pH 7 are those of pK_a of

TABLE 2.2. *Influence of α and state of ionization on general acid catalysis*

α	$\dfrac{\text{Rate in 1 M solution of acid}}{\text{Rate in water}}$			$\dfrac{\text{Rate in 1 M solution at pH 7}^a}{\text{Rate in water}}$		
	$pK_a = 5$	7	9	5	7	9
0	1	1	1	1	1	1
0·3	31	8·6	2·9	1·3	4·8	2·9
0·5	$4·3 \times 10^3$	427	44	42	214	43·2
0·7	6×10^5	$2·4 \times 10^4$	951	6×10^3	$1·2 \times 10^4$	940
0·85	$2·5 \times 10^7$	$4·9 \times 10^5$	$9·7 \times 10^3$	$2·4 \times 10^5$	$2·4 \times 10^5$	$9·7 \times 10^3$
1	10^9	10^7	10^5	10^7	5×10^6	$9·9 \times 10^4$

[a] The rate of a 1 M solution of both the acid and base forms compared with the uncatalysed water reaction. It should be noted that the proton becomes an efficient catalyst at higher values of α. For $\alpha = 0·3$, 0·5, 0·7, 0·85, and 1·0, the reaction rate increases at pH 7 by factors of 1·0003, 2, 3×10^3, $1·3 \times 10^6$, and $5·5 \times 10^8$ respectively, thus swamping out the catalysis by other acids at the higher values

about 7. This accounts for the widespread involvement of histidine, with an imidazole pK_a of 6–7, in enzyme catalysis.

3. Intramolecular catalysis: the 'effective concentration' of a group on an enzyme

Acid–base catalysis is seen to be an effective way of catalysing reactions. We would now like to know the contribution of this to enzyme catalysis, but there is a fundamental problem in directly applying the results of the last section to an enzyme. The crux of the matter is that the rate constants for the solution catalysis are second order, the rate increasing with increasing concentration of catalyst, whilst reactions in an enzyme–substrate complex are first order, the acids and bases being an integral part of the molecule. So what is the concentration of the acid or base that is to be used in the calculations? The experimental approach is to synthesize model compounds with the catalytic group part of the substrate molecule, and to compare the reaction rates with the corresponding intermolecular reactions.

A typical example of an *intramolecularly catalysed* reaction is the hydrolysis of aspirin.[12] The hydrolysis of the ester bond is catalysed by intramolecular general base catalysis. Comparison with the uncatalysed hydrolysis rate of similar compounds gives a rate enhancement of some

100-fold due to the catalysis.[13] This may be extrapolated to a figure of 5000-fold if the pK_a of the base is 7 rather than the value of 3·7 in aspirin.

The intermolecular general base catalysis of the hydrolysis may also be measured. Comparing the rate constants for this with the intramolecular reaction shows that a 13 M solution of an external base is required to give the same first-order rate as the intramolecular reaction.[12] The 'effective concentration' of the carboxylate ion in aspirin is therefore 13 M. This is a typical value for intramolecular general acid–base catalysis.

The effective concentrations of nucleophiles in intramolecular reactions are often far higher than this. The examples that follow are for 'unstrained' systems. The chemist can synthesize compounds that are strained. The relief of strain in the reaction then gives a large rate enhancement. In the succinate and aspirin derivatives below, the attacking nucleophile can rotate away from the ester bond to relieve any strain.

The observed rate enhancements are due entirely to the high effective concentration of the neighbouring group.

(a) Rates of acyl transfer in succinates[14]

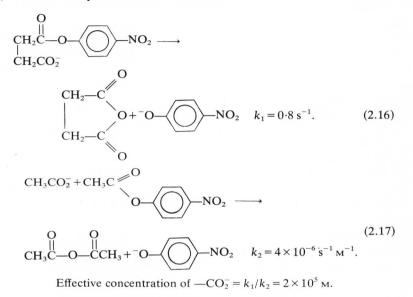

$$k_1 = 0.8 \text{ s}^{-1}. \quad (2.16)$$

$$(2.17)$$

$$k_2 = 4 \times 10^{-6} \text{ s}^{-1} \text{ M}^{-1}.$$

Effective concentration of $-CO_2^- = k_1/k_2 = 2 \times 10^5$ M.

(b) Rate of acyl transfer in aspirin derivatives[15]

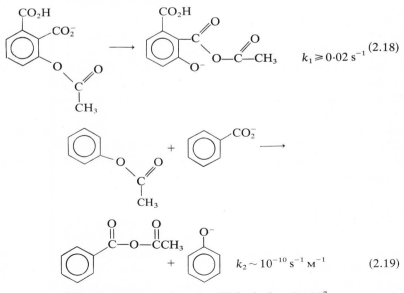

$$k_1 \geqslant 0.02 \text{ s}^{-1} \quad (2.18)$$

$$k_2 \sim 10^{-10} \text{ s}^{-1} \text{ M}^{-1} \quad (2.19)$$

Effective concentration of $-CO_2^- = k_1/k_2 > 2 \times 10^7$ M.

(c) Equilibria for acyl transfer in succinates[16,17]

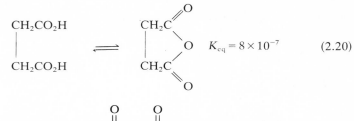

$$2CH_3CO_2H \rightleftharpoons CH_3\overset{O}{\overset{||}{C}}-O-\overset{O}{\overset{||}{C}}CH_3 \qquad K_{eq} = 3 \times 10^{-12}\,\text{M} \qquad (2.21)$$

Effective concentration of $-CO_2H = 3 \times 10^5$ M.

These examples show that enormous rate enhancements come from intramolecular nucleophilic catalysis.

4. Entropy: the theoretical basis of intramolecular catalysis and effective concentration[18,19]

The high effective concentration of intramolecular groups is one of the most important reasons for the efficiency of enzyme catalysis. This can be explained theoretically using transition state theory and examining the entropy term in the rate eqn (2.7). It will be seen that effective concentrations may be calculated by substituting certain entropy contributions into the $\exp(\Delta S^{\ddagger}/R)$ term of eqn (2.7).

a. Meaning of entropy

The most naive explanation of entropy, much frowned upon by purists but adequate for this discussion, is that entropy is a measure of the degree of randomness or disorder of a system. The more disordered the system, the more it is favoured and the higher its entropy. Entropy is similarly associated with the spatial freedom of atoms and molecules.

The catalytic advantage of an intramolecular reaction over its intermolecular counterpart is due to entropy. The intermolecular reaction involves two or more molecules associating to form one. This leads to an increase in 'order' and a consequent loss of entropy. An 'effective concentration' may be calculated from the entropy loss.

b. Magnitude of entropy

The entropy of a molecule is composed of the sum of its translational, rotational, and internal entropies. The translational and rotational entropies may be precisely calculated for the molecule in the gas phase from its mass and geometry. The entropy of the vibrations may be calculated from their frequencies, and the entropy of the internal rotations from the energy barriers to rotation.

TABLE 2.3 *Entropy of translation, rotation, and vibration* (298 K)

Motion	Entropy	
	(J/deg/mol)	(cal/deg/mol)
3 degrees of *translational freedom* for molecular weight = 20–200 (Standard state = 1 M)	120–150	29–36
3 degrees of *rotational freedom*		
Water	44	10·5
n-Propane	90	21·5
endo-Dicyclopentadiene	114	27·2
Internal rotation	13–21	3–5
Vibrations cm^{-1}		
1000	0·4	0·1
800	0·8	0·2
400	4·2	1·0
200	9·2	2·2
100	14·2	3·4

From M. I. Page and W. P. Jencks, *Proc. natn. Acad. Sci. U.S.A.* **68**, 1678 (1971)

The translational entropy is high, about 120 J/deg/mol (30 cal/deg/mol) for a 1 M solution of a small molecule. This is equivalent to about 40 kJ/mol (9 kcal/mol) at 25° (298 K). It is proportional to the volume occupied by the molecule; the smaller the volume, the more it is restricted, and the lower the entropy. Similarly the entropy decreases with increasing concentration since the average volume occupied by a molecule is inversely proportional to its concentration. It is important to note that the dependence on mass is low (Table 2.3). A tenfold increase in mass on going from a molecular weight of 20 to 200 leads to only a small increase in translational entropy.

The rotational entropy is again high, up to 120 J/deg/mol (30 cal/deg/mol) for a large organic molecule. It also increases only slowly with increasing mass, but is independent of concentration.

Stiff vibrations, as found in most covalent bonds, make very low individual contributions to the entropy. Low frequency vibrations, where the atoms are less constrained, can contribute a few entropy units. Internal rotations have entropies in the range of 13–21 J/deg/mol (3–5 cal/deg/mol) (Table 2.3).

c. The loss of entropy on two molecules condensing to form one
The combination of two molecules to form one leads to the loss of one set of rotational and translational entropies. The rotational and translational entropies of the adduct of the two molecules are only slightly larger than those of one of the original molecules since these entropies increase only slightly with size (Table 2.3). The entropy loss is of the order of

190 J/deg/mol (45 cal/deg/mol) or 55–9 kJ/mol (13–14 kcal/mol) at 25°. This may be offset somewhat by an increase in internal entropy due to new modes of internal rotation and vibration.

This loss is for a standard state of 1 M. If the solutions are more dilute the loss will be correspondingly greater since the translational entropy is concentration dependent.

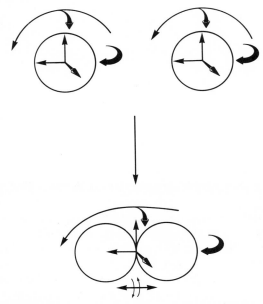

FIG. 2.5. A free molecule has three degrees of translational entropy and three degrees of overall rotational entropy. When two molecules condense to form one, the resulting adduct has only three degrees of translational and three degrees of overall rotational entropy, a loss of three degrees of each. There is a compensating gain of internal vibrational and rotational entropy which partly offsets this loss

d. Entropic advantage of a unimolecular over a bimolecular reaction
Let us compare the reaction of two molecules A and B combining together to form a third, with its intramolecular counterpart.

$$A + B \rightarrow AB^{\ddagger} \rightarrow AB \qquad (2.22)$$

$$A \smile B \rightarrow AB^{\ddagger} \rightarrow AB \qquad (2.23)$$

The formation of the transition state $AB^{\ddagger}$ leads to the loss of translational and rotational entropy as described above. The intramolecular cyclization in eqn (2.23) involves the loss of only some entropy of internal rotation. However, the entropy loss on the formation of $AB^{\ddagger}$ is compensated slightly by some increased internal rotation.

Depending on the relative gains and losses in internal rotation, the intramolecular reaction is favoured entropically by up to 190 J/deg/mol (45 cal/deg/mol) or 55–59 kJ/mol (13–14 kcal/mol) at 25°. This entropy advantage can be considered as equivalent to the effective concentration of the neighbouring group. Substituting 190 J/deg/mol (45 cal/deg/mol) into the $\exp(\Delta S^{\ddagger}/R)$ term of eqn (2.7) gives a value of 6×10^9 M for the maximum effective concentration of a neighbouring group.

The loss of internal rotation lowers the effective concentration quite considerably. In the succinate case there are three internal rotations lost. Page and Jencks suggest that the loss of the rotation about the methylene

group lowers the entropy by about 13 kJ/deg/mol (3 kcal/deg/mol), and that about each methylene carboxyl bond by 25 kJ/deg/mol (6 kcal/deg/mol). This is equivalent to a factor of 2×10^3. If the free rotations in the succinate compounds are 'frozen out', the neighbouring carboxyl group would have an effective concentration of about 5×10^8 M. This may be increased by a further factor of 10 allowing for the unfavourable energy of eclipsing the methylene hydrogens when forming the five-membered ring.

The theoretical basis of these calculations is completely valid for the gas phase. However, it was not generally realized before the work of Page and Jencks that reactions in solution involve similar entropy changes. Before this, the effective concentration of a neighbouring group was thought to be up to 55 M, the concentration of water in water. This is the figure that is normally found for intramolecular general base catalysis.

e. Dependence of effective concentration on 'tightness' of transition state
The lower effective concentrations found in intramolecular base catalysis are due to these reactions having loose transition states. In nucleophilic reactions, the nucleophile and electrophile are fairly rigidly aligned so that there is a large entropy loss. In general base or acid catalysis, there is considerable spatial freedom in the transition state. The position of the catalyst is not as closely defined as in nucleophilic catalysis. There is consequently a smaller loss in entropy in general base catalysis so that the intramolecular reactions are not favoured as much as their nucleophilic counterparts.

Koshland's original treatment of the problem of calculating effective concentrations was to consider that the concentration of an intramolecular group is approximately the same as that of water in aqueous solution, since a molecule in solution is completely surrounded by water.[20] This gives an upper limit of 55 M for effective concentration, equivalent to 34 J/deg/mol (8 cal/deg/mol) of entropy. This figure does represent the probability of two molecules being next to each other in solution. But, as soon as the two molecules are tightly linked, there is a large loss of entropy. A loose transition state may, perhaps, be considered as two molecules in close juxtaposition but retaining considerable entropic freedom.

In summary, one of the most important factors in enzyme catalysis is entropy. Catalysed reactions in solution are slow because the bringing together of the catalysts and substrate involves a considerable loss of entropy. Enzymic reactions take place in the confines of the enzyme–substrate complex. The catalytic groups are part of the same molecule as the substrate so there is no loss of translational or rotational entropy in the transition state. One way of looking at this is that the catalytic groups have very high effective concentrations compared with bimolecular reactions in solution. This gain in entropy is 'paid for' by the enzyme–substrate binding energy; the rotational and translational entropies of the substrate are lost on formation of the enzyme–substrate complex and not during the chemical steps. (The loss of entropy on forming the enzyme–substrate complex increases its dissociation constant).

5. 'Orbital steering'[21]

Attempts have been made to account for the rate enhancements in intramolecular catalysis on the basis of an effective concentration of 55 M combined with the requirement of very precise alignment of the electronic orbitals of the reacting atoms—'orbital steering'. Although this treatment does have the merit of emphasizing the importance of correct orientation in the enzyme-substrate complex, it overestimates its importance because, as we now know, the value of 55 M is a severe underestimate of the contribution of translational entropy to effective concentration. The consensus of opinion is that while there are requirements for the satisfactory overlap of orbitals in the transition state, these amount to an accuracy of only 10° or so.[22,23] The distortion of even a fully formed carbon–carbon bond by 10° causes a strain of only 11 kJ/mol (2·7 kcal/mol). A distortion of 5° costs only 2·8 kJ/mol (0·68 kcal/mol).[19]

6. Electrostatic catalysis

Chemical studies on model compounds do not show large effects due to electrostatic catalysis. This has led some chemists to reject the idea of electrostatic catalysis, but this is based on a misunderstanding of the electrostatic forces involved. The electrostatic interaction energy between two point charges e_1 and e_2 separated by a distance r in a medium of dielectric constant D is given by

$$E = e_1 e_2 / Dr. \qquad (2.24)$$

For a proton and an electron separated by $3 \cdot 3$ Å *in vacuo* this gives -418 kJ/mol (-100 kcal/mol). But in water of dielectric constant 79, this drops to $-5 \cdot 4$ kJ/mol ($-1 \cdot 3$ kcal/mol).

It should be noted that the water does not have to be inserted between the two charges to lower the interaction energy. This may be illustrated from the pK_as of the two following amine bases in water:

$$H_2NNH_2 \underset{H^+}{\overset{pK_a=8}{\rightleftharpoons}} H_2NNH_3^+ \underset{H^+}{\overset{pK_a=-1}{\rightleftharpoons}} {}^+H_3NNH_3^+ \qquad (2.25)$$

$$\text{N}\underset{}{\text{N}} \underset{H^+}{\overset{pK_a=8.8}{\rightleftharpoons}} \text{N}\underset{}{\text{NH}^+} \underset{H^+}{\overset{pK_a=3}{\rightleftharpoons}} {}^+\text{HN}\underset{}{\text{NH}^+} \qquad (2.26)$$

In the protonation of hydrazine (2.25), the juxtaposition of the two positive charges in the dication should destabilize it by 920 kJ (220 kcal/mol) according to two positive charges separated by $1 \cdot 5$ Å *in vacuo*. Instead, the two pK_as differ by only 9 units, a difference of only $50 \cdot 2$ kJ/mol (12 kcal/mol). In eqn (2.26), the two nitrogen atoms of triethylenediamine are separated by $2 \cdot 6$ Å. The two positive charges in the dication would be expected to have an unfavourable interaction energy of 546 kJ/mol (130 kcal/mol) *in vacuo*, but the second pK_a is perturbed by only $33 \cdot 4$ kJ (8 kcal). In both cases there appears to be an effective dielectric constant of about 17 between the two nitrogen atoms. This is due to the positively charged ions polarizing the solvent and inducing dipoles. The electrostatic field from these dipoles and any counter ions partially neutralize the positive field from the cations. For this reason, the surrounding of ions by a dielectric without interposing it between them lowers the interaction energy between the ions. Model studies in water severely underestimate the importance of electrostatic catalysis in proteins.

Electrostatic interactions are much stronger in organic solvents than in water due to the lower dielectric constants. This has been used by synthetic chemists to increase reaction rates. The same is true of enzymes. In Perutz's words 'The non-polar interiors of enzymes provide the living cell with the equivalent of the organic solvent used by the chemists'.[24]

It has recently been calculated that the carbonium ion in the lysozyme reaction is stabilized by $37 \cdot 6$ kJ (9 kcal) by the interaction with carboxylate ion of Asp-52 (relative to the un-ionized carboxyl).[25] This is equivalent to a rate enhancement of 4×10^6.

7. Metal ion catalysis[26]

a. Electrophilic catalysis

One obvious role for metals in metalloenzymes is to function as electrophilic catalysts, stabilizing the negative charges that are formed. In carboxypeptidase (Chapter 12), the carbonyl oxygen of the amide substrate is coordinated to the Zn^{2+} of the enzyme. This polarizes the amide to nucleophilic attack and strongly stabilizes the tetrahedral intermediate.

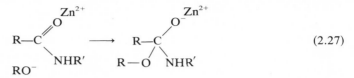

$$(2.27)$$

This has been mimicked in model compounds to give rate enhancements of 10^4–10^6.[27,28] For example, the base catalysed hydrolysis of glycine ethyl ester, is increased 2×10^6-fold when coordinated to $(\text{ethylenediamine})_2\text{Co}^{3+}$.

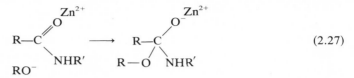

$$(2.28)$$

$$H_2NCH_2CO_2Et \xrightarrow[\text{HO}^-]{k_2 = 0.6\ s^{-1}M^{-1}}.$$

$$(2.29)$$

b. A source of hydroxyl ions at neutral pH

Another very important result that has recently come from kinetic studies in inorganic chemistry is that metal-bound hydroxyl ions are potent nucleophiles.[29–31] The Co-bound water molecule in (2.30) ionizes with a

$$(NH_3)_5Co^{3+}OH_2 \rightleftharpoons (NH_3)_5Co^{2+}OH + H^+$$

$$(2.30)$$

pK_a of 6·6. This is 9 units below the pK_a of free H_2O, yet the Co-bound hydroxyl group is only 40 times less reactive than the free hydroxide ion in catalysing the hydration of carbon dioxide.[29] This insensitivity of the high reactivity to pK_a is quite general and independent of the metal involved.[31] Thus, metal-bound water molecules provide a source of nucleophilic hydroxyl groups at neutral pH. Just as a base of pK_a 7 is most effective in general base catalysis (Section B2b), so a metal-bound water molecule of pK_a 7 is most effective for nucleophilic attack because it combines a high reactivity with a high fraction in the correct ionization state.

This is of relevance to the mechanism of carbonic anhydrase. This enzyme, which catalyses the hydration of CO_2, has at its active site a Zn^{2+} ion ligated to the imidazole rings of three of the histidines. The classical mechanism for the reaction is that the fourth ligand is a water molecule which ionizes with a pK_a of 7.[32] The reactive species is considered to be the zinc-bound hydroxyl. Chemical studies show that zinc-bound hydroxyls are no exception to the rule of high reactivity. The H_2O in (2.31) ionizes with a pK_a of 8·7 and catalyses the hydration of carbon dioxide and acetaldehyde.[33]

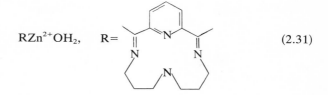

$$RZn^{2+}OH_2, \quad R= \qquad\qquad\qquad (2.31)$$

The carbonic anhydrase mechanism probably involves the following step:

$$E-Zn^{2+}-O\overset{H}{\underset{}{}}C\overset{O}{\underset{O}{}} \rightleftharpoons E-Zn^{2+}+HCO_3^-. \qquad (2.32)$$

The combination of a metal-bound hydroxyl group and an intramolecular reaction has provided one of the largest rate enhancements found in a strain-free system. The complex of glycylglycine with ethylenediamine)$_2$Co^{3+} and a *cis*-hydroxyl (2.33) hydrolyses at pH 7 nearly 10^{10} times faster than the free glycylglycine.[34]

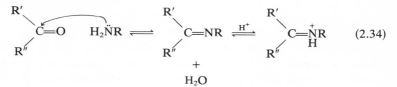

$$NH_2CH_2CONHCH_2CO_2^-$$

$$Co^{2+} \qquad\qquad \overset{k_1=5\cdot5\times10^{-3}\,s^{-1}}{\underset{pH\,7,\,25°}{\longrightarrow}} \qquad (2.33)$$

$$OH$$

There are more complex examples of metal ion catalysis. Cobalt in vitamin B_{12} reactions forms covalent bonds with carbons of substrates.[35,36] Metals can also act as electron conduits in redox reactions. For example, in cytochrome c the iron in the haem is reversibly oxidized and reduced.

C. Covalent catalysis

1. Electrophilic catalysis by Schiff base formation[37]

A good example of how transient chemical modification can activate a substrate is Schiff base formation from the condensation of an amine with a carbonyl compound.

$$\overset{R'}{\underset{R''}{}}C=O \quad H_2\ddot{N}R \rightleftharpoons \overset{R'}{\underset{R''}{}}C=NR \overset{H^+}{\rightleftharpoons} \overset{R'}{\underset{R''}{}}C=\overset{+}{\underset{H}{N}}R \qquad (2.34)$$

$$+$$

$$H_2O$$

The Schiff base may be protonated at neutral pH. This acts as an *electron sink* to stabilize the formation of a negative charge on one of the α-carbons:

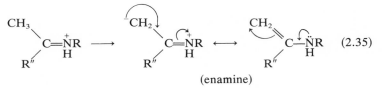

$$(2.35)$$

(enamine)

After tautomerization to form the enamine, the methylene carbon is activated as a nucleophile. Another consequence of Schiff base formation is that the carbonyl carbon is activated towards nucleophilic attack because of the strong electron withdrawal of the protonated nitrogen.

a. Acetoacetate decarboxylase[38,39]

This enzyme catalyses the decarboxylation of acetoacetate to acetone and carbon dioxide. The non-enzymic reaction involves the expulsion of a highly basic enolate ion at neutral pH (2.36), but the enzymic reaction

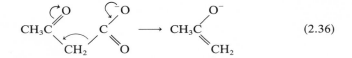

$$(2.36)$$

circumvents this by the prior formation of a Schiff base with a lysine residue. The protonated imine is then readily expelled. This process may be mimicked in solution using aniline as a catalyst.

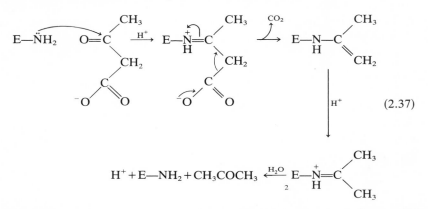

$$(2.37)$$

The evidence for the intermediate is that the enzyme is irreversibly inhibited when sodium borohydride is added to the complex with the substrate. Borohydride is known to reduce Schiff bases and the hydrolysate of the inhibited protein is found to contain isopropyllysine.

$$E\!-\!NH_2 + CH_3COCH_2CO_2^- \longrightarrow E\!-\!\overset{+}{N}H\!=\!C\!\!\begin{array}{c}CH_3\\ \\CH_3\end{array} \quad \xrightarrow{\;BH_4^-\;} \quad E\!-\!\overset{+}{N}H_2C\!\!\begin{array}{c}CH_3\\ \\CH_3\\ H\end{array}$$

$$\text{(2.38)}$$

$$CH_3\!-\!\underset{H}{\overset{CH_3}{C}}\!-\!\overset{+}{N}H_2(CH_2)_4\underset{\overset{+}{N}H_3}{\overset{CO_2^-}{CH}}$$

The carbon in the Schiff base is activated to the attack of an H^- ion from the borohydride.

$$-\underset{H}{\overset{+}{N}}\!=\!C\!\!\overset{\frown}{}\quad H\!-\!\bar{B}H_3 \qquad \text{(2.39)}$$

b. Aldolase[40,41]

The aldol condensation and the reverse cleavage reaction catalysed by this enzyme both involve a Schiff base. The cleavage reaction is similar to the acetoacetate decarboxylase mechanism, the protonated imine being expelled. The condensation reaction illustrates the other function of a Schiff base, the activation of carbon via an enamine:

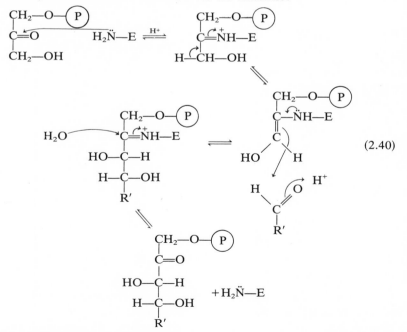

$$\text{(2.40)}$$

The intermediate may be trapped as before.

2. Pyridoxal phosphate—electrophilic catalysis[37,42]

The principles of the above reactions form the basis of a series of important metabolic interconversions involving the coenzyme pyridoxal phosphate (2.41). This condenses with amino acids to form a Schiff base (2.42).

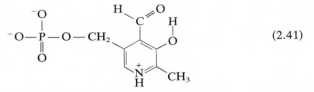

(2.41)

The pyridine ring in the Schiff base acts as an electron sink which very effectively stabilizes a negative charge.

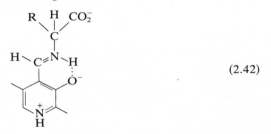

(2.42)

Each one of the groups around the chiral carbon of the amino acid may be cleaved, forming an anion which is stabilized by the Schiff base and pyridine ring.

a. Removal of the α-hydrogen

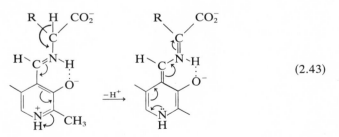

(2.43)

The removal of the α-hydrogen gives a key intermediate which may react in several different ways.

(i) *Racemization.* Addition of the proton back to the amino acid will lead to racemization unless done stereospecifically.

(ii) *Transamination.* Addition of a proton to the carbonyl carbon of the pyridoxal leads to a compound which is the Schiff base of an α-keto acid and pyridoxamine. Hydrolysis of the Schiff base gives the α-keto acid and pyridoxamine, which may react with a different α-keto acid to reverse the sequence.

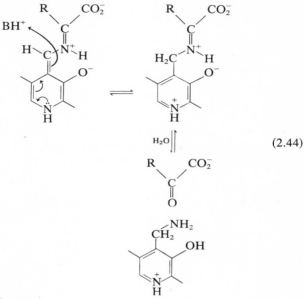

$$(2.44)$$

The overall reaction is

$$R'CH(NH_3^+)CO_2^- + R''COCO_2^- \rightleftharpoons R'COCO_2^- + R''CH(NH_3^+)CO_2^- \quad (2.45)$$

(iii) *β-decarboxylation.* When the amino acid is aspartate, the second compound in (2.44) is analogous to the Schiff base in the acetoacetate decarboxylase reaction and may readily decarboxylate.

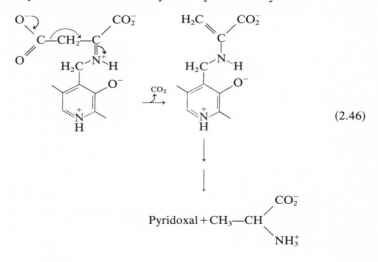

$$(2.46)$$

(iv) *Interconversion of side chains.* When, in (2.47), RX— is a good
leaving group, it may be expelled thus:

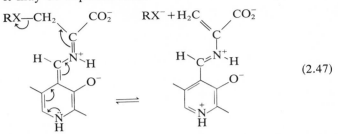

(2.47)

RX may be a thiol, hydroxyl, or indole group. In this way, serine,
threonine, cysteine, tryptophan, cystathionine, and serine and threonine
phosphates may be interconverted or degraded.

b. α-Decarboxylation
The 'electron sink' allows facile decarboxylation,

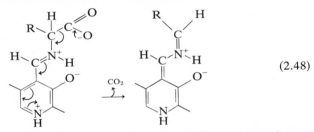

(2.48)

The decarboxylated adduct will add a proton to either the amino acid
carbonyl carbon, and then hydrolyse to give the amine and pyridoxal, or
will add the proton to the pyridoxal carbonyl carbon and then hydrolyse
to give the aldehyde and pyridoxamine:

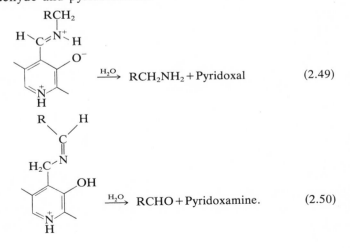

$\xrightarrow{\text{H}_2\text{O}}$ RCH$_2$NH$_2$ + Pyridoxal   (2.49)

$\xrightarrow{\text{H}_2\text{O}}$ RCHO + Pyridoxamine.   (2.50)

3. Thiamine pyrophosphate—electrophilic catalysis

Thiamine pyrophosphate (2.51) is another coenzyme that covalently bonds to a substrate and stabilizes a negative charge. The positive charge

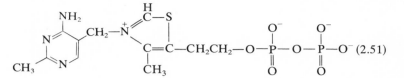

on the nitrogen promotes the ionization of the C-2 carbon by electrostatic stabilization. The ionized carbon is a potent nucleophile.

$$\text{H}-\text{C} \overset{+|}{\underset{\text{S}}{\text{N}}} \quad \rightleftharpoons \quad {}^{-}\text{C} \overset{+|}{\underset{\text{S}}{\text{N}}} + \text{H}^{+} \qquad (2.52)$$

The nitrogen atom can also stabilize by delocalization a negative charge on the adduct of thiamine with many compounds, for example, as in *hydroxyethylthiamine pyrophosphate* (2.53), a form in which much of the coenzyme is found *in vivo*.

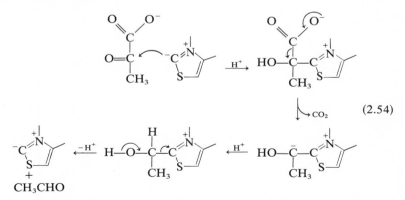

The combination of these reactions allows the decarboxylation of pyruvate by the route (2.54).

Other carbon–carbon bonds adjacent to a carbonyl group may be cleaved in the same manner.

The hydroxyethylthiamine pyrophosphates are potent nucleophiles and may add to carbonyl compounds to form carbon–carbon bonds. A good illustration of carbon–carbon bond making and breaking occurs in the reactions of transketolase. The enzyme contains tightly bound thiamine pyrophosphate and shuttles a dihydroxyethyl group between D-xylulose-5-phosphate and D-ribose-5-phosphate to form D-sedoheptulose-7-phosphate and D-glyceraldehyde-3-phosphate (2.55 and 2.56).

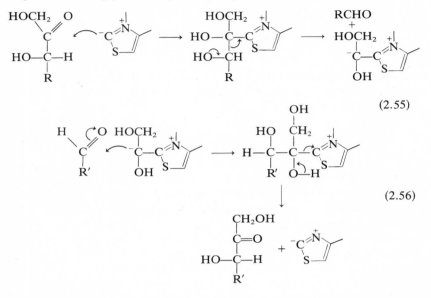

$$(2.55)$$

$$(2.56)$$

Hydroxyethylthiamine pyrophosphate is also nucleophilic towards a thiol of reduced lipoic acid. A hemithioacetal is formed which decomposes to give a thioester.

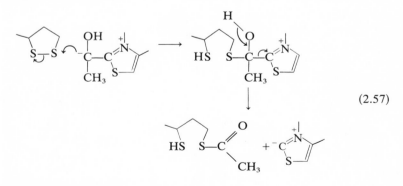

$$(2.57)$$

4. Nucleophilic catalysis

The most common nucleophilic groups in enzymes that are functional in catalysis are the serine hydroxyl, occurring in the serine proteases, cholinesterases, esterases, lipases, and acid and alkaline phosphatases,

TABLE 2.4. *Nucleophilic groups in enzymes*

Nucleophile	Enzyme	Intermediate
—OH (Serine)	Serine proteases	Acylenzyme
	Acid and alkaline phosphatases, phosphoglucomutase	Phosphorylenzyme
OH^- (Zinc-Bound)	Carbonic anhydrase	—
—SH (Cysteine)	Thiol proteases, glyceraldehyde-3-phosphate dehydrogenase	Acylenzyme
—CO_2^- (Aspartate)	Pepsin	Acylenzyme
	ATPase (K^+/Na^+, Ca^{2+})	Phosphorylenzyme
—NH_2 (Lysine)	Acetoacetate decarboxylase, aldolase, pyridoxal enzymes	Schiff Base
Imidazole (Histidine)	Phosphoglycerate mutase, succinyl-CoA synthetase, nucleoside diphosphokinase, histone phosphokinase	Phosphorylenzyme
—OH (Tyrosine)	Glutamine synthetase	Adenylenzyme

and the cysteine thiol, which occurs in the thiol proteases (papain, ficin, and bromelain), glyceraldehyde-3-phosphate dehydrogenase etc. The imidazole of histidine usually functions as an acid–base catalyst, enhancing the nucleophilicity of hydroxyl and thiol groups, but it sometimes acts as a nucleophile with the phosphoryl group in phosphate transfer.

The hydrolysis of peptides by these proteases represents classic nucleophilic catalysis. The relatively inert peptide is converted to the far more reactive ester or thioester acylenzyme, which is rapidly hydrolysed. The use of the serine hydroxyl rather than the direct attack of a water molecule on the substrate is favoured in several ways: alcohols are often better nucleophiles than the water molecule in both general-base-catalysed and direct nucleophilic attack; the serine reaction is intramolecular and hence favoured entropically; the arrangement of groups is more 'rigid' and defined for the serine hydroxyl compared with a bound water molecule.

D. Structure–reactivity relationships

One of the most fruitful approaches in studying organic reaction mechanisms has been that of noting the change in reactivity with changing structure of the reagents. These studies have given considerable information on the electronic structures of transition states and the features that determine reactivity and whether a group is a good nucleophile or leaving group. Structure–reactivity studies with enzymes tend to measure the effects of changes in the structure of the substrate on its interaction with the enzyme rather than the effects of these on the electronic changes in the transition state. In general, it is difficult to obtain useful data on the electronic requirements of enzymic reactions from structure–reactivity studies because of the restricted range of changes that can be made to the substrate and because of the inductive effects of substituents often being obscured by the effects on binding. However, the lessons learned from the chemical studies have been invaluable to our understanding of the mechanisms of enzymic reactions.

1. Nucleophilic attack at the carbonyl group

Structure–reactivity studies have been used to give information concerning the charge distribution in the transition state by noting the effects of electron withdrawing and electron donating substituents on the reaction rate. For example, it has been found that the rate of nucleophilic attack on esters increases with (a) electron withdrawal in the acyl portion ($CHCl_2CO_2Et$ is far more reactive than CH_3CO_2Et); (b) electron withdrawal in the leaving group (p-nitrophenyl acetate is more reactive than phenyl acetate); and (c) increasing basic strength of the nucleophile, that is, electron donation in the nucleophile (the hydroxide ion is far more reactive than the acetate ion). Using the idea from transition state theory that the reaction rate depends on the energy difference between the transition state and the ground state, we can deduce that the reaction involves an increase of negative charge on the substrate since the reaction rate is increased by electron withdrawal, and a decrease in charge on the nucleophile since the rate is increased by electron donation (i.e. electron repulsion). This is consistent with either mechanism (2.58), where the

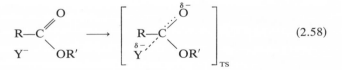

$$\text{(2.58)}$$

rate-determining step is the formation of the tetrahedral intermediate, or (2.59), where the rate-determining step is the breakdown.

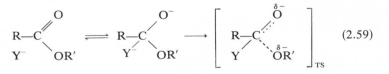

$$\text{(2.59)}$$

a. Linear free energy relationships and the Brönsted and Hammett equations

A quantitative assessment of the sensitivity of the reaction to electron withdrawal and donation in the attacking nucleophile may be made by measuring the second-order rate constants for the attack of a series of nucleophiles on a particular ester. A plot of the logarithms of the rate constants against the pK_as of the nucleophiles may be made in the same way as for general base catalysis. Generally, when the measurements are restricted to nucleophiles of a similar chemical nature and to a not too wide range of pK_a, a straight line relationship is found. The slope of the line is termed β as for general base catalysis.

These linear relationships between the logarithm of a rate constant and a pK_a are known as *linear free energy relationships* since the logarithm of a rate constant is proportional to its Gibbs energy of activation and the logarithm of an equilibrium constant (such as a pK_a) is proportional to the Gibbs energy change of a reaction (Gibbs energy was formerly called free energy). The relationship between the nucleophilicity of a nucleophile and its basic strength shows that the Gibbs energy of activation of bond formation with the carbonyl carbon is proportional to the Gibbs energy of transfer of a proton to the nucleophile.

A similar Brönsted relationship is found when the rate constants for the attack of a particular nucleophile on a series of esters with differing leaving groups are plotted against the pK_a of the leaving group. An alternative way of plotting the data for aromatic compounds is to use the *Hammett* equation and the Hammett substituent constants. The substituent constants, which measure the electron donating- or withdrawing-power of a substituent in a benzene ring, are derived empirically from the pK_as of substituted benzoic acids using the equation

$$(pK_a)_X = (pK_a)_0 - \sigma_X, \qquad (2.60)$$

where X is a substituent in the meta or para position of benzoic acid, $(pK_a)_X$ the pK_a of the substituted acid, $(pK_a)_0$ that of the unsubstituted acid, and σ_X the substituent constant for X (different for the para and meta positions). It is found that other reactions of benzoic acid derivatives, such as the rate of hydrolysis of benzoate esters, and also the equilibria of other reactions involving the benzene ring, such as the ionization of phenols, follow similar Hammett equations except that a constant of proportionality ρ, the reaction constant, is invoked to measure the sensitivity of the reaction towards changes in σ. For example, for the ionization of phenols,

$$(pK_a)_X = (pK_a)_0 - \rho\sigma_X \qquad (2.61)$$

where $\rho = 2 \cdot 1$. And for the alkaline hydrolysis of phenyl acetates,

$$\log k_X = \log k_0 + \rho\sigma_X \qquad (2.62)$$

where $\rho = 0 \cdot 8$.

The Brönsted and Hammett plots are equivalent. In the Hammett treatment the logarithms of the rate constants for, say, the alkaline

TABLE 2.5. *Substituent constants*

Substituent	Aliphatic[a] σ_I	Aromatic[b] σ_m	σ_p
—NH$_2$	0·10	−0·16	−0·66 (−0·17)[c]
—NH$_3^+$	0·60	0·63	
—NHAc	0·28	0·21	0·00
—OH	0·25	0·121	−0·37 (−0·11)[c]
—OAc	0·39	0·39	0·31
—N(Me)$_3^+$	0·73	0·88	0·82
—CO$_2^-$	−0·17	−0·10	0·00
—CO$_2$H	0·34	0·37	0·45
—COCH$_3$	0·29	0·38	0·50 (0·87)[d]
—F	0·52	0·337	0·062
—Cl	0·47	0·373	0·227
—Br	0·45	0·391	0·232
—I	0·39	0·352	0·18
—NO$_2$	0·63	0·710	0·778 (1·24)[d]
—CN	0·58	0·56	0·66 (0·90)[d]
—CH$_3$	−0·05	−0·069	−0·17
—OCH$_3$	0·25	0·115	−0·268 (−0·11)[c]

[a] Based on $(\sigma_I)_X = [(pK_a)_{CH_3CO_2H} - (pK_a)_{X—CH_2CO_2H}]/3·95$
[b] Based on $\sigma = (pK_a)_{PhCO_2H} - (pK_a)_{X—PhCO_2H}$
[c] For use with compounds other than benzoic acids and carbonyl compounds
[d] For use with phenols, anilines, and thiophenols
From M. Charton, *J. org. Chem.* **29**, 1222 (1964); C. D. Ritchie and W. F. Sager, *Progress in Physical Organic Chemistry* **2**, 323 (1964)

hydrolysis of phenyl acetates are proportional to σ, and so are the pK_as of the parent phenols. Hence the logarithms of the rate constants for the hydrolysis are proportional to the pK_as, as is directly found from the Brönsted plot of logarithm of rate constant against pK_a.

Brönsted and Hammett relationships are also found for the attack of a particular nucleophile on an ester of a particular leaving group as the acyl portion is varied.

b. Interpretation of Brönsted β-values

The important parameter that is derived from the Brönsted plots is the β-value. The sign and magnitude of this are an indication of the charge developed in the transition state. Consider, for example, the attack of nucleophiles on esters. The β-value for the equilibrium constants for the transfer of acetyl groups between oxyanions and also tertiary amines is 1·6–1·7. The value is greater than 1·0 because the acetyl group is more electron-withdrawing than the proton (for which $\beta = 1·0$ by definition) and is more sensitive to the pK_a of the alcohol or amine. Now it is found that the value of β for the attack of tertiary amines on esters of very basic alcohols is $+1\cdot5$ for the variation of the nucleophile and $-1\cdot5$ for the variation of the alcohol.[43] This shows that the transition state for the reaction is very close to the structure of the products, that is, close to complete acyl transfer from the alcohol to the amine (2.63).

$$R_3N + CH_3C\begin{matrix} O \\ \diagup\diagdown \\ OEt \end{matrix} \longrightarrow \begin{bmatrix} & O \\ CH_3{-}C \diagdown \\ \overset{+}{R_3N} & \bar{O}Et \end{bmatrix}_{TS} \qquad (2.63)$$

At the other extreme, the reaction of basic nucleophiles with esters containing activated leaving groups exhibits β-values of only $+0\cdot1{-}0\cdot2$ for the variation of the nucleophile and $-0\cdot1{-}0\cdot2$ for the variation of leaving group pK_a. This shows that the transition state involves little bond making and breaking and is close to the starting materials (2.64).

$$RO^- + CH_3C\begin{matrix} O \\ \diagup\diagdown \\ OAr \end{matrix} \longrightarrow \begin{bmatrix} & O \\ CH_3C \diagdown \\ RO^- & OAr \end{bmatrix}_{TS} \qquad (2.64)$$

β is a measure of the charge formed in the transition state rather than the extent of bond formation. However, in (2.63) and (2.64) and other examples where there is no acid–base catalysis, charge and bond formation are linked, so β does also give a measure of the extent of bond formation. But, when there is also acid–base catalysis partly neutralizing the charges formed in the transition state, there is no relation between β and the extent of bond formation.[44]

2. Factors determining nucleophilicity and leaving group ability

The magnitude of general acid–base catalysis by oxygen and nitrogen bases depends only on their pK_as and is independent of their chemical nature (apart from an enhanced activity of oximes in general acid catalysis). Nucleophilic reactivity depends markedly on the nature of the reagents. These reactions may be divided into two broad classes, nucleophilic attack on *soft* electrophilic centres, and on *hard* centres.[45]

a. Nucleophilic reactions with the carbonyl, phosphoryl, sulphuryl, and other hard groups

The attack of a nucleophile on an amide, ester, or carbonyl carbon involves the formation of a 'real' chemical intermediate, and the valency of carbon is not extended beyond its normal value of 4. The attack on a phosphate ester is similar; a pentacovalent phosphate intermediate is formed. The transition state of the reaction involves the formation of a normal bond. This is a characteristic of hard centres. The dominant feature controlling nucleophilicity in these reactions is the basic strength of the nucleophile; the stronger the basic strength, the greater the nucleophilicity. There are, however, differences in reactivity between different classes of nucleophiles: amines and thiolate anions tend to be more nucleophilic than oxyanions.[46,47] Also, certain nucleophiles which have two electronegative atoms next to each other, such as NH_2OH, NH_2NH_2, $NH_2CONHNH_2$, HOO^-, and $MeOO^-$, are more reactive than expected from their pK_as. This is known as the α-effect. The relative reactivities of the different classes are often reflected in increased equilibrium constants for their addition to aldehydes and ketones.[48]

The ease of expulsion of a group depends on both its pK_a and state of protonation. Basically, a 'good leaving group' is one that is stable in solution. For example, the p-nitrophenolate ion is a good leaving group because it is weakly basic; the pK_a of p-nitrophenol is 7·0. The chemical cause of the stability of the ion is that the negative charge is delocalized around the aromatic ring and on to the nitro group (2.65). The ion is readily and directly expelled from a tetrahedral intermediate (2.66).

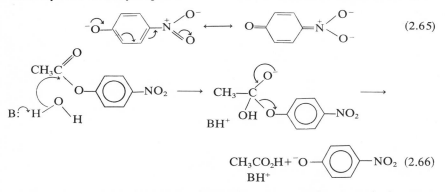

$$(2.65)$$

$$\text{CH}_3\text{CO}_2\text{H} + {}^-\text{O}\!\!-\!\!\!\langle\bigcirc\rangle\!\!-\!\!\text{NO}_2 \quad (2.66)$$
$$\text{BH}^+$$

In this class of leaving groups, the lower the basic strength the greater the ease of expulsion. Acetate (pK_a of acetic acid $= 4·76$) is a better leaving group than p-nitrophenol and phosphate (pK_a about 7), which are better leaving groups than OH^- ($pK_a = 15·8$).

Alcoholate ions are difficult to expel because they are strongly basic; the pK_as of simple alcohols are about 16. The expulsion of alcohols is aided by general acid catalysis (2.67).

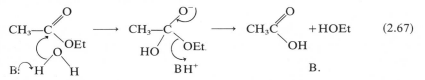

$$(2.67)$$

Nitrophenyl esters are often used as leaving groups in synthetic substrates for two reasons: (a) the nitrophenolate ion has a characteristic

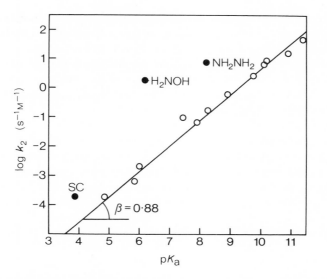

FIG. 2.6. Brönsted plot for the nucleophilic attack of primary and secondary amines on *p*-nitrophenyl acetate (W. P. Jencks and M. Gilchrist, *J. Amer. chem. Soc.* **90**, 2622 (1968)). Note that the '*α*'-effect nucleophiles, semicarbazide (SC), hydroxylamine, and hydrazine, are more reactive than expected from their pK_as

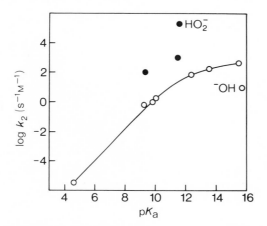

FIG. 2.7. Brönsted plot for the attack of oxyanion nucleophiles on *p*-nitrophenyl acetate. As in Fig. 2.6, the '*α*'-effect nucleophiles (●) are unusually reactive. Note how the linear plot breaks down with increasing pK_a for the more reactive nucleophiles. In general, the Brönsted relationships hold only over a limited range of pK_a in these reactions. The curvature is often not seen in practice because of the limited range of bases used

absorption at 400 nm and is thus easily assayed spectrophotometrically; (b) it is a very good leaving group so it forms a reactive substrate. Both these factors are caused by the delocalization shown in (2.65).

Amines have to be protonated to be expelled from a molecule since the amide ions, RNH^-, are far too unstable to be directly released into solution. (The exceptions to this rule are the highly activated derivatives such as 2,4-dinitroaniline.)

b. Nucleophilic reactions with saturated carbon

The attack of a nucleophile on saturated carbon, such as the bimolecular attack of a thiol on the methyl carbon of S-adenosylmethionine, involves a transition state in which 5 groups surround the normally tetravalent carbon (2.68). This is not a 'normal' bond with carbon and is peculiar to

$$\text{>S:} \quad \overset{\diagup}{\underset{\diagdown}{C}} \overset{+}{S} \diagup \longrightarrow \left[\text{>S} \cdots \overset{|}{\underset{\diagdown}{C}} \cdots \text{S<} \right]_{TS} \qquad (2.68)$$

the transition state. This reaction is typical of a 'soft' centre. Large, polarizable, atoms, such as sulphur and iodine (i.e. 'soft' ligands), react more rapidly in these reactions, whilst the small atoms of low polarizability, oxygen and nitrogen (i.e. 'hard' ligands), are less reactive. The dominant factor in nucleophilicity towards alkyl groups and other soft centres is polarizability. Within any particular class of compounds, increasing basic strength increases the nucleophilicity, but between classes, polarizability is all important (Table 2.6).

TABLE 2.6. *Nucleophilic reactivity towards saturated carbon*

Nucleophile	pK_a	Relative reactivity towards CH_3Br
H_2O	$-1 \cdot 74$	$1 \cdot 00$
NO_3^-	$-1 \cdot 3$	11
F^-	$3 \cdot 17$	100
$CH_3CO_2^-$	$4 \cdot 76$	525
Cl^-	$-7 \cdot 0$	$1 \cdot 1 \times 10^3$
C_5H_5N	$5 \cdot 17$	$4 \cdot 0 \times 10^3$
$HPO_4^=$	$7 \cdot 21$	$6 \cdot 3 \times 10^3$
Br^-	$-9 \cdot 0$	$7 \cdot 8 \times 10^3$
OH^-	$15 \cdot 74$	$1 \cdot 6 \times 10^4$
$C_6H_5NH_2$	$4 \cdot 62$	$3 \cdot 1 \times 10^4$
I^-	$-10 \cdot 0$	$1 \cdot 1 \times 10^5$
CN^-	$9 \cdot 40$	$1 \cdot 3 \times 10^5$
SH^-	$7 \cdot 00$	$1 \cdot 3 \times 10^5$

C. G. Swain and C. B. Scott, *J. Am. chem. Soc.* **75**, 141 (1953)

As with reactions at the carbonyl group, weakly basic leaving groups are more readily displaced than strongly basic ones.

c. Leaving group activation

It was seen in the last section that highly basic groups are not readily displaced from carbonyl compounds and from saturated carbon. An extreme example of this is the esterification of an alcohol by a carboxylate ion. This would require the formation of a tetrahedral intermediate with two negatively charged oxygens and the subsequent expulsion of $O^=$.

$$RCO_2^- + R'OH \; \xrightleftharpoons \; R-\underset{\substack{|\\OR'\\+H^+}}{\overset{\substack{O^-\\|}}{C}}-O^- \;\xrightarrow{\;\;\times\;\;}\; RCO_2R' + O^= \qquad (2.69)$$

In the esterification of tRNA with amino acids catalysed by the amino-acyl-tRNA synthetases, the amino acid is activated by forming an enzyme-bound mixed anhydride with AMP (2.70) in the same way that an organic chemist activates a carboxylic acid by forming an acyl chloride or mixed anhydride. (Note that in (2.70) the substrate is the magnesium

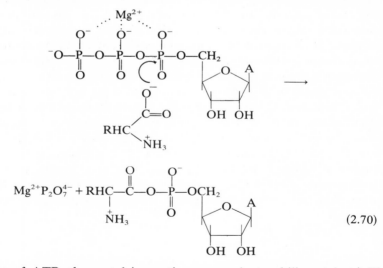

complex of ATP, the metal ion acting as an electrophilic catalyst.) The carbonyl group of the amino acid is activated by (2.70) because it is bound to a good leaving group, the phosphate of AMP having a pK_a of about 6 or 7. It may be readily attacked by one of the hydroxyl groups of the ribosyl ring of the terminal adenosine of tRNA.

Another example of leaving group activation is the utilization of S-adenosylmethionine rather than methionine in methylation reactions. A

relatively basic thiolate anion has to be expelled from methionine, whilst
the non-basic neutral sulphur is displaced from the activated derivative.

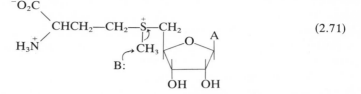

$$(2.71)$$

3. Application of linear free energy relationships to enzyme reactions

Although β-values have been successfully obtained for some reactions,
there are some theoretical difficulties. There is first the problem of
whether to examine the effects of substituents on k_{cat} or k_{cat}/K_M. The
difficulty with k_{cat}/K_M is that the binding energy terms come directly into
this, and so the electronic effects of a substituent may be masked by its
contribution to binding. On the other hand, k_{cat} is complicated in that all
the effects such as the accumulation of intermediates, non-productive
binding, strain, and induced fit alter this term (see Chapters 3 and 10).
There is at least one example in which the effect of substituents on
non-productive binding has been mistakenly attributed to an inductive
effect on the reaction rate.[49] In many ways k_{cat}/K_M is the safer parameter
to use since all the above artefacts that obscure k_{cat} cancel out in this
(Chapter 10). Nevertheless, some good structure–reactivity studies have
been performed with enzymes. Some earlier studies on the acylation of
chymotrypsin by non-specific substrates[50] and the deacylation of non-
specific acylenzymes[51] showed that the reactions are very sensitive to
electron withdrawal, indicating that there is the attack of a basic group on
the substrate. A more recent study on the hydrolysis of substituted
anilides by papain (Chapter 12B2) shows that there is a high negative
value of ρ, that is, the values of k_{cat} and k_{cat}/K_M are increased by electron
donating substituents in the aniline ring.[52] This is consistent with the
rate-determining breakdown of a tetrahedral intermediate in which the
aniline ring is protonated (2.72). These examples are relatively

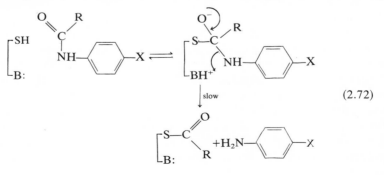

$$(2.72)$$

straightforward to interpret as the ρ-values are large. Small values are more difficult to interpret since they can be caused by a variety of different reasons; for example, a transition state in which there is very little bond making and breaking will be very insensitive to electron withdrawal, as will be a transition state in which extensive bond making and breaking are compensated by extensive proton transfer from acid–base catalysis.[44]

The reactions of yeast alcohol dehydrogenase have been analysed by more sophisticated techniques.[53] Multiple linear regression analysis has been applied to separate the results of both the electronic effects of substituents and the contributions of their hydrophobic binding energy (see Chapter 9A). A similar analysis has been applied to the reactions of acetylcholinesterase.[54] In both examples there is a feature that will be discussed at length in Chapter 10—binding energy is used to increase k_{cat}.

Although it is sometimes thought that structure–reactivity relationships are more useful in teaching us the principles of catalysis in simple model systems than in being directly applicable to enzymes, there is no doubt that these examples from enzymology have provided useful information and ideas.

E. Principle of microscopic reversibility or detailed balance

The principle of microscopic reversibility or detailed balance is used in thermodynamics to place limitations on the nature of transitions between different quantum or other states. It applies also to chemical and enzymic reactions, considering each chemical intermediate or conformation as a 'state'. The principle requires that the transitions between any two states take place with equal frequency in either direction at equilibrium.[55] That is, the process $A \rightarrow B$ is exactly balanced by $B \rightarrow A$ and so equilibrium cannot be maintained by a cyclic process, with the reaction being $A \rightarrow B$ in one direction and $B \rightarrow C \rightarrow A$ in the opposite. A useful way of restating the principle for reaction kinetics is that the reaction pathway for the reverse of a reaction at equilibrium is the exact opposite of the pathway for the forward direction. In other words, the transition states for the forward and reverse reactions are identical. This also holds for (non-chain) reactions in the steady state, under a given set of reaction conditions.[56]

The principle of microscopic reversibility is very useful for elucidating the nature of a transition state from a knowledge of that for the reverse reaction. For example, as the attack of ethanol on acetic acid is general-base-catalysed at low pH, the reverse reaction must involve the general-acid-catalysed expulsion of ethoxide ion from the tetrahedral intermediate (2.73).

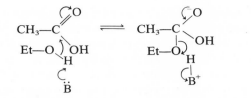

General base catalysis   General acid catalysis

Similarly, since *p*-nitrophenol is ionized above pH 7, its attack on a carbonyl compound cannot be general-base-catalysed: the *p*-nitrophenolate ion carrying a full negative charge is both a more powerful nucleophile and is present at a higher concentration than *p*-nitrophenol. Therefore the ion directly attacks a carbonyl compound and, by the principle of microscopic reversibility, the expulsion of *p*-nitrophenolate ion from a tetrahedral intermediate is uncatalysed (2.74).

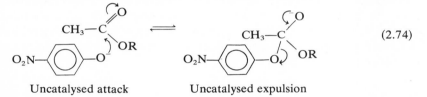

Uncatalysed attack   Uncatalysed expulsion

Similar arguments have been used to show that the attack of thiols upon esters is not general-base-catalysed but involves the direct attack of the thiolate ion even at ten pH units below the pK_a of the thiol where only one part in 10^{10} is ionized.[44]

Care must always be taken that a proposed reaction mechanism satisfies the principle of microscopic reversibility. Periodically, mechanisms are published that contravene the principle by the reverse reaction using a different pathway from the forward reaction under the same set of reaction conditions.

F. Principle of kinetic equivalence

There is an inherent ambiguity, known as the principle of kinetic equivalence, in interpreting the pH dependence of chemical reactions. When a rate law shows, for example, that the reaction rate is proportional to the concentration of an acid HA, it means that the net ionic charge of the acid appears in the transition state of the reaction; that is, either the undissociated HA *or* the two ions H^+ and A^-. Similarly, if the reaction rate varies as the concentration of A^-, the transition state contains either A^- *or* HA and OH^-. This may be shown algebraically as follows.

Rearranging the equation for the ionization of an acid gives

$$[HA] = [A^-][H^+]/K_a. \tag{2.75}$$

The concentration of the acid is related to the product of the concentrations of its conjugate base and the proton. Because of this, it is not possible to tell whether a reaction which depends on the concentration of [HA] really involves the undissociated acid or involves the combination of an H^+ with the conjugate base in the transition state. Similarly, since the concentration of the basic form is related to that of the hydroxide ion and the acid by

$$[A^-] = [HA][OH^-]K_a/K_w \qquad (2.76)$$

(where K_w is the ionic product of water), it is not possible to distinguish a reaction involving A^- from that involving a combination of HA and OH^- by examining the concentration dependence. This is the principle of kinetic equivalence. For example, the two following mechanisms for general base catalysis follow the same rate law:

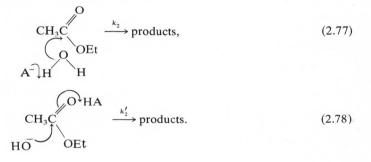

$$\qquad (2.77)$$

$$\qquad (2.78)$$

For (2.77),

$$v = k_2[A^-][\text{Ester}]. \qquad (2.79)$$

For (2.78),

$$v = k_2'[HA][OH^-][\text{Ester}] \qquad (2.80)$$

$$= k_2'[A^-][\text{Ester}]K_w/K_a \qquad (2.81)$$

$$= k_2''[A^-][\text{Ester}]. \qquad (2.82)$$

It is seen that eqns (2.79) and (2.82) are equivalent and the two mechanisms cannot be distinguished by the concentration or pH dependence. The two mechanisms cannot be distinguished by the sign of the Brönsted β for the variation of the pK_a of the catalyst. One can see this intuitively from transition state theory since both reactions involve a dispersion of negative charge in the transition state. Alternatively, this can be derived more mathematically. Mechanism (2.77) involves a positive value of β since the catalysis will be stronger for stronger bases. Mechanism (2.78) also involves a positive β-value, since although the general acid component involves a negative value of β, the term $1/K_a$ in eqn (2.81) means that a component of $1 \times pK_a$ must be added to log v which more than compensates for the fractionally negative β-value for the chemical step.

The only time that kinetically equivalent mechanisms can be distinguished is when one of the mechanisms involves an 'impossible' step, such as a second-order rate constant that is faster than diffusion controlled reaction or, as in the case of mechanism (2.78) for aspirin hydrolysis, a negative energy of activation.[12]

G. Kinetic isotope effects[57]

Information about the extent and nature of the bond-making and bond-breaking steps in the transition state may sometimes be obtained by studying the effects of isotopic substitution on the reaction rates. The effects may be divided into two classes depending on the position of the substitution.

1. Primary isotope effects

A primary isotope effect results from the cleavage of a bond to the substituted atom. For example, it is often found that the cleavage of a C—D bond is several times slower than that of a C—H bond. Smaller decreases in rate, up to a few per cent, are sometimes found on the substitution of ^{15}N for ^{14}N, or ^{18}O for ^{16}O. The magnitude of the change in rate gives some idea of the extent of the breaking of the bond in the transition state.

A simple way of analysing isotope effects is to compare an enzymic reaction with a simple chemical model whose chemistry has been established by other procedures. In this section we are interested primarily in the empirical results of the model experiments, but the following oversimplified account of the theoretical origins of the effects is helpful in understanding their nature. The plot of the energy of a carbon–hydrogen bond against interatomic distance gives the characteristic curve shown in Fig. 2.8. The carbon–deuterium bond gives an identical plot since the shape is determined by the electrons in the orbitals. According to quantum theory, the lowest energy level is at a value of $\frac{1}{2}h\nu$ above the bottom of the well, i.e. the *zero-point energy*, where ν is the frequency of vibration. ν may be found from the infrared stretching frequencies to give values of 17·4 and 12·5 kJ/mol (4·15 and 3 kcal/mol) for the zero-point energies of the C—H and C—D bonds respectively. If, during the transition state for the reaction, the hydrogen or deuterium atom is at a potential-energy maximum, rather than in a well, there will be no zero-point energy. It is therefore easier to break the C—H bond by 17·4–12·5 kJ/mol (4·15–3·0 kcal/mol), a factor of 7 at 25°. In practice, the kinetic isotope effect may be higher than this because of quantum

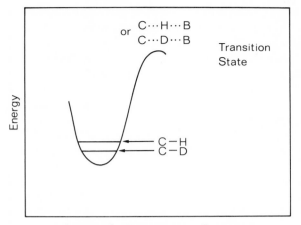

C—H or C—D Interatomic Distance

Fɪɢ. 2.8. The energy changes during the transfer of hydrogen or deuterium from carbon. The energy of the transition state is the same for both (subject to the provisos in the text) but the hydrogen is at a higher energy in the starting materials because of its higher zero-point energy. The activation energy for the transfer of hydrogen is therefore less than that for deuterium

mechanical tunnelling, or lower because there are compensating bending motions in the transition state. As a general rule, values of 2 to 15 are good evidence that a C—H bond is being broken in the transition state. It will be seen in Chapter 12A1 that values of 3 to 5 are found for k_H/k_D for hydride transfer between the substrate and NAD^+ in the reactions of alcohol dehydrogenase.

Oxygen and nitrogen kinetic isotope effects have been used to probe the nature of the rate-determining step in the reactions of chymotrypsin and papain. In a model system, the hydrolysis of methyl formate $(HC(=O)^{16,18}OCH_3)$, alkaline and general-base-catalysed hydrolyses are characterized by $k_{16_O}/k_{18_O} = 1 \cdot 01$. Hydrazinolysis, where the rate-determining step is breakdown of the tetrahedral intermediate, has an isotope

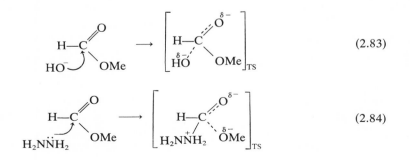

(2.83)

(2.84)

effect of 1·062. These values may be compared with a value of 1·052 calculated for the complete loss of the zero-point energies of the bonds.[58] The kinetic isotope effects on the chymotrypsin-catalysed hydrolysis of ester substrates are close to those found for the general-base-catalysed hydrolysis of methyl formate.[59] That is, there is probably the rate-determining formation of a tetrahedral intermediate. The papain-catalysed hydrolysis of benzoyl-L-arginine amide has an $^{14}N/^{15}N$ kinetic isotope effect of about 1·024, close to the upper limit of 1·01 to 1·025 found for C—N bond cleavage in model reactions.[60] This indicates a considerable degree of C—N bond cleavage in the transition state in the enzymic reaction and is almost certainly caused by the rate-determining breakdown of a tetrahedral intermediate (see Chapter 12B2). A kinetic nitrogen isotope effect of 1·006 to 1·01 found for the chymotrypsin-catalysed hydrolysis of acetyltryptophan amide has been interpreted as evidence for a tetrahedral intermediate in which both formation and breakdown contribute to the rate-determining step.[61]

2. Secondary isotope effects

These result from bond cleavage between atoms *adjacent* to the isotopically substituted atom. Secondary isotope effects are caused by a change in the electronic hybridization of the bond linking the isotope rather than by cleavage of the bond. Perhaps the best-known example of this in enzymic reactions is the substitution of deuterium or tritium for hydrogen on the C-1 carbon of substrates for lysozyme (Chapter 12E). A carbonium ion is formed in the reaction and the C-1 carbon changes from sp^3 to sp^2. Model reactions give a value of 1·14 for k_H/k_D, compared with 1·11 found for the enzymic reaction.[62,63]

$$\underset{R}{\overset{R'}{\diagdown}}C\diagup\overset{OPh}{\underset{D}{\diagdown}} \longrightarrow \underset{R}{\overset{R'}{\diagdown}}C^+\!\!-\!D \qquad\qquad (2.85)$$

3. Solvent isotope effects

These are found from comparing the rates of a reaction in H_2O and D_2O. They are usually a result of proton transfers between electronegative atoms *accompanying* the bond-making and -breaking steps between the heavier atoms in a reaction such as (2.86).

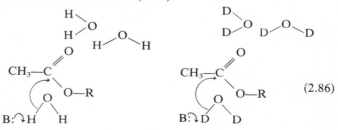

(2.86)

Solvent isotope effects differ in origin from those caused by the cleavage of a C—H bond and their cause is not fully understood. Proton transfers between electronegative atoms are very rapid compared with normal bond-making and -breaking steps and the proton probably sits at the bottom of a potential-energy well during the reaction. There is not the loss of zero-point energy that occurs in the transfer of a proton from carbon, where the actual C—H bond-breaking step is slow. Solvation of the reagents and secondary isotope effects caused by the exchange of deuterium for hydrogen in the reagents may make contributions to the solvent isotope effect.

Solvent isotope effects are a useful diagnostic tool in simple chemical reactions although they can be variable. For example, it is found that general-base-catalysed reactions such as (2.86) have k_H/k_D about 2, whilst the nucleophilic attack on an ester has k_H/k_D about 1. The isotope effects in enzymic reactions are more difficult to analyse because there are so many protons in the protein that may exchange with deuterons from D_2O.[64] Also there may be slight changes in the structure of the protein on the change of solvent.

H. Stereochemistry[65]

The idea that enzymes may be specific for only one enantiomer of a pair of optically active substrates is as old as the study of stereochemistry itself. Pasteur, that towering genius of chemistry and biochemistry, reported in 1858 a form of yeast which fermented dextrorotatory tartaric acid but not laevorotatory. Early work on the proteases showed that derivatives of L amino acids and not those of D amino acids are hydrolysed.

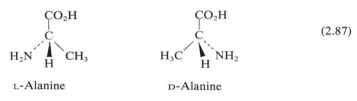

(2.87)

L-Alanine   D-Alanine

A compound is optically active, rotating the plane of polarization of plane polarized light, if it is not superimposable with its mirror image, i.e. with its enantiomer. A simple diagnostic test for superimposability is the presence of a plane or centre of symmetry. The presence of such symmetry indicates a lack of activity; the absence indicates activity. As with many organic molecules of biological interest, the optical activity in the amino acids is caused by an asymmetric carbon atom with four different groups around it. This type of carbon is now called a *chiral* centre (from the Greek for hand). A carbon atom of the form $CR_2R'R''$ is called *prochiral*. Although it is not optically active because it is bound to two identical groups (and thus has a plane of symmetry and is superimposable with its mirror image), it is *potentially* chiral since one of the R groups could be chemically replaced by another group (not = R', R'').

It was pointed out some 40 or 50 years ago that the recognition of a chiral carbon by an enzyme implies that at least three of the groups surrounding the chiral carbon must be bound to the enzyme (the three-point attachment theory).[66] If only two of the groups are bound by the enzyme, the other two may be interchanged without affecting the binding (2.88). The cause of the stereospecificity of the proteases (on which the

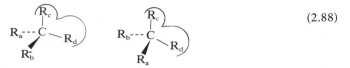

(2.88)

early stereochemical experiments were performed) is immediately obvious on examining the crystal structures of the enzymes (Chapters 1 and 12). D-amino acid derivatives differ from those of L amino acids by having the H atom and side chain attached to the chiral carbon interchanged. The D derivatives cannot bind because of steric hindrance between the side chain and the walls of enzyme around the position normally occupied by the H atom of L derivatives.

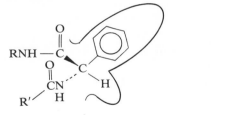

(2.89)

1. An asymmetric enzyme can confer asymmetry on a symmetric substrate

Although the two hydrogen atoms in CH_2RR' are equivalent in simple chemical reactions, the equivalence may be lost on binding to an asymmetric active site of an enzyme.[67] For example, the attachment to the enzyme by R and R' in (2.90) causes the two hydrogen atoms to be

(2.90)

exposed to different environments. H_a may be next to a catalytic base whilst H_b may be in an inert position.

Another example of this is found in the reactions of carbonyl compounds and carbon–carbon double bonds. An optically active compound is formed if a reagent attacks just one side of a planar trigonal carbon. For example, if in (2.91) the nucleophile attacks acetaldehyde from the

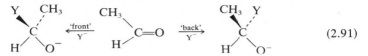

$$(2.91)$$

'front' side, the product on the left is formed, whereas attack from the 'back' of the page gives the enantiomer in the right. (Note: attack on trigonal carbon always occurs perpendicular to the plane of the double bond.) In a simple chemical reaction in solution, there is an equal probability of attack at either face of the trigonal carbon in (2.91) so that a racemic mixture of 50% of each enantiomer is formed. But in an enzymic reaction, attack may occur on one face only because the substrate is firmly held in an asymmetric active site with only one face exposed to the attacking group.

2. Notation[65]

The terms D and L that are usually used to denote configuration have the drawback of not being absolute but are relative to a reference compound. A more useful notation, denoting *absolute* molecular asymmetry, is the *RS* convention. In this, the groups around the chiral carbon are assigned an order of 'priority' based on a series of rules that depend on atomic number and mass. The atom directly attached to the chiral carbon is considered first; the higher its atomic number, the higher its priority. For isotopes, the higher mass number has priority. For groups that have the same type of atom attached to the chiral carbon, the atomic and mass numbers of the next atoms out are considered. This is best illustrated by the following list of the most commonly found groups: —SH > —OR > —OH > —NHCOR > —NH$_2$ > —CO$_2$R > —CO$_2$H > —CHO > —CH$_2$OH > —C$_6$H$_5$ > —CH$_3$ > —T > —D > —H. (Note: —CHO has priority over —CH$_2$OH because a C=O carbon is counted as being bonded to *two* oxygen atoms.)

A chiral carbon is designated as being *R* or *S* as follows. The carbon is viewed from the direction opposite to the ligand of lowest priority. If the priority order of the remaining three ligands decreases in a clockwise direction, the absolute configuration of the molecule is said to be *R* (= *rectus*, right). If it decreases in the anticlockwise direction it is said to be *S* (= *sinister*, left). For example:

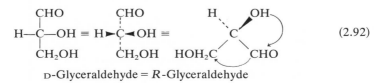

$$(2.92)$$

D-Glyceraldehyde = *R*-Glyceraldehyde

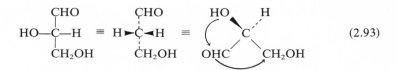

$$\text{L-Glyceraldehyde} = S\text{-Glyceraldehyde}$$ (2.93)

(Note: the formulae for glyceraldehyde on the left-hand side of the page are written according to the Fischer projection notation; bonds in the 'east–west' direction come forwards and those going 'north–south' go into the paper, as represented in the formulae in the centres of (2.92) and (2.93).)

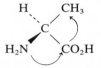

(2.94)

$$\text{L-Alanine} = S\text{-Alanine}$$

There is also a *prochirality* rule. For example, in ethanol, if we label the two protons by subscripts a and b, and *arbitrarily* give H_a priority over H_b as in (2.95), H_a is said to be *pro-R* because of the clockwise order of

(2.95)

priority. Conversely, H_b is *pro-S*. Note that if one repeats the treatment by giving H_b priority over H_a, H_a is still found to be *pro-R* and H_b *pro-S*. Prochirality is thus absolute and does not depend on whether H_a or H_b is given priority.

The two faces of a compound containing a trigonal carbon atom are described as *re* (rectus) and *si* (sinister) by a complicated set of rules. Two simple cases are illustrated by the faces presented towards the reader by the following compounds:

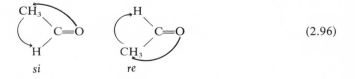

(2.96)

3. Some examples of enzyme stereospecificity

a. Nicotinamide adenine dinucleotide-dependent oxidation and reduction
NAD^+ (2·97, R = H) and $NADP^+$ (2·97, R = PO_3^-) function as coenzymes

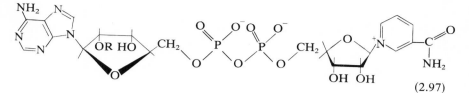

$$(2.97)$$

in redox reactions by reversibly accepting hydrogen at the 4 position of the nicotinamide ring.

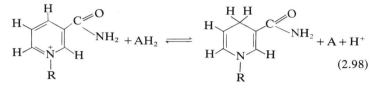

$$(2.98)$$

Position 4 in the dihydronicotinamide ring is prochiral. The faces of the nicotinamide ring and the C-4 protons of the dihydronicotinamide rings may be labelled according to the *RS* convention by giving the portion of the ring containing the —CONH₂ group priority over the other portion.

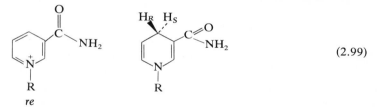

$$(2.99)$$

re

It was discovered in a historically important series of experiments that there is direct and stereospecific transfer between the substrate and NAD⁺.[68,69] Yeast alcohol dehydrogenase transfers one mole of deuterium from CH_3CD_2OH to NAD⁺. On incubating the NADD formed with the enzyme and unlabelled acetaldehyde, *all* the deuterium is lost from the NADD and incorporated in the alcohol that is formed. The deuterium or hydrogen is transferred stereospecifically to one face of the NAD⁺ and then transferred back from the same face. A non-specific transfer to both faces would lead to a transfer back to the acetaldehyde of 50% of the deuterium (or considerably less than this because of the kinetic isotope effect slowing down deuterium transfer). Some dehydrogenases transfer to the same face as does the alcohol dehydrogenase (class A dehydrogenases), others to the opposite face (class B); these are listed in Chapter 12. It has subsequently been found that the class A enzymes transfer to the *re* face of NAD⁺ (or (NADP⁺) and use the pro-*R* hydrogen of NADH.

$$NAD^+ + CH_3CD_2OH \rightarrow NADD + CH_3CDO + H^+ \qquad (2.100)$$

$$NADD + CH_3CHO + H^+ \rightarrow NAD^+ + CH_3CDHOH \qquad (2.101)$$

The transfer to the aldehyde is also stereospecific. The alcohol formed in (2.102) is the S enantiomer and is optically active with a specific rotation of $-0.28 \pm 0.03°$.[70]

$$(2.102)$$

b. Fumarase-catalysed hydration of fumarate

Fumarase catalyses the addition of the elements of water across the double bond of fumarate (2.103). The addition of D_2O is found from the

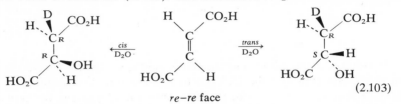

$$(2.103)$$

$re-re$ face

analysis of the stereochemistry of the malic acid that is formed to be *trans* (D^+ adds to the re face at the top and OD^- adds to the si face at the bottom) rather than *cis* (both adding re).

c. cis Enediol intermediate in aldose–ketose isomerases

It is discussed in Chapter 12G that the catalysis of reaction by aldose–ketose isomerases involves an enediol intermediate in which the transferred proton (T in eqn (2.104)) remains on the *same* face of the inter-

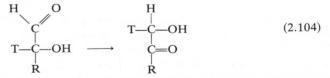

$$(2.104)$$

mediate. The stereochemistry of the products shows that the intermediate is *cis* rather than *trans*.[71]

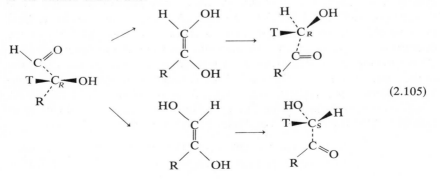

$$(2.105)$$

Aldoses that are R at C-2, as in (2.105), always form ketoses that are R at C-1. This implies the cisenediol in the upper branch of (2.105) rather than the *trans* intermediate.

References

1 H. Pelzer and E. Wigner, *Z. phys. Chem.* **B15,** 445 (1932).
2 H. Eyring, *Chem. Rev.* **17,** 65 (1935).
3 M. G. Evans and M. Polanyi, *Trans. Faraday Soc.* **31,** 875 (1935).
4 A. A. Frost and R. G. Pearson, *Kinetics and mechanism*, John Wiley & Sons Inc. (1961).
5 G. S. Hammond, *J. Am. chem. Soc.* **77,** 334 (1955).
6 J. N. Brönsted and K. Pedersen, *Z. phys. Chem.* **A108,** 185 (1923).
7 W. P. Jencks and J. Carriuolo, *J. Am. chem. Soc.* **83,** 1743 (1961).
8 T. H. Fife, *Accs. chem. Res.* **5,** 264 (1972).
9 G. A. Craze and A. J. Kirby, *J. chem. Soc. Perk. II*, 61 (1974).
10 R. F. Atkinson and T. C. Bruice, *J. Am. chem. Soc.* **96,** 819 (1974).
11 F. G. Bordwell and W. J. Boyle, Jr, *J. Am. chem. Soc.* **94,** 3907 (1972).
12 A. R. Fersht and A. J. Kirby, *J. Am. chem. Soc.* **89,** 4853, 4857 (1967); **90,** 5818, 5826 (1968).
13 T. St. Pierre and W. P. Jencks, *J. Am. chem. Soc.* **90,** 3817 (1968).
14 Extrapolated from the data of E. Gaetjens and H. Morawetz, *J. Am. chem. Soc.* **82,** 5328 (1960) and V. Gold, D. G. Oakenfull, and T. Riley, *J. chem. Soc.* **1968B,** 515 (1968).
15 A. R. Fersht and A. J. Kirby, *J. Am. chem. Soc.* **90,** 5833 (1968).
16 T. Higuchi, L. Eberson, and J. D. Macrae, *J. Am. chem. Soc.* **89,** 3001 (1967).
17 W. P. Jencks, F. Barley, R. Barnett, and M. Gilchrist, *J. Am. chem. Soc.* **88,** 4464 (1966).
18 M. I. Page and W. P. Jencks, *Proc. natn. Acad. Sci. U.S.A.* **68,** 1678 (1971).
19 M. I. Page, *Chem. Soc. Revs.* **2,** 295 (1973).
20 D. E. Koshland, Jr., *J. theorèt. Biol.* **2,** 75 (1962).
21 D. R. Storm and D. E. Koshland, Jr, *Proc. natn. Acad. Sci. U.S.A.* **66,** 445 (1970); *J. Am. chem. Soc.* **94,** 5805 (1972).
22 T. C. Bruice, A. Brown, and D. C. Harris, *Proc. natn. Acad. Sci. U.S.A.* **68,** 658 (1971).
23 W. P. Jencks and M. I. Page, *Biochem. biophys. Res. Comm.* **57,** 887 (1974).
24 M. F. Perutz, *Proc. Ry. Soc.* **B167,** 448 (1967).
25 A. Warshel and M. Levitt, *J. molec. Biol.* **103,** 227 (1976).
26 A. S. Mildvan, *The Enzymes* **2,** 246 (1970).
27 D. A. Buckingham, C. E. Davis, D. M. Foster, and A. M. Sargeson, *J. Am. chem. Soc.* **92,** 5571 (1970).
28 D. A. Buckingham, J. MacB. Harrowfield, and A. M. Sargeson, *J. Am. chem. Soc.* **96,** 1726 (1974).
29 E. Chaffee, T. P. Dasgupta, and G. M. Harris, *J. Am. chem. Soc.* **95,** 4169 (1973).
30 D. A. Palmer and G. M. Harris, *Inorg. Chem.* **13,** 965 (1974).

31 D. A. Buckingham and L. M. Engelhardt, *J. Am. chem. Soc.* **97,** 5915 (1975).

32 J. E. Coleman, *Prog. bioorg. Chem.* **1,** 159 (1971), and references therein.

33 P. Woolley, *Nature, Lond.* **258,** 677 (1975).

34 D. A. Buckingham, F. R. Keene, and A. M. Sargeson, *J. Am. chem. Soc.* **96,** 4981 (1974).

35 T. H. Finlay, J. Valinsky, K. Sato, and R. H. Abeles, *J. biol. Chem.* **247,** 4197 (1974).

36 B. T. Golding and L. Radom, *J. Am. chem. Soc.* **98,** 6331 (1976).

37 E. E. Snell and S. J. di Mari, *The Enzymes* **2,** 335 (1976).

38 G. A. Hamilton and F. H. Westheimer, *J. Am. chem. Soc.* **81,** 6332 (1959).

39 S. G. Warren, B. Zerner, and F. H. Westheimer, *Biochemistry* **5,** 817 (1966).

40 I. A. Rose, E. L. O'Connell, and A. H. Mehler, *J. biol. Chem.* **240,** 1758 (1965).

41 I. A. Rose and E. L. O'Connell, *J. biol. Chem.* **244,** 126 (1969).

42 L. Davis and D. E. Metzler, *The Enzymes* **7,** 33 (1972).

43 A. R. Fersht and W. P. Jencks, *J. Am. chem. Soc.* **92,** 5442 (1970).

44 A. R. Fersht, *J. Am. chem. Soc.* **93,** 3504 (1971).

45 R. G. Pearson, *J. chem. Educ.* **45,** 103, 643 (1968).

46 W. P. Jencks and J. Carriuolo, *J. Am. chem. Soc.* **82,** 1778 (1960).

47 W. P. Jencks and M. Gilchrist, *J. Am. chem. Soc.* **90,** 2622 (1968).

48 E. G. Sander and W. P. Jencks, *J. Am. chem. Soc.* **90,** 6154 (1968).

49 J. Fastrez and A. R. Fersht, *Biochemistry* **12,** 1067 (1973).

50 M. L. Bender and K. Nakamura, *J. Am. chem. Soc.* **84,** 2577 (1962).

51 M. Caplow and W. P. Jencks, *Fedn Proc.* **21,** 248 (1962).

52 G. Lowe and Y. Yuthavong, *Biochem. J.* **124,** 107 (1971).

53 J. Klinman, *Biochemistry* **15,** 2018 (1976).

54 J. Järv, T. Kesvatera, and A. Aaviksaar, *Eur. J. Biochem.* **67,** 315 (1976).

55 J. S. Thomsen, *Phys. Rev.* **91,** 1263 (1953).

56 R. M. Krupka, H. Kaplan, and K. J. Laidler, *Trans. Farad. Soc.* **62,** 2754 (1966).

57 J. H. Richards, *The Enzymes* **2,** 321 (1970).

58 C. B. Sawyer and J. F. Kirsch, *J. Am. chem. Soc.* **95,** 7375 (1973).

59 C. B. Sawyer and J. F. Kirsch, *J. Am. chem. Soc.* **97,** 1963 (1975).

60 M. H. O'Leary, M. Urberg, and A. P. Young, *Biochemistry* **13,** 2077 (1974).

61 M. H. O'Leary and M. D. Kluetz, *J. Am. chem. Soc.* **94,** 3584 (1972).

62 F. W. Dahlquist, T. Rand-Meir, and M. A. Raftery, *Proc. natn. Acad. U.S.A.* **61,** 1194 (1968).

63 L. E. H. Smith, L. H. Mohr, and M. A. Raftery, *J. Am. chem. Soc.* **95,** 7497 (1973).

64 A. J. Kresge, *J. Am. chem. Soc.* **95,** 3065 (1972).

65 G. Popjak, *The Enzymes* **2,** 115 (1970).

66 M. Bergman and J. S. Fruton, *J. biol. Chem.* **117,** 193 (1937) .

67 A. Ogston, *Nature, Lond.* **162,** 963 (1948).

68 F. H. Westheimer, H. Fisher, E. E. Conn, and B. Vennesland, *J. Am. chem. Soc.* **73,** 2043 (1951).

69 H. F. Fisher, E. E. Conn, B. Vennesland, and F. H. Westheimer, *J. biol. Chem.* **202,** 687 (1953).

70 H. R. Levy, F. A. Loewus, and B. Vennesland, *J. Am. chem. Soc.* **79,** 2949 (1957).
71 I. A. Rose and E. L. O'Connell, *Biochim. biophys. Acta* **42,** 159 (1960).

Further reading

W. P. Jencks, *Catalysis in chemistry and enzymology*, McGraw-Hill Inc. (1969). (A scholarly and detailed account of most of the topics of this chapter).
M. I. Page, 'The energetics of neighbouring group participation', *Chem. Soc. Revs* **2,** 295 (1973). (Deals with entropy, strain etc. in intramolecularly catalysed reactions.)
T. C. Bruice and S. J. Benkovic, *Biooganic mechanisms*, W. A. Benjamin (1966). (A comprehensive account of the chemical mechanisms of many biologically important reactions.)

Chapter 3

The basic equations of enzyme kinetics

A. Steady-state kinetics

The concept of the steady state is used widely in dynamic systems. It generally refers to the value of a particular quantity being constant, or in the steady state, when its rate of formation is balanced by its rate of destruction. For example, the population of a country is in a steady state when the birth and immigration rates equal those of death and emigration. Similarly, the concentration of a metabolite in the cell is at a steady-state level when it is being produced as rapidly as it is being degraded. In enzyme kinetics, the concept is applied to the concentrations of enzyme-bound intermediates. On mixing an enzyme with a large excess of substrate there is an initial period, known as the *pre-steady state*, during which the concentrations of these intermediates build up to their steady-state levels. After this period, the reaction rate changes relatively slowly with time and the intermediates are at the steady-state concentrations. It is during this steady-state period that the rates of enzymic reactions are traditionally measured. The steady state is an approximation since the substrate is gradually depleted during the course of an experiment. But, provided the rate measurements are restricted to a short time interval over which the concentration of the substrate does not greatly change, it is a very good approximation.

Whilst the use of pre-steady-state kinetics is undoubtedly superior as a means of analysing the chemical mechanisms of enzyme catalysis (Chapters 4 and 7), steady-state kinetics are more important for the understanding of metabolism since they measure the catalytic activity of an enzyme in the steady-state conditions in the cell.

1. The experimental basis: the Michaelis–Menten equation[1]

In deriving the following kinetic expressions, it is assumed that the concentration of the enzyme is negligible compared with that of the substrate. Apart from the rapid reaction measurements described in the next chapter, this is generally true in practice because of the high catalytic efficiency of enzymes.

It is found experimentally in most cases that the initial rate (v) of formation of products or destruction of substrate by an enzyme is directly proportional to the concentration of enzyme $[E_0]$. However, v generally follows *saturation kinetics* with respect to the concentration of substrate, $[S]$, in the following way (see Fig. 3.1). At sufficiently low $[S]$, v increases linearly with $[S]$. But as $[S]$ is increased this relationship begins to break

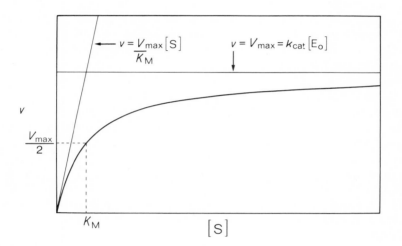

FIG. 3.1. Reaction rate plotted against substrate concentration for a reaction obeying Michaelis–Menten (or saturation) kinetics

down and v increases less rapidly than $[S]$ until at sufficiently high or *saturating* $[S]$, v reaches a limiting value termed V_{max}. This is expressed quantitatively in (3.1), the Michaelis–Menten equation, the basic equation of enzyme kinetics.

$$v = \frac{[E_0][S]k_{cat}}{K_M + [S]} \qquad (3.1)$$

where

$$k_{cat}[E_0] = V_{max}. \qquad (3.2)$$

The concentration of substrate at which $v = \frac{1}{2}V_{max}$ is termed K_M, the Michaelis constant. Note that at low $[S]$, where $[S] \ll K_M$,

$$v = (k_{cat}/K_M)[E_0][S]. \qquad (3.3)$$

2. Interpretation of the kinetic phenomena for single substrate reactions: the Michaelis–Menten mechanism

In 1913 Michaelis and Menten developed the theories of earlier workers and proposed the following scheme:

$$E + S \underset{}{\overset{K_S}{\rightleftharpoons}} ES \overset{k_{cat}}{\longrightarrow} EP. \qquad (3.4)$$

The catalytic reaction is divided into two processes. The enzyme and substrate first combine to give an enzyme–substrate complex, ES. This step is assumed to be rapid and reversible with no chemical changes taking place; the enzyme and substrate are held together by physical forces. The chemical processes then take place in a second step with a first-order rate constant k_{cat} (the turnover number). The rate equations are solved in the following manner:

From eqn (3.4),

$$\frac{[E][S]}{[ES]} = K_S \qquad (3.5)$$

and

$$v = k_{cat}[ES]. \qquad (3.6)$$

Also the total enzyme concentration $[E_0]$ and the free enzyme $[E]$ are related by

$$[E] = [E_0] - [ES] \qquad (3.7)$$

Thus

$$[ES] = \frac{[E_0][S]}{K_S + [S]} \qquad (3.8)$$

and

$$v = \frac{[E_0][S]k_{cat}}{K_S + [S]}. \qquad (3.9)$$

This is identical to eqn (3.1) where K_M is equal to the dissociation constant of the enzyme–substrate complex.

The concept of the enzyme–substrate complex is the foundation stone of enzyme kinetics and our understanding of the mechanism of enzyme catalysis. In honour of its introduction, this non-covalently bound complex is often termed the *Michaelis* complex.

3. Extensions and modifications of the Michaelis–Menten mechanism

A distinction must be drawn between the equation and the mechanism proposed by Michaelis and Menten. Their equation holds for many mechanisms, but their mechanism is not always appropriate.

The Michaelis–Menten mechanism assumes that the enzyme–substrate complex is in thermodynamic equilibrium with free enzyme and substrate. This is true only if, in the following scheme, $k_2 \ll k_{-1}$

$$E + S \underset{k_{-1}}{\overset{k_1}{\rightleftharpoons}} ES \overset{k_2}{\longrightarrow} EP. \qquad (3.10)$$

The case when k_2 is comparable to k_{-1} was first analysed by Briggs and Haldane in 1925.

a. Briggs–Haldane kinetics:[2] $K_M > K_S$

The method of solution of (3.10) is somewhat more complicated than for the Michaelis–Menten scheme; the steady-state approximation is applied to the concentration of ES. That is, if the reaction rate measured is approximately constant over the time interval concerned, then [ES] is also constant:

$$\frac{d[ES]}{dt} = 0 = k_1[E][S] - k_2[ES] - k_{-1}[ES]. \qquad (3.11)$$

Substituting eqn (3.7) gives

$$[ES] = \frac{[E_0][S]}{[S] + (k_2 + k_{-1})/k_1} \qquad (3.12)$$

and, as $v = k_2[ES]$,

$$v = \frac{[E_0][S]k_2}{[S] + (k_2 + k_{-1})/k_1}. \qquad (3.13)$$

This is identical to the Michaelis–Menten eqn (3.1) where now

$$K_M = (k_2 + k_{-1})/k_1. \qquad (3.14)$$

Since K_S for the dissociation of [ES] is equal to k_{-1}/k_1,

$$K_M = K_S + k_2/k_1 \qquad (3.15)$$

Of course, when $k_{-1} \gg k_2$, eqn (3.14) simplifies to $K_M = K_S$.

b. Intermediates occurring after ES: $K_M < K_S$

The Michaelis–Menten scheme may be extended to cover a variety of cases where additional intermediates, covalently or non-covalently bound, occur on the reaction pathway. It is found in all examples that the Michaelis–Menten equation still applies, although K_M and k_{cat} are now combinations of various rate and equilibrium constants. K_M is always $\leqslant K_S$ in these cases. Suppose, as for example in scheme (3.16), there are several intermediates and the final catalytic step is slow:

$$E + S \underset{}{\overset{K_S}{\rightleftharpoons}} ES \underset{}{\overset{K}{\rightleftharpoons}} ES' \underset{}{\overset{K'}{\rightleftharpoons}} ES'' \underset{slow}{\overset{k_4}{\longrightarrow}} EP \qquad (3.16)$$

where $[ES'] = K[ES]$, and $[ES''] = K'[ES]$. Then

$$K_M = K_S/(1 + K + KK') \qquad (3.17)$$

and

$$k_{cat} = k_4 KK'/(1 + K + KK'). \qquad (3.18)$$

The chymotrypsin-catalysed hydrolysis of esters and amides proceeds through (3.19):

$$E + S \underset{}{\overset{K_S}{\rightleftharpoons}} ES \underset{\underset{P_1}{\searrow}}{\overset{k_2}{\longrightarrow}} EA \overset{k_3}{\longrightarrow} EP_2 \qquad (3.19)$$

where EA is an 'acylenzyme'.

Applying the steady-state assumption to $[EA]$, it may be shown that

$$v = [E_0][S] \left\{ \frac{k_2 k_3/(k_2 + k_3)}{K_S k_3/(k_2 + k_3) + [S]} \right\}, \qquad (3.20)$$

i.e. a Michaelis–Menten equation where

$$K_M = K_S \frac{k_3}{k_2 + k_3} \qquad (3.21)$$

and

$$k_{cat} = \frac{k_2 k_3}{k_2 + k_3}, \qquad (3.22)$$

or, alternatively,

$$\frac{1}{k_{cat}} = \frac{1}{k_2} + \frac{1}{k_3}. \qquad (3.23)$$

In more complex reactions, the generalization that $K_M < K_S$ can break down.

4. All three mechanisms occur in practice

In the Briggs–Haldane mechanism where k_2 is much greater than k_{-1}, k_{cat}/K_M is equal to k_1, the rate constant for the association of enzyme and substrate. It is shown in Chapter 4 that association rate constants should be of the order of $10^8 \, s^{-1} \, M^{-1}$. This leads to a diagnostic test for the Briggs–Haldane mechanism: k_{cat}/K_M is about 10^7–$10^8 \, s^{-1} \, M^{-1}$. Catalase, acetylcholinesterase, carbonic anhydrase, crotonase, fumarase, and triosephosphate isomerase all exhibit Briggs–Haldane kinetics by this criterion (see Table 4.4). Another example is one of the first enzymes studied by rapid reaction techniques, horse-radish peroxidase.[3] It first forms a Michaelis complex with hydrogen peroxide and then undergoes a second-order reaction with a hydrogen donor. At sufficiently high donor concentrations the rate of the second reaction is much faster than the dissociation of the Michaelis complex.

It is extremely common for intermediates to occur after the initial enzyme–substrate complex as in eqn (3.19). However, it is often found for the physiological substrates that these intermediates do not accumulate and the slow step in eqn (3.19) is k_2. (There is a theoretical reason for this that is discussed in Chapter 10, where examples are given). Under these conditions, K_M is equal to K_S, the dissociation constant, and the original Michaelis–Menten mechanism is obeyed to all intents and purposes. The opposite occurs in many laboratory experiments. The enzyme kineticist often uses synthetic, highly reactive, substrates to assay enzymes, and covalent intermediates frequently accumulate.

B. The significance of the Michaelis–Menten parameters

1. Meaning of k_{cat}

In the simple Michaelis–Menten mechanism where there is only one enzyme–substrate complex and all binding steps are fast, k_{cat} is simply the first-order rate constant for the chemical conversion of the ES complex to the EP complex. For more complex reactions, k_{cat} is a function of all the first-order rate constants and cannot be assigned to any particular process except when simplifying features occur. For example, in the Briggs–Haldane mechanism where the dissociation of the EP complex is fast, k_{cat} is equal to k_2 (eqn (3.10)). But if dissociation of the EP complex is slow, the rate constant for this process contributes to k_{cat} and, in the extreme case where EP dissociation is far slower than the chemical steps, k_{cat} will be equal to the dissociation rate constant. In the example of eqn (3.19), k_{cat} was seen to be a function of k_2 and k_3. But if one of these constants is much smaller than the other, it becomes equal to k_{cat}. For example, if $k_3 \ll k_2$, then, from eqn (3.22), $k_{cat} = k_3$. An extension of this is that k_{cat} cannot be greater than any first-order rate constant on the forward reaction pathway.[4] It thus sets a *lower* limit on the chemical rate constants.

k_{cat} is often called the *turnover number* of the enzyme because it represents the maximum number of substrate molecules converted to products per active site per unit time, or the number of times the enzyme 'turns over' per unit time.

2. Meaning of K_M: real and apparent equilibrium constants

Although it is only for the simple Michaelis–Menten mechanism or in similar cases that $K_M = K_S$, the true dissociation constant of the enzyme–substrate complex, K_M may be treated for *some* purposes as an *apparent* dissociation constant. For example, the concentration of free enzyme in solution may be calculated from the relationship

$$\frac{[E][S]}{\sum [ES]} = K_M \tag{3.24}$$

where $\sum [ES]$ is the sum of *all* the bound enzyme species.†

The concept of *apparent* values is very useful and appears in other phenomena, such as pK_a values. Quite often pK_a values do not represent the microscopic ionization of a particular group but are a combination of this value with various equilibrium constants between different conformational states of the molecule. The result is an *apparent* pK_a which may be handled titrimetrically as a simple pK_a. This simple-minded approach must not be taken too far and, when considering the effects of temperature, pH etc. on an apparent K_M, it must be realized that the rate constant components of this term are also affected. The same applies to k_{cat} values. There are examples in the literature where breaks in the temperature dependence of k_{cat} have been interpreted as indicative of conformational changes in the enzyme when, in fact, they are due to a different temperature dependence of the individual rate constants in k_{cat}, e.g. k_2 and k_3 in eqn (3.22).

An illustration of the way that K_M is a measure of the amount of enzyme that is bound in any form whatsoever to the substrate is given by the following (cf. the chymotrypsin mechanism):

$$E + S \underset{}{\overset{K_S}{\rightleftharpoons}} ES \xrightarrow{k_2} ES' \xrightarrow{k_3} E + P. \tag{3.25}$$

Application of the steady-state approximation to $[ES']$ gives

$$[ES'] = [ES]k_2/k_3. \tag{3.26}$$

When k_2 is much greater than k_3, $[ES']$ is much greater than $[ES]$, so that ES' makes a more important contribution to K_M than does ES and is the

---

† Kinetic curiosities may be devised that give Michaelis–Menten kinetics without the enzyme being saturated with the substrate. For example, in the following scheme where the active form of the enzyme reacts with the substrate in a second-order reaction to give the products and an inactive form of the enzyme E', that slowly reverts to the active form, apparent saturation kinetics are followed with $k_{cat} = k_2$ and $K_M = k_2/k_1$. Eqn (3.24) applies to this example by treating E' as a 'bound' form of the enzyme.

$$E + S \xrightarrow{k_1} E' + P$$
$$\downarrow{\scriptstyle k_2}$$
$$E$$

predominant enzyme-bound species. Without solving the equations for the reaction, we can say intuitively that K_M must be smaller than K_S by a factor of about k_3/k_2, i.e.

$$K_M \simeq K_S(k_3/k_2). \tag{3.27}$$

3. Meaning of k_{cat}/K_M

It was pointed out earlier that the reaction rate for low substrate concentrations is given by $v = (k_{cat}/K_M)[E_0][S]$ (eqn (3.3)), that is, k_{cat}/K_M is an *apparent* second-order rate constant. It is not a true microscopic rate constant except in the extreme case where the rate-determining step in the reaction is the encounter of enzyme and substrate.

The importance of k_{cat}/K_M is that it relates the reaction rate to the concentration of free, rather than total, enzyme. This is readily seen from eqn (3.3), mentioned above, since at low substrate concentration the enzyme is largely unbound and $[E] \sim [E_0]$. At low substrate concentration, the reaction rate is thus given by:

$$v = [E][S]k_{cat}/K_M.$$

It is shown later (eqn (3.41)) that this result holds at *any* substrate concentration. It will also be shown later (eqn (3.44)) that k_{cat}/K_M determines the specificity for competing substrates.

k_{cat}/K_M cannot be greater than any second-order rate constant on the forward reaction pathway.[4] It thus sets a *lower* limit on the rate constant for the association of enzyme and substrate.

C. Graphical representation of data

It is very useful to transform the Michaelis–Menten equation into a linear form for analysing data graphically and detecting deviations from the ideal behaviour. One of the best known methods is the double reciprocal or Lineweaver–Burk plot. Inverting both sides of eqn (3.1) and substituting eqn (3.2) gives eqn (3.28), the Lineweaver–Burk plot.[5]

$$1/v = 1/V_{max} + K_M/V_{max}[S]. \tag{3.28}$$

Plotting $1/v$ against $1/[S]$ (Fig. 3.2) gives an intercept of $1/V_{max}$ on the y-axis as $1/[S]$ tends to zero, and $1/[S] = -1/K_M$ on the x-axis. The slope of the line is K_M/V_{max}.

Another common plot is that of Eadie and Hofstee (eqn (3.29)).[6,7] Eqns (3.1) and (3.2) may be rearranged to give:

$$v = V_{max} - K_M v/[S]. \tag{3.29}$$

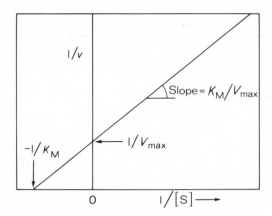

FIG. 3.2. The Lineweaver–Burk plot

Plotting v against $v/[S]$ (Fig. 3.3) gives an intercept of V_{max} on the y-axis as $v/[S]$ tends to zero. The slope of the line is equal to $-K_M$. The intercept on the x-axis is at $v/[S] = V_{max}/K_M$.

The Lineweaver–Burk plot has the disadvantage of compressing the data points at high substrate concentrations into a small region and emphasizing the points at lower concentrations. It does have the advantage that it is easy to read from it values of v for a given value of $[S]$.

The Eadie plot does not compress the higher values, but it is more difficult to determine rapidly from it the values of v against $[S]$. The Eadie plot is considered more accurate and generally superior.[8,9]

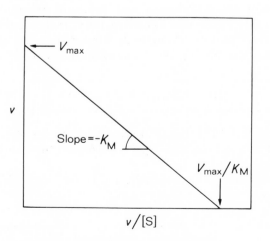

FIG. 3.3. The Eadie–Hofstee plot

D. Inhibition

As well as being irreversibly inactivated by heat or chemical reagents, enzymes may be *reversibly* inhibited by the non-covalent binding of inhibitors. There are four main types of inhibition.

1. Competitive inhibition

If an inhibitor I reversibly binds to the active site of the enzyme and prevents S binding and vice versa, I and S compete for the active site and I is said to be a *competitive* inhibitor. In the case of the simple Michaelis–Menten mechanism (eqn (3.4), $K_M = K_S$), an additional equilibrium must be considered, i.e.

$$E \xrightleftharpoons{S, K_M} ES \xrightarrow{k_{cat}} E + P.$$

$$\underset{I, K_I}{\Big\Updownarrow}$$

$$EI$$

(3.30)

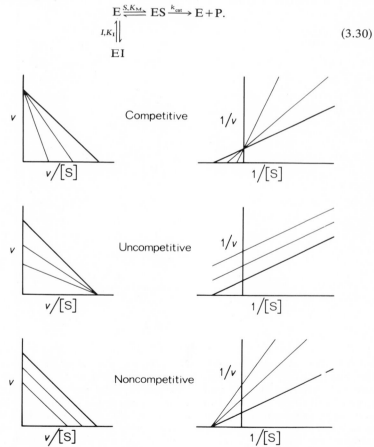

FIG. 3.4. Eadie–Hofstee (left) and Lineweaver–Burk (right) plots of different types of inhibition. Heavy lines are for the reaction in the absence of inhibitor, faint lines are for the presence

Solving the equilibrium and rate equations using

$$[E_0] = [ES] + [EI] + [E] \qquad (3.31)$$

gives

$$v = \frac{[E_0][S]k_{cat}}{[S] + K_M(1 + [I]/K_I)}. \qquad (3.32)$$

K_M is apparently increased by a factor of $(1 + [I]/K_I)$. This equation holds for all mechanisms obeying the Michaelis–Menten equation. Competitive inhibition affects K_M only, and not V_{max}, since infinitely high concentrations of S displace I from the enzyme.

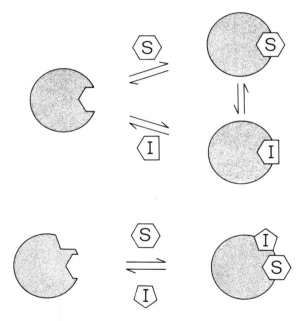

Fig. 3.5. Top: competitive inhibition; inhibitor and substrate compete for the same binding site. For example, indole, phenol, and benzene bind in the binding pocket of chymotrypsin and inhibit the hydrolysis of derivatives of tryptophan, tyrosine, and phenylalanine. Bottom: non-competitive inhibition; inhibitor and substrate bind simultaneously to the enzyme. An example is the inhibition of fructose 1,6-diphosphatase by AMP. This type of inhibition is very common with multisubstrate enzymes. A rare example of uncompetitive inhibition of a single substrate enzyme is the inhibition of alkaline phosphatase by L-phenylalanine (N. K. Ghosh and W. H. Fishman, *J. biol. Chem.* **241,** 2516 (1966)). This enzyme is composed of two identical sub-units so the phenylalanine presumably binds at one site and the substrate at the other

2. Non-competitive, uncompetitive, and mixed inhibition

Different inhibition patterns occur if I and S bind simultaneously to the enzyme instead of competing for the same binding site.

$$
\begin{array}{ccc}
\text{E} & \xrightarrow{S, K'_M} \text{ES} \xrightarrow{k_{cat}} & \\
\big\updownarrow {\scriptstyle I, K_I} & \big\updownarrow {\scriptstyle I, K'_i} & \\
\text{EI} & \xrightarrow{S, K_M} \text{ESI} \xrightarrow{k'} &
\end{array}
\tag{3.33}
$$

It may be shown for the simplifying case of the Michaelis–Menten mechanism where the dissociation constant of S from EIS is the same as that from ES (i.e. $K_M = K'_M$) but ESI does not react (i.e. $k' = 0$), that

$$
v = \frac{k_{cat}/(1+[I]/K_I)}{[S]+K_M}.
\tag{3.34}
$$

This is termed *non-competitive inhibition*: K_M is unaffected, but k_{cat} is lowered by a factor of $(1+[I]/K_I)$. More commonly, the dissociation constant of S from EIS is different from that from ES. In this case both K_M and k_{cat} are altered and the inhibition is termed *mixed*. A further type of inhibition, *uncompetitive*, occurs when I binds to ES but not E.

E. Non-productive binding

There are examples known where a substrate binds in an alternative unreactive mode at the active site of the enzyme, in competition with the productive mode of binding. This is known as non-productive binding.

$$
\begin{array}{c}
\text{E} \underset{K'_S}{\overset{K_S}{\big\langle}} \begin{array}{l} \text{ES} \xrightarrow{k_2} \text{EP} \\ \text{ES}' \end{array}
\end{array}
\tag{3.35}
$$

The effect of this on the Michaelis–Menten mechanism is to lower both k_{cat} and K_M. k_{cat} is lowered since, at saturation, only a fraction of the substrate is productively bound. K_M is lower than K_S since the existence of additional binding modes must lead to apparently tighter binding.

Solving eqn (3.35) by the usual procedures gives

$$
v = \frac{[E_0][S]k_2}{K_S+[S](1+K_S/K'_S)}.
\tag{3.36}
$$

By comparison with the Michaelis–Menten eqn (3.1) this gives:

$$
k_{cat} = k_2/(1+K_S/K'_S)
\tag{3.37}
$$

and

$$
K_M = K_S/(1+K_S/K'_S).
\tag{3.38}
$$

It should be noted that

$$k_{cat}/K_M = k_2/K_S. \tag{3.39}$$

That is, k_{cat}/K_M is unaffected by the presence of the additional binding mode since k_{cat} and K_M are altered in a compensating manner. For example, if the non-productive site binds a thousand times more strongly than the productive site, K_M will be a thousand times lower than K_S, but since only one molecule in a thousand is productively bound, k_{cat} is a thousand times lower than k_2 (strictly speaking the factor is 1001).

F. $k_{cat}/K_M = k_2/K_S$

The result of eqn (3.39) for non-productive binding is quite general. It applies to cases where intermediates occur on the reaction pathway as well as in the non-productive modes. For example, in eqn (3.19) for the action of chymotrypsin on esters where an acylenzyme accumulates, it is seen from the ratios of eqns (3.21) and (3.22) that $k_{cat}/K_M = k_2/K_S$. This relationship clearly breaks down for the Briggs–Haldane mechanism in which the enzyme–substrate complex is not in thermodynamic equilibrium with the free enzyme and substrates. It should be borne in mind that K_M might be a complex function when there are several enzyme-bound intermediates in rapid equilibrium, as in eqn (3.16). Here k_{cat}/K_M is a function of all the bound species.

G. Competing substrates

1. An alternative formulation of the Michaelis–Menten equation

Suppose two substrates compete for the active site of the enzyme.

$$\tag{3.40}$$

The reaction rates may be calculated by the usual steady-state or Michaelis–Menten assumptions. However, there is an alternative approach for rapidly calculating the ratio of the reaction rates. Substitution of eqn (3.24) into the Michaelis–Menten eqn (3.1) gives

$$v = (k_{cat}/K_M)[E][S], \tag{3.41}$$

where $[E]$ is the concentration of free or unbound enzyme. This is a useful equation since it is based on $[E]$ rather than $[E_0]$. Several important relationships may be inferred directly from this equation without the need for a detailed mechanistic analysis as shown below.

2. Specificity for competing substrates

If two substrates A and B compete for the enzyme, then as

$$\mathrm{d}[A]/\mathrm{d}t = v_A = (k_{cat}/K_M)_A[E][A] \tag{3.42}$$

and

$$\mathrm{d}[B]/\mathrm{d}t = v_B = (k_{cat}/K_M)_B[E][B], \tag{3.43}$$

$$v_A/v_B = (k_{cat}/K_M)_A[A]/(k_{cat}/K_M)_B[B]. \tag{3.44}$$

The important conclusion is that specificity, in the sense of discrimination between two competing substrates, is determined by the ratios of k_{cat}/K_M and not by K_M alone. Since k_{cat}/K_M is unaffected by nonproductive binding (E) or the accumulation of intermediates (F), these phenomena do not affect specificity (see Chapter 11).

H. Reversibility: the Haldane equation[10]

1. Equilibria in solution

$$\mathrm{S} \underset{k_r}{\overset{k_f}{\rightleftharpoons}} \mathrm{P} \tag{3.45}$$

$$K_{eq} = [P]/[S] = k_f/k_r \tag{3.46}$$

An enzyme cannot alter the equilibrium constant between the free solution concentrations of S and P. This places constraints on the relative values of k_{cat}/K_M for the forward and reverse reactions. Specifically, as the rates of formation of P and S are equal at equilibrium, application of eqn (3.44) gives

$$(k_{cat}/K_M)_S[E][S] = (k_{cat}/K_M)_P[E][P] \tag{3.47}$$

so that

$$(k_{cat}/K_M)_S/(k_{cat}/K_M)_P = K_{eq}. \tag{3.48}$$

This relationship is known as the Haldane equation, after Haldane who derived it in 1930.

2. Equilibria on the enzyme surface

The Haldane equation does not relate the equilibrium constant between ES and EP to that between S and P in solution. The equilibrium constant for the enzyme-bound reagents is often very different from that in solution for several reasons.

(i) '*Strain*'. The geometry of the active site may be such that, for example, P is bound more tightly than S. The equilibrium on the enzyme surface will favour P more than in the equilibrium in solution.

(ii) '*Non-productive binding*'. If the enzyme has alternative binding modes for S apart from the catalytically productive mode, these will favour [ES] in the equilibrium.

(iii) '*Entropy*'. In the case where 'P' is two separate molecules, that is

$$S \rightleftharpoons P + P', \tag{3.49}$$

the equilibrium constant in solution has a term reflecting the favourable gain in entropy on forming two molecules from one. However, if both P and P' are bound on the enzyme surface,

$$ES \rightleftharpoons EPP', \tag{3.50}$$

the relevant equilibrium constant will not have this entropic contribution.

One example where an enzyme bound equilibrium is vastly different from that in solution is given in Chapter 7E. The hydrolysis constant for ATP in equilibrium with ADP and orthophosphate bound to myosin (S_1) is only 9, compared with a value of about 10^5 to 10^6 for the equilibrium in solution.

I. Breakdown of the Michaelis–Menten equation

Apart from essentially trivial reasons, such as the experimental inability to measure initial rates, there are two main reasons for the failure of the equation.

The first is substrate inhibition. A second molecule of substrate binds to give an ES_2 complex which is catalytically inactive. If, in a simple Michaelis–Menten mechanism, the second dissociation constant is K'_S, then

$$v = \frac{[E_0][S]k_{cat}}{K_S + [S] + [S]^2/K'_S}. \tag{3.51}$$

At low concentrations of [S] the rate is given by $v = [E_0][S]k_{cat}/K_S$ as usual. But as [S] increases, there is first a maximum value of v followed by a decrease.

The second case is substrate activation: an ES_2 complex is formed which is more active than ES.

J. Multisubstrate systems

We have dealt so far with enzymes that react with a single substrate only. The majority of enzymes, however, involve two substrates. The dehydrogenases, for example, bind both NAD^+ and the substrate that is to be oxidized. Many of the principles developed for the single-substrate systems may be extended to multisubstrate systems. However, the general solution of the equations for such systems is complicated and well beyond the scope of this book. Four books devoted almost solely to the detailed analysis of the steady-state kinetics of multisubstrate systems have been recently published, and the reader is referred to these for advanced study.[11–14] The excellent short accounts by Cleland[15] and Dalziel[16] are highly recommended.

From the point of view of this book, the most important experimental observation is that most reactions obey Michaelis–Menten kinetics when the concentration of one substrate is held constant and the other is varied. Furthermore, in practice, only a limited range of mechanisms is commonly observed. In this section we shall just list some common pathways and give a glossary of terms.

Reactions in which all the substrates bind to the enzyme before the first product is formed are called *sequential*. Reactions in which one or more products are released before all the substrates are added are called *ping pong*. Sequential mechanisms are called *ordered* if the substrates combine with the enzyme and the products dissociate in an obligatory order. A *random* mechanism implies no obligatory order of combination or release. The term *rapid equilibrium* is applied when the chemical steps are slower than those for the binding of reagents. Some examples follow.

1. Random sequential mechanism

$$\text{(3.52)}$$

The complex EAB is called a *ternary* or *central* complex.

2. Ordered mechanism

$$E \xrightleftharpoons{A} EA \xrightleftharpoons{B} EAB \longrightarrow \qquad (3.53)$$

Ordered mechanisms often occur in the reactions of the NAD^+-linked dehydrogenases, the coenzyme binding first. The molecular explanation for this is that the binding of the dinucleotide causes a conformational change which increases the affinity of the enzyme for the other substrate (see Chapter 12).

3. Theorell–Chance mechanism

The Theorell–Chance mechanism is an example of the above mechanism in which the ternary complex does not accumulate under the reaction conditions, as is found for horse liver alcohol dehydrogenase.

$$E \xrightleftharpoons{A} EA \xrightarrow{\quad B \quad P \quad} EQ \tag{3.54}$$

(P is one product and Q the other—acetaldehyde and NADH respectively for the liver alcohol dehydrogenase).

4. Ping-pong (or substituted-enzyme or double-displacement) mechanism

The following type of reaction, in which the enzyme reacts with one substrate to give a covalently modified enzyme and release one product, and then reacts with the second substrate, gives rise to the characteristic family of Lineweaver–Burk plots illustrated in Fig. 3.6.

$$E + A \rightleftharpoons E.A \rightleftharpoons E{-}P + Q \tag{3.55}$$

$$E{-}P + B \rightleftharpoons E{-}P.B \rightarrow E + P{-}B. \tag{3.56}$$

An example of a ping-pong reaction is a phosphate-transferring enzyme, such as phosphoglycerate mutase, which is phosphorylated by one

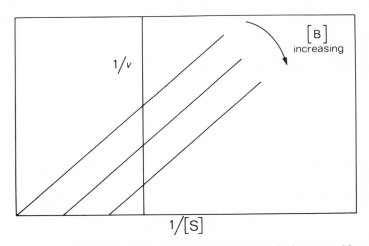

Fig. 3.6. The characteristic parallel line reciprocal plots of ping-pong kinetics. As the concentration of the second substrate in the sequence increases, V_{max} increases as does the K_M for the first substrate. V_{max}/K_M, the reciprocal of the slope of the plot, remains constant

substrate to form a phosphorylenzyme (E—P in eqn (3.55)), and then transfers the phosphoryl group to a second substrate (3.57), where, in the

$$\text{E—N} + \text{ROPO}_3^= \rightleftharpoons \text{E—N.ROPO}_3^= \underset{\text{ROH}}{\searrow} \text{E—N—PO}_3^= \overset{\text{R'OH}}{\searrow} \text{E—N} + \text{R'OPO}_3^= \qquad (3.57)$$

case of phosphoglycerate mutase, N is the imidazole side chain of a histidine residue—Chapter 12G3). Another example occurs in the transfer of an acyl group from acetyl coenzyme A to sulphanilimide or other amines catalysed by acetyl coenzyme A: arylamine acetyltransferase

$$\text{E—SH} + \text{CH}_3\text{COSCoA} \rightleftharpoons \text{E—SH.CH}_3\text{COSCoA} \underset{\text{HSCoA}}{\searrow} \text{E—S—COCH}_3 \overset{\text{RNH}_2}{\searrow}$$

$$\text{E—SH} + \text{RNHCOCH}_3$$

$$(3.58)$$

(3.58). This reaction almost certainly involves the formation of an acyl-thioenzyme in which the —SH of a cysteine residue is acylated.[17]

In many ways, ping-pong kinetics are the most mechanistically informative of all the types of steady-state kinetics, since information is given about the occurrence of a covalent intermediate. The finding of ping-pong kinetics is often used as evidence for such an intermediate, but as other kinetic pathways can give rise to the characteristic parallel double-reciprocal plots of Fig. 3.6, the evidence must always be treated with caution and confirmative evidence sought.

Steady-state kinetics may be used to distinguish between the various mechanisms mentioned above. Under the appropriate conditions, their application can determine the order of addition of substrates and the order of release of products from the enzyme during the reaction. For this reason, the term 'mechanism' when used in steady-state kinetics often refers just to the sequence of substrate addition and product release.

References

1   L. Michaelis and M. L. Menten, *Biochem. Z.* **49**, 333 (1913).
2   G. E. Briggs and J. B. S. Haldane, *Biochem. J.* **19**, 338 (1925).
3   B. Chance, *J. biol. Chem.* **151**, 553 (1943).
4   L. Peller and R. A. Alberty, *J. Am. chem. Soc.* **81**, 5907 (1959).
5   H. Lineweaver and D. Burk, *J. Am. chem. Soc.* **56**, 658 (1934).
6   G. S. Eadie, *J. biol. Chem.* **146**, 85 (1942).
7   B. H. J. Hofstee, *Nature, Lond.* **184**, 1296 (1959).
8   J. E. Dowd and D. S. Riggs, *J. biol. Chem.* **249**, 863 (1965).
9   G. L. Atkins and I. A. Nimmo, *Biochem. J.* **149**, 775 (1975).
10 J. B. S. Haldane, *Enzymes* Longmans (1930).

11 H. J. Fromm, *Initial rate enzyme kinetics*, Springer (1975).
12 J. T.-F. Wong, *Kinetics of enzyme mechanisms*, Academic Press (1975).
13 I. H. Segel, *Enzyme kinetics*, Wiley (1975).
14 A. Cornish-Bowden, *Principles of enzyme kinetics*, Butterworths (1975).
15 W. W. Cleland, *The Enzymes* **2,** 1 (1970).
16 K. Dalziel, *The Enzymes* **10,** 2 (1975).
17 W. P. Jencks, M. Gresser, M. S. Valenzuela, and F. C. Huneeus, *J. biol. Chem.* **247,** 3756 (1972).

Chapter 4

Measurement and magnitude of enzymic rate constants

Part 1 Methods for measurement: an introduction to pre-steady-state kinetics

Steady-state kinetic measurements on an enzyme usually give only two pieces of kinetic data, the K_M value, which may or may not be the dissociation constant of the enzyme-substrate complex, and the k_{cat} value, which may be a microscopic rate constant but may also be the combination of the rate constants for several steps. The kineticist does have a few tricks that he may be able to use on occasion to detect intermediates and even measure individual rate constants but these are not general and depend on mechanistic interpretations. (Some examples of these methods will be discussed in Chapter 7.) In order to measure the rate constants of the individual steps on the reaction pathway and detect transient intermediates, it is necessary to measure the rate of approach to the steady state. It is during the time period in which the steady state is set up that the individual rate constants may be observed.

Since values of k_{cat} lie between 1 and $10^7 \, s^{-1}$, measurements must be made in the time range $1–10^{-7} \, s$. This requires either techniques for rapidly mixing and then observing the enzyme and substrate, or totally new methods. Also, since the events that are to be observed occur on the enzyme itself, the enzyme must be available in substrate quantities. It is the development of apparatus for measuring these rapid reactions and of techniques for isolating large quantities of pure proteins that has revolutionized enzyme kinetics.

In the following we shall discuss two types of techniques. The first is *rapid mixing*. This is extremely useful since fortunately it is possible to mix two solutions in a fraction of a millisecond and the majority of enzyme turnover numbers are less than $1000 \, s^{-1}$. The second is *relaxation kinetics*. The time barrier due to mixing is overcome by using premixed solutions.

A. Rapid mixing and sampling techniques

1. Continuous-flow method

In 1923, Hartridge and Roughton introduced the continuous-flow method to solution kinetics in order to study the combination of deoxyhaemoglobin

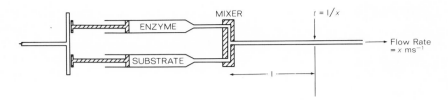

FIG. 4.1. Continuous-flow apparatus

with ligands.[1] The principle of the method is illustrated in Fig. 4.1. Two syringes are connected by a mixing chamber to a flow tube. One syringe is filled with the enzyme, the other with substrate, and the two are compressed at a constant rate. The two solutions mix thoroughly in the mixing chamber, pass down the flow tube and 'age'. At a constant flow rate the age of the solution is linearly proportional to the distance down the flow tube and the flow rate; e.g. if the flow rate is 10 m s^{-1}, then 1 cm after the mixing chamber the solution is 1 ms old, 10 cm after mixing it is 10 ms old and so on. The flow rate of the liquid must be kept above a critical velocity in order to ensure 'turbulent flow'. Below this value, about 2 m s^{-1} for a tube of 1 mm diameter, the flow may be laminar, the liquid at the centre travelling faster than that near the wall. This places an upper time limit on the apparatus for a particular length of tube.

2. Stopped-flow method

This was introduced by Roughton in 1934[2] and greatly improved by Chance some six years later.[3] The principle is illustrated in Fig. 4.2.[4] Instead of the continuous-flow system above, the two driving syringes are

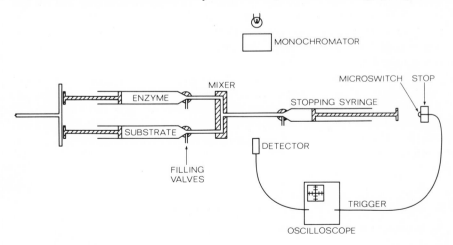

FIG. 4.2. Stopped-flow apparatus

compressed to express about 50–200 μl from each and then mechanically stopped. Suppose there is an observation point 1 cm after the mixing chamber. If the flow rate is 10 m s^{-1} during the period of compression, then during this *continuous-flow* period the detector sees a solution which is 1 ms old. When the flow is stopped, the solution ages normally with time and the detector then sees the events occurring after one millisecond. The age of the solution at the initial observation is known as the 'dead time' of the apparatus.

The stopped-flow method is the routine laboratory tool whilst the continuous-flow apparatus is used in a few specialized cases only. The stopped-flow technique requires only 100–400 μl of solution or less for a complete time course of a reaction, has a dead time as low as 0·5 ms or so and observations may be extended to several minutes. It does, however, require a rapid detection and recording system. The continuous-flow system requires very large reaction volumes and readings may be taken only up to about 100 ms or so due to the impracticabilities of using longer observation tubes. Also there are difficulties in that the whole length of the flow tube must be scanned by the detector. Apart from mechanical problems, this may lead to systematic errors if the tube is not uniform in dimensions, thermostatting etc. It is, however, a slightly faster technique, with dead times as low as 100–200 μs since there are no mechanical problems of stopping the flow—the stopping takes a fraction of a millisecond and can set up shock waves if it is too vigorous. A second advantage is that slowly responding detectors may be used since at a particular point on the flow tube the age of the solution is constant. This was particularly important in the original experiments of Hartridge and Roughton who, before the introduction of photomultipliers, used a hand (reversion) spectroscope as a detector!

3. Rapid quenching techniques

Instead of using a photomultiplier or other detector in the flow systems, the solutions may be quenched by, say, the addition of an acid, such as trichloroacetic acid, and the reaction products directly analysed by chromatographic or other techniques.

a. Quenched flow

The simplest form of the method is to submerge the end of the observation tube of a continuous-flow apparatus in a beaker of acid. A somewhat more sophisticated version is illustrated in Fig. 4.3. A third syringe is added which mixes the quenching acid with the reagent solutions via a second mixing chamber. Such an apparatus may have a dead time of only 4 or 5 ms. However, the maximum practical reaction time that may be measured using small volumes of reagents is about 100–150 ms, otherwise excessively long reaction tubes are required.

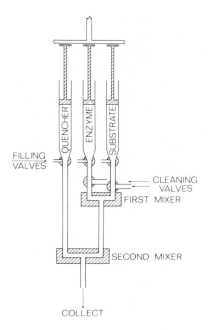

FIG. 4.3. Quenched-flow apparatus

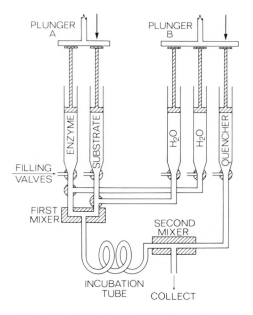

FIG. 4.4. Pulsed quenched-flow apparatus

b. Pulsed quenched-flow technique
The time range of the quenched-flow technique may be extended by a procedure similar to that of stopped-flow.[5] As illustrated in Fig. 4.4, the enzyme and substrate are first mixed and driven into an incubation tube by a plunger actuated by compressed air, and after the desired time interval, 150 ms or greater, a second plunger is actuated which drives the incubated mixture with a pulse of distilled water into a second mixer where it is quenched.

B. Relaxation methods

1. Temperature jump technique
The time involved in mixing places a limit on the dead time of flow techniques. The only way to increase the time resolution is to cut out the mixing and have an equilibrium mixture of the reagents already incubated together, and then somehow measure the desired rate constant.

The most common method of doing this is to use a relaxation technique. The system is perturbed from its equilibrium position and its rate of relaxation to a new equilibrium is measured. For example, in the temperature jump method[6] (Fig. 4.5), a solution is incubated in an absorbance or fluorescence cell and its temperature is raised through 5–10° in less than a microsecond by the discharge of a capacitor (or, in recent developments, in 10–100 ns by the discharge of an infrared laser). If the equilibrium involves an enthalpy change, the equilibrium position will change. The system will proceed to its new equilibrium position via a series of *relaxation times*, τ ($\equiv$ reciprocal of rate constant).

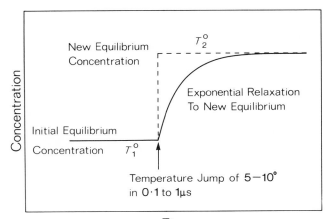

Fig. 4.5. Illustration of temperature jump

Clearly, this method cannot be applied to systems where there are irreversible chemical processes involved. It is most suitable for situations involving simple ligand binding, such as NAD$^+$ with a dehydrogenase, inhibitor binding, or conformational changes in the protein. There have been some attempts to combine the temperature-jump with the stopped-flow method.

2. Nuclear magnetic resonance[7–10]

NMR may sometimes be used to measure the *dissociation* rate constants of enzyme–inhibitor complexes. These may then be combined with the binding constants for the reaction to give the association rate constants.

C. Analysis of pre-steady-state and relaxation kinetics

It is relatively straightforward to solve the differential equations for the time dependence of the transients in simple cases. However, it is important to understand the physical meaning of why a particular case gives rise to a particular form of solution. In this section we will concentrate on an intuitive approach in order to do this. Once a feel for the subject has been developed, algebraic mistakes will not be made and also some complex kinetic schemes may be solved by inspection.

1. Simple exponentials

a. Irreversible reactions

Suppose a compound A transforms into B with a first-order rate constant k_f, and the reaction proceeds to completion:

$$A \xrightarrow{k_f} B. \tag{4.1}$$

Then

$$\frac{d[A]}{dt} = -k_f[A]. \tag{4.2}$$

This is solved by integration to give

$$[A_t] = [A_0] \exp(-k_f t) \tag{4.3}$$

where $[A_0]$ is the initial concentration of A. Since

$$[A_t] + [B_t] = [A_0], \tag{4.4}$$

$$[B_t] = [A_0]\{1 - \exp(-k_f t)\}. \tag{4.5}$$

Both [A] and [B] follow simple exponentials.

It should be noted that the half life of the reaction, $t_{\frac{1}{2}}$, where $[A] = [B] = ([A_0]/2)$ is given by

$$\exp(-k_f t) = \tfrac{1}{2} \tag{4.6}$$

i.e.

$$t_{\frac{1}{2}} = \frac{0 \cdot 6931}{k_f} = 0 \cdot 6931 \tau. \tag{4.7}$$

b. Method of initial rates

When discussing more complex examples we shall use an adaptation of the *method of initial rates* to illustrate the physical meaning of some of the expressions. This method is often used in experiments which are too slow to follow over a complete time course or when there are complicating side reactions.

The initial rate v_0 of eqn (4.1) is given by

$$v_0 = k_f[A_0]. \tag{4.8}$$

k_f is determined by dividing v_0 by the expected change in reagent concentrations thus:

$$k_f = v_0 / \Delta[A_0] \tag{4.9}$$

c. Reversible reactions

In this case A does not completely transform to B but there is an equilibrium concentration of A:

$$A \underset{k_r}{\overset{k_f}{\rightleftharpoons}} B \tag{4.10}$$

and

$$K_e = [B]/[A] = k_f/k_r. \tag{4.11}$$

Here,

$$\frac{d[A]}{dt} = -k_f[A] + k_r[B]. \tag{4.12}$$

Substitution of eqn (4.4) gives

$$\frac{d[A]}{dt} = -k_f[A] + k_r([A_0] - [A]). \tag{4.13}$$

This equation may be integrated by separating the variables and multiplying each side of the equation by an exponential factor:

$$\frac{d[A]}{dt} + [A](k_f + k_r) = k_r[A_0] \qquad (4.14)$$

$$\frac{d[A]}{dt} \exp(k_f + k_r)t + [A](k_f + k_r)\exp(k_f + k_r)t = k_r[A_0]\exp(k_f + k_r)t$$

$$(4.15)$$

$$\therefore \quad \frac{d}{dt}([A]\exp(k_f + k_r)t) = k_r[A_0]\exp(k_f + k_r)t$$

$$\therefore \quad [A]\exp(k_f + k_r)t = \frac{k_r}{k_f + k_r}[A]\exp(k_f + k_r)t + \text{constant}.$$

$$(4.16)$$

Using the boundary conditions that at $t = 0$, $[A] = [A_0]$ and at $t = \infty$ the equilibrium concentration of $[A]$, $[A]_{eq}$, is given by

$$[A_{eq}] = [A_0]k_r/(k_f + k_r) \qquad (4.17)$$

(from eqns (4.4) and (4.11)), the solution is

$$[A_t] = \frac{[A_0]}{k_f + k_r}[k_f \exp\{-(k_f + k_r)t\} + k_r]. \qquad (4.18)$$

The expression for the time course (4.18) may be divided into various factors. There is first the exponential term with rate constant, or in terms of relaxation kinetics, a reciprocal relaxation time $1/\tau$ given by

$$1/\tau = k_f + k_r, \qquad (4.19)$$

and second an *amplitude* factor given by

$$k_f/(k_f + k_r). \qquad (4.20)$$

Now, supposing that the reaction is started from the other direction, with B initially present but not A. Then

$$[B_t] = \frac{[B_0]}{k_f + k_r}[k_r \exp\{-(k_f + k_r)t\} + k_f]. \qquad (4.21)$$

This expression has the same relaxation time as in eqn (4.18), but a different amplitude factor.

The first important point to be noted is that the rate constant for the approach to equilibrium is greater than either of the individual first-order rate constants, k_f and k_r, and is equal to their sum. The reason why this is so is readily understood on applying the principle of initial rates. The initial velocity in the reversible case (4.10) is the same as for the irreversible eqn (4.1), but the reaction does not have to proceed so far. For example, before B accumulates only A is present, so again

$$v_0 = k_f[A_0],$$

but the total change in [A] is given by

$$\Delta[A_0] = [A_0] - [A_{eq}]. \tag{4.22}$$

Substitution of eqn (4.17) gives

$$\Delta[A_0] = [A_0]k_f/(k_f + k_r) \tag{4.23}$$

$$\therefore \quad 1/\tau = v_0/\Delta[A_0] = k_f + k_r \tag{4.24}$$

The second point to be noted is that k_f and k_r cannot be assigned without a knowledge of the amplitude factor. This basic symmetry in the relaxation times occurs in many cases, and, in general, the rate constants for unimolecular reactions cannot be assigned unless the concentrations of A and B at equilibrium may be determined. It will be seen later that when the reactions are not unimolecular, but pseudo-unimolecular because of the presence of a second reagent, the relaxation time will have a concentration dependence which removes this ambiguity.

2. Association of enzyme and substrate

$$[E] + [S] \rightleftharpoons [ES]. \tag{4.25}$$

If $[S] \gg [E]$, the reaction is effectively first order since the concentration of [S] is hardly affected by the reaction. If the second-order rate constant for the association is k_{on}, and that for dissociation is k_{off}, then the system reduces to

$$[E] \underset{k_{off}}{\overset{k_{on}[S]}{\rightleftharpoons}} [ES]. \tag{4.26}$$

The relaxation time for this reaction is, from eqn (4.19),

$$\boxed{1/\tau = k_{off} + k_{on}[S]. \tag{4.27}}$$

Two points should be noted; (a) as the rate constants are *pseudo-unimolecular*, there is a concentration dependence so that k_{on} and k_{off} may be resolved without the amplitude factor; (b) there is a lower limit to $1/\tau$, that is, $1/\tau$ cannot be less than k_{off}. This sets a limit on the measurement of these rate constants. A good stopped-flow spectrophotometer can cope only with rate constants of $1000 \, s^{-1}$ or less, and many enzyme–substrate dissociation constants are faster than this.

A favourable example is given in Fig. 4.6. The dissociation rate constant of tyrosine from its complex with tyrosyl-tRNA synthetase is low so that the association and dissociation rate constants can be measured by

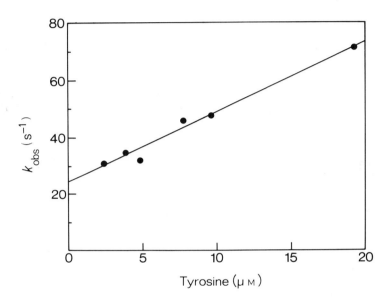

Fig. 4.6. Binding of tyrosine to the tyrosyl–tRNA synthetase from *B. stearothermophilus* (A. R. Fersht, R. S. Mulvey, and G. L. E. Koch, *Biochemistry* **14**, 13 (1975))

stopped flow. (Note that sometimes a two-step process may be mistaken for the above single-step reaction—see Section 6).

Where there is no subsequent turnover of a substrate, such as occurs on the omission of a co-substrate in a multisubstrate reaction, or on inhibitor binding, the T-jump technique is generally the most useful tool for the determination of these constants.

3. Consecutive reactions

a. Irreversible reactions

The simplest case of consecutive reactions is

$$A \xrightarrow{k_1} B \xrightarrow{k_2} C. \qquad (4.28)$$

This is simply solved using the conservation equation and the integration procedures above to give:

$$[A] = [A_0] \exp - k_1 t$$

$$[B] = \frac{[A_0]k_1}{k_2 - k_1} \{\exp(-k_1 t) - \exp(-k_2 t)\} \qquad (4.29)$$

$$[C] = [A_0] \left[1 + \frac{1}{k_1 - k_2} \{k_2 \exp(-k_1 t) - k_1 \exp(-k_2 t)\} \right].$$

[B] is a transient intermediate that appears and then disappears. If $k_1 \gg k_2$, it is formed with rate constant k_1 and then slowly decomposes

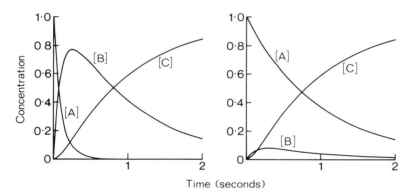

FIG. 4.7. Plots of the concentrations of A, B, and C in the reaction $A \to B \to C$ (eqn (4.28)). LHS, $k_1 = 10 \text{ s}^{-1}$, $k_2 = 1 \text{ s}^{-1}$. RHS, $k_1 = 1 \text{ s}^{-1}$, $k_2 = 10 \text{ s}^{-1}$. Note: (a) the progress curves for C are identical; (b) the *shapes* of the two curves for B are identical—they differ only in amplitude; (c) A decreases ten times more rapidly on the LHS. Thus, unless the concentration of A is monitored, the two examples cannot be distinguished on the basis of measured rate constants only

with rate constant k_2. However, if $k_2 \gg k_1$, [B] reaches a steady-state concentration with a rate constant k_2 and decays slowly with rate constant k_1. The apparently paradoxical situation is that the intermediate appears to be formed with its decomposition rate constant and decomposes with its formation rate constant! This is readily understood for the initial rate treatment. When $k_1 \ll k_2$, [B] reaches a steady-state concentration given by

$$\frac{d[B]}{dt} = 0 = k_1[A] - k_2[B]. \qquad (4.30)$$

The initial concentration of B on setting up the steady state is given by

$$[B]_{SS} \simeq \frac{k_1}{k_2}[A_0] \qquad (i.e. \ll [A_0]) \qquad (4.31)$$

and

$$v_0 = k_1[A_0]. \qquad (4.32)$$

$$\therefore \quad 1/\tau = v_0/[B_{SS}] = k_2. \qquad (4.33)$$

In this latter case, where $k_2 > k_1$, [B] is at a very low concentration and in the former, where $k_1 > k_2$, it accumulates. The two cases can be resolved if the concentration of [B] can be monitored.

This type of kinetic situation sometimes occurs in protein renaturation experiments where the kinetics are often monitored by fluorescence changes. The biphasic traces cannot be resolved in such circumstances unless the quantum yield of the transient intermediate is known so that its absolute concentration can be determined.

An example of the application of these equations is found in Chapter 7D. The aminoacyl-tRNA synthetase that specifically esterifies the tRNA molecule that accepts valine, $tRNA^{Val}$, 'corrects' the error of mistakenly forming an aminoacyl adenylate with threonine by the following scheme:

$$E.Thr \sim AMP.tRNA \xrightarrow[AMP]{transfer} E.Thr\text{-}tRNA \xrightarrow{hydrolysis} E + Thr + tRNA \qquad (4.34)$$

It will be seen that the rate of disappearance of E.Thr $\sim$ AMP.tRNA was directly measured by the formation of AMP, the intermediate E.Thr-tRNA was directly measured, and the second step (hydrolysis) measured directly and independently by isolating the mischarged tRNA and adding it to the enzyme.

b. Quasi-reversible: steady state

A more common situation in enzyme kinetics is the following:

$$E \xrightarrow{k_1 S} ES' \xrightarrow{k_2} E \qquad (4.35)$$
$$+P_1 \qquad +P_2$$

For example, chymotrypsin reacts with p-nitrophenyl acetate (AcONp) according to the above scheme (when $[AcONp] \ll K_S$ for the first step) to give an intermediate acylenzyme, EAc:

$$E \xrightarrow[HONp]{k_1[AcONp]} EAc \xrightarrow[AcOH]{H_2O} E \qquad (4.36)$$

(where $k_1 = k_{cat}/K_M$ for this step).

Since the acylenzyme is continuously being formed and turning over, its concentration is in the steady state (providing $[AcONp] \gg [E]$). The steady-state concentration of the acylenzyme is given by:

$$\frac{d}{dt}[EAc] = 0 = k_1[AcONp][E] - k_2[EAc]_{ss}. \qquad (4.37)$$

Now since

$$[E] + [EAc] = [E_0] \qquad (4.38)$$

$$0 = k_1[AcONp]([E_0] - [EAc]) - k_2[EAc] \qquad (4.39)$$

and

$$[EAc]_{ss} = \frac{k_1[AcONp][E_0]}{k_2 + k_1[AcONp]}. \qquad (4.40)$$

Applying the initial rate treatment,

$$v_0 = k_1[E_0][AcONp], \qquad (4.41)$$

$$1/\tau = v_0/[EAc]_{ss} \qquad (4.42)$$

$$= k_2 + k_1[AcONp]. \qquad (4.43)$$

Just as in the case of reversible reactions, the intermediate is formed with a rate constant that is greater than the rate constant for the transformation of the preceding intermediate.

The analytical solution for the rate constant is

$$[HONp] = [E_0]\left(\frac{k_1'}{k_1' + k_2}\right)\left(\frac{k_1'}{k_1' + k_2}[1 - \exp{-(k_1' + k_2)t}] + k_2 t\right)$$

$$(4.44)$$

where $k_1' = k_1[AcONp]$. If saturation kinetics are observed for the acylation step, k_1' is of the form $k_{cat}[S]/(K_M + [S])$. There is an initial exponential phase, which dies out after t is about 5 times greater than τ, and a linear term which eventually predominates.

c. Consecutive reversible reactions

The general solution for these reactions is given later in Section 6. We shall deal here with cases where one step is fast compared with the other. Under these circumstances the relaxation times are on different time scales and do not 'mix' with each other.

(i) *First step fast.* (pre-equilibrium)

$$E \underset{k_{-1}}{\overset{k_1[S]}{\rightleftharpoons}} ES \underset{k_{-2}}{\overset{k_2}{\rightleftharpoons}} ES' \qquad (4.45)$$

This example may be readily solved by inspection on applying two simple rules. (1) There must be two sets of relaxation times since there are two sets of reactions involved. (2) As the reactions occur on different time scales, each may be dealt with separately.

The first relaxation time is for the binding step. By analogy with eqn (4.27) this is given by

$$1/\tau_1 = k_{-1} + k_1[S]. \qquad (4.46)$$

The second relaxation time is for the slow step. This is an example of a reversible reaction and, by analogy with eqn (4.19), the reciprocal relaxation time is given by the sum of the forward and reverse rate constants for the step. However, the effective forward rate constant for this step is given by k_2 multiplied by the fraction of the enzyme that is in the form of ES, i.e.

$$1/\tau_2 = k_{-2} + \frac{k_2[S]}{[S] + K_S} \qquad (4.47)$$

where

$$K_S = k_{-1}/k_1. \qquad (4.48)$$

(ii) *Second step fast* The above reaction could be an example of a substrate-induced conformational change in the enzyme, where ES′ is just a different conformational state, or alternatively, the accumulation of an intermediate on the pathway. The following reaction is an example of the displacement of an equilibrium between two conformational states of an enzyme caused by the binding of a substrate to one form only.

$$E' \underset{k_{-1}}{\overset{k_1}{\rightleftharpoons}} E \underset{\text{fast}}{\overset{S, K_S}{\rightleftharpoons}} ES. \qquad (4.49)$$

This situation is found for the binding of ligands to chymotrypsin: this exists in two conformational states and only one binds aromatic substrates. It may be shown from the formal analysis given later in Section 6 that

$$1/\tau_2 = k_1 + k_{-1}\left(\frac{K_S}{[S] + K_S}\right) \qquad (4.50)$$

This type of situation may be distinguished from case (i) since $1/\tau_2$ *decreases* with increasing [S]. This may be understood by analogy with the examples of irreversible and reversible reactions (eqns (4.1) and (4.10)). Clearly, when [S] is very high the reaction is essentially irreversible since E′ is transformed completely to ES and so $1/\tau_2$ tends to k_1.

Similarly as [S] tends to zero, there is very little [ES] and $1/\tau_2$ tends to $k_1 + k_{-1}$. Hence the concentration dependence.

4. Parallel reactions

Parallel reactions are said to arise when a compound undergoes two or more reactions simultaneously. This situation often occurs in enzymic reactions where an activated intermediate is formed which may react with several competing acceptors.

$$
\begin{array}{c}
\overset{k_\mathrm{B}}{\nearrow}\ \mathrm{B} \\
\mathrm{A} \\
\underset{k_\mathrm{C}}{\searrow}\ \mathrm{C}
\end{array}
\tag{4.51}
$$

The kinetic equations are easily solved by integration, but it is instructive to solve them intuitively. It is obvious that [A] decreases with a rate constant that is the sum of k_B and k_C, and also that B and C are formed in the ratio of the rate constants. Since the rates of formation of B and C depend on the concentration of A, they must both be formed with a rate constant that is the same as for its disappearance. Therefore,

$$[A] = [A_0] \exp -(k_\mathrm{B} + k_\mathrm{C})t, \tag{4.52}$$

$$[B] = \frac{[A_0]k_\mathrm{B}}{k_\mathrm{B} + k_\mathrm{C}} \{1 - \exp -(k_\mathrm{B} + k_\mathrm{C})t\}, \tag{4.53}$$

$$[C] = \frac{[A_0]k_\mathrm{C}}{k_\mathrm{B} + k_\mathrm{C}} \{1 - \exp -(k_\mathrm{B} + k_\mathrm{C})t\}, \tag{4.54}$$

The situation is similar to that for reversible reactions (eqn (4.10)) where the relaxation time is composed of the sums of those for two reactions.

Examples of parallel reactions are given in Chapter 7, e.g. the attack of various nucleophiles on acylchymotrypsins measured by steady-state and pre-steady-state kinetics.

5. Derivation of equations for temperature jump

As an illustration, consider the association of an enzyme and substrate in a one-step reaction:

$$\mathrm{E + S} \underset{k_{-1}}{\overset{k_1}{\rightleftharpoons}} \mathrm{ES} \tag{4.55}$$

Suppose that because of the change of temperature the equilibrium moves to a new position so that

$$[E] = [E_e] + e \tag{4.56}$$

$$[S] = [S_e] + s \tag{4.57}$$

$$[ES] = [ES_e] + es \tag{4.58}$$

where $[E_e]$, $[S_e]$, and $[ES_e]$ are the equilibrium concentrations at the new temperature.

$$d[ES]/dt = k_1([E_e]+e)([S_e]+s) - k_{-1}([ES_e]+es) \qquad (4.59)$$
$$= k_1[E_e][S_e] - k_{-1}[ES_e] + k_1([E_e]s + [S_e]e + e \cdot s) - k_{-1}es. \qquad (4.60)$$

Eqn (4.60) may be simplified since the first two terms on the right-hand side cancel out as they are equal at equilibrium. Also since the reagents are conserved, $e = s = -es$. And since $[ES_e]$ is a constant, $d[ES]/dt = des/dt$. Therefore,

$$-des/dt = k_1([E]es + [S]es + e \cdot s) + k_{-1}es. \qquad (4.61)$$

Now, if the perturbation from equilibrium is small, the second-order term $e \cdot s$ may be ignored. Eqn (4.61) may then be integrated to give the relaxation time:

$$1/\tau = k_1([E]+[S]) + k_{-1}. \qquad (4.62)$$

If the equilibrium is perturbed only slightly, the return to equilibrium is always a first-order process even though the reagents may be present at similar concentrations.

6. General solution of two-step consecutive reversible reactions

The solution of eqn (4.63) involves simultaneous linear differential equations.

$$A \underset{k_{-1}}{\overset{k_1}{\rightleftharpoons}} B \underset{k_{-2}}{\overset{k_2}{\rightleftharpoons}} C. \qquad (4.63)$$

Two relaxation times are obtained:

$$1/\tau_1 = \tfrac{1}{2}(p+q) \qquad (4.64)$$
$$1/\tau_2 = \tfrac{1}{2}(p-q) \qquad (4.65)$$

where

$$p = k_1 + k_{-1} + k_2 + k_{-2}, \quad \text{and} \quad q = [p^2 - 4(k_1 k_2 + k_{-1} k_{-2} + k_1 k_{-2})]^{\frac{1}{2}}. \qquad (4.66)$$

These basic equations may be manipulated to cover many cases. A useful trick is to express the rate constants as the sums and products of the relaxation times:

$$1/\tau_1 + 1/\tau_2 = k_1 + k_{-1} + k_2 + k_{-2} \qquad (4.67)$$
$$1/\tau_1 \tau_2 = k_1 k_2 + k_{-1} k_{-2} + k_1 k_{-2} \qquad (4.68)$$

The equations are easy to solve if a concentration dependence is involved. For example, if the sequence is the pseudo first-order series,

$$E \underset{k_{-1}}{\overset{k_1'[S]}{\rightleftharpoons}} ES \underset{k_{-2}}{\overset{k_2}{\rightleftharpoons}} ES', \qquad (4.69)$$

$k_1'[S]$ may be substituted for k_1 in (4.67) and (4.68). Also, by analogy with eqn (4.62) for temperature-jump experiments, $k_1'([S]+[E])$ may be substituted for k_1 in a relaxation experiment.

The equations may be simplified if one of the relaxation times is much faster than the other. For example, if in (4.69) the first step is fast, $1/\tau_2$ and $k_2 + k_{-2}$ may be ignored in (4.67). τ_2 may then be solved from substituting (4.67) into (4.68). In the case of a temperature-jump experiment, this gives

$$1/\tau_1 = k_1'([E]+[S]) + k_{-1} \qquad (4.70)$$

$$1/\tau_2 = k_{-2} + \frac{k_2([E]+[S])}{k_{-1}/k_1' + ([E]+[S])}. \qquad (4.71)$$

The same manipulations may be performed for

$$E \underset{k_{-1}}{\overset{k_1}{\rightleftharpoons}} E' \underset{k_{-2}}{\overset{k_2'[S]}{\rightleftharpoons}} ES', \qquad (4.72)$$

where the first step is slow, to give

$$1/\tau_1 = k_{-2} + k_2'([E]+[S]) \qquad (4.73)$$

$$1/\tau_2 = k_1 + \frac{k_{-1}\{(k_{-2}/k_2') + [E']\}}{k_{-2}/k_2' + ([E']+[S])} \qquad (4.74)$$

Two practical points should be noted. The kinetic mechanisms in eqns (4.69) and (4.72) may be distinguished by the concentration dependence of $1/\tau_2$. For (4.69) this increases with increasing [S], for (4.72) this decreases. But there are situations that are difficult to resolve. For example, in eqn (4.72) if $[E'] \gg [E]$, there will be a burst of formation of ES' with relaxation time τ_1 followed by a small increase, as E converts to E', at relaxation time τ_2. The concentration dependence of τ_2 will be small since $k_1 \gg k_{-1}$ for $[E'] \gg [E]$ (see Fig. 4.8). This can be mistaken for the scheme in eqn (4.69) where only a little ES' is formed. In this case also the concentration dependence of $1/\tau_2$ is small as $k_{-2} \gg k_2$. In both cases the amplitudes of the changes will often be small and the rate constants difficult to measure precisely.

A more common situation that leads to difficulties is the two-step combination of an enzyme and substrate, as in (4.69), where the dissociation constant for the first step, k_{-1}/k_1', is high. If measurements are made

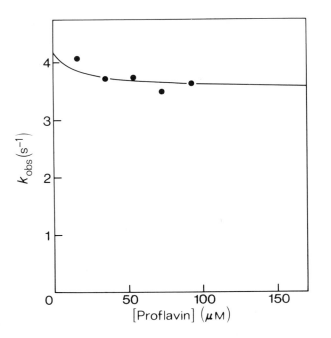

FIG. 4.8. Illustration of the difficulty of distinguishing between mechanisms (4.69) and (4.72) by eqns (4.71) and (4.74). The rate constant for the binding of proflavin to α-chymotrypsin at pH 6·84 and 25° is plotted against the concentration of proflavin. The reaction scheme is

$$\mathrm{E} \underset{0 \cdot 6\,\mathrm{s}^{-1}}{\overset{3 \cdot 1\,\mathrm{s}^{-1}}{\rightleftharpoons}} \mathrm{E}' \underset{\mathrm{PF}}{\overset{\mathrm{Fast}}{\rightleftharpoons}} \mathrm{E}'\mathrm{PF}.$$

(A. R. Fersht and Y. Requena, *J. mol. Biol.* **60,** 279 (1971)). The rate constant decreases with increasing proflavin as predicted by eqn (4.19), but the decrease is small; the maximum possible change is only 19%

only in the region where $k_{-1}/k'_1 > ([\mathrm{E}]+[\mathrm{S}])$, eqn (4.71) reduces to†

$$1/\tau_2 = k_{-2} + k'_1(k_2/k_{-1})([\mathrm{E}]+[\mathrm{S}]). \qquad (4.75)$$

This has the form of a simple one-step association of an enzyme and substrate, as in (4.62), and may mistakenly be interpreted as this. In this case, the association rate constant would appear to be $k'_1(k_2/k_{-1})$, a value lower than the true rate constant of k'_1. Some of the low values seen later in Table 4.3 are undoubtedly caused by this. Measurements should always

---

† Eqn (4.71) was derived on the assumption that $k_{-1} \gg k_2$. If k_2 is appreciable, eqn (4.75) should be modified to

$$1/\tau_2 = k_{-2} + \frac{k'_1 k_2}{k_{-1}+k_2}\,([\mathrm{E}]+[\mathrm{S}]).$$

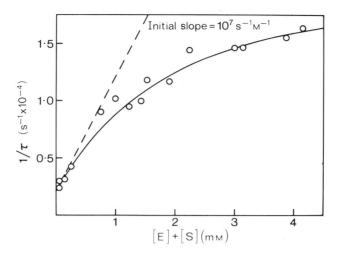

FIG. 4.9. The two-step binding of (N-acetylglucosamine)$_2$ to lysozyme. Solutions of enzyme (0·03 to 0·2 mM) and ligand (0·02 to 4.1 mM) were temperature-jumped from 29° to 38° at pH 6. Experiments at low concentration are in the linear region of the curve and give an apparent second-order rate constant of $10^7 \, \text{s}^{-1} \, \text{M}^{-1}$ for binding. However, measurements at higher concentrations reveal that the rate reaches a plateau, indicating a two-step process (ref. 13).

be extended to high substrate concentrations to search for a levelling off of rate as predicted by eqn (4.71) (Fig. 4.9; see also Fig. 7.10).

7. Experimental application of pre-steady-state kinetics

Later, in Chapter 7, we shall discuss several examples of the successful application of transient kinetics to the solution of enzyme mechanisms. Here, we briefly discuss some of the strategies and tactics used by the kineticist to initiate a transient kinetic study. On many occasions, steady-state kinetics and other studies have set the kineticist a well-defined and specific question to answer. At other times, he just wishes to study a particular system to gather information. In both cases there is no substitute for imagination and insight in designing the incisive experiment. But there is a systematic approach that can be used.

The analysis of a relaxation time may be divided into two basic steps:

(1) The relaxation time must be assigned to a specific physical event.
(2) This physical event must be fitted into the overall kinetic mechanism of the process under study.

Let us now consider two examples. The first is the binding of a ligand to a protein under non-catalytic conditions. Two likely physical events are the initial binding step and a ligand-induced conformational change. The first experiment is the measurement of the concentration dependence and number of relaxation times in order to determine the number of intermediate states and the rate constants for their interconversion. Under ideal circumstances, the number of relaxation times will be equal to the number of steps in the reaction. Even if only one relaxation time is found, its concentration dependence may indicate a two-step process by being non-linear (e.g. eqns (4.71) and (4.74)). Additional information may be obtained by using more than one physical probe, for example fluorescence and absorbance, and studying both the ligand and the protein, since some steps may show up in one of these and not the other. Further physical processes may occur, such as proton release or uptake during the reaction, or a change in the state of aggregation of the protein. The former provides an additional convenient probe since changes of pH may be measured by a chromophoric pH indicator. Aggregation complicates the kinetics, but may be detected and measured to provide additional information. Relaxation techniques are often more powerful than flow methods for these simple reactions because they can measure faster processes. However, there are times when stopped flow is more useful: for example, processes that are too slow for detection by temperature jump may be measured by stopped flow; and also certain experiments, such as the effect of a large change in pH, can be performed only by mixing experiments (although small changes in pH can be made in a temperature-jump experiment by using a buffer whose pK_a is temperature dependent).

The second example is the reaction mechanism of an enzyme under catalytic conditions. In addition to determining the binding steps and conformational changes described above, it is even more important to measure the bond-making and -breaking steps and detect the chemical intermediates in the reaction. The chemical steps are usually best studied by the methods which directly measure the concentration of chemical species, for example stopped-flow spectrophotometry and quenched flow. Indirect measurements of chemical steps, such as a change in protein fluorescence, must somehow be assigned. The ideal situation for study is when an intermediate accumulates and may be detected and measured. In general, as many different probes as possible should be used in order to confirm existing information and add further details. Good examples of the application of such methods are to be found in Chapter 7.

D. The absolute concentration of enzymes

1. Active-site titration and the magnitudes of 'bursts'

The calculation of rate constants from steady-state kinetics and the determination of binding stoichiometries requires the knowledge of the concentration of active sites of the enzyme. It is not sufficient to calculate this from the molecular weight of the protein and its concentration since enzymes are not always isolated 100% pure. This problem has been overcome by the introduction of the technique of active-site titration, a combination of steady-state and pre-steady-state kinetics whereby the concentration of active enzyme is related to an initial burst of product formation. This type of situation occurs when an enzyme-bound inter-mediate accumulates during the reaction. The first mole of substrate rapidly reacts with the enzyme to form stoichiometric amounts of the enzyme-bound intermediate and product, whilst the subsequent reaction is slow since it depends on the slow breakdown of the intermediate to release free enzyme.

$$E + S \xrightarrow{k_1'} EI \xrightarrow{k_2} E \qquad (4.76)$$
$$+ \qquad +$$
$$P_1 \qquad P_2$$

Clearly, if in (4.76) k_1' is very fast and k_2 is negligibly slow, the release of P_1 is easily measured and related to the concentration of enzyme. However, in practice k_2 is generally not negligible, so that there is an initial burst of formation of P_1 followed by a progressive increase as the intermediate turns over. The mathematics of this situation were described previously (eqns (4.35) to (4.44)). It was shown that the overall release of products was linear with time after an initial transient. From eqn (4.44) it is seen that the linear portion extrapolates backs to a burst π, given by

$$\pi = [E_0]\{k_1'/(k_1' + k_2)\}^2. \qquad (4.77)$$

It should be noted that the burst depends on a 'squared' relationship with the rate constants. If the ratio of $k_1' : k_2$ is high, the square term is close to

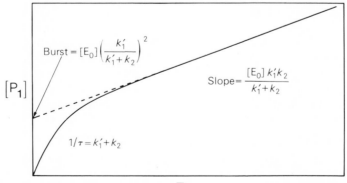

Time

FIG. 4.10. The principle of active-site titration

1 so that the burst is equal to the enzyme concentration. If this condition does not hold, the concentration will be underestimated unless both rate constants are measured and substituted into (4.77).

2. Dependence of burst on substrate concentration

In (4.76) the term k'_1 is the apparent first-order rate constant for the formation of the intermediate under the particular reaction conditions. In general this will follow the Michaelis–Menten equation, that is,

$$k'_1 = k_{cat}[S]/([S] + K_M), \tag{4.78}$$

where k_{cat} and K_M refer to the first step. At sufficiently low concentrations of S there will be no burst, but, provided k_{cat} is greater than k_2, one

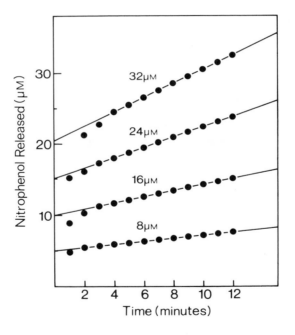

FIG. 4.11. The 'original' active-site titration experiment (B. S. Hartley and B. A. Kilby, *Biochem. J.* **56,** 288 (1954)). The indicated concentrations of chymotrypsin were mixed with *p*-nitrophenyl ethyl carbonate, EtOCO₂—⟨◯⟩ NO₂. The acylenzyme E—O—COOEt is rapidly formed but hydrolyses slowly. Note that about 0·63 moles of *p*-nitrophenol are released per mole of enzyme in the 'burst'. The enzyme is either only 63% pure (active) or the rate constant for the formation of the acylenzyme is not sufficiently greater than that for deacylation for the acylenzyme to accumulate fully

will occur at higher concentrations. Substituting (4.78) into (4.77) gives

$$\frac{1}{\sqrt{\pi}} = \frac{1}{\sqrt{E_0}} \left[1 + \frac{k_2}{k_{cat}} + \frac{K_M k_2}{[S] k_{cat}} \right]. \tag{4.79}$$

If k_{cat} is much greater than k_2, (4.79) may be used to extrapolate the burst from measurements at various substrate concentrations. It is obvious that when k_2 is not negligible, care must be taken not to underestimate the concentration of the enzyme.

3. Active-site titration versus rate assay

Active-site titration is not always applicable since it requires the accumulation of an intermediate in the reaction. The more usual procedure is to determine the concentration of an enzyme from a rate assay. This has the disadvantage that is does not give the absolute concentration of the enzyme unless it has been calibrated against an active-site titration. Further, rate measurements are sensitive to the reaction conditions. Whereas these may be controlled with some precision in a particular laboratory, they often vary from laboratory to laboratory. Active-site titration suffers from the disadvantage that several mg of enzyme are required for a spectrophotometric assay, but this may be reduced a thousand times using radioactive methods.[11] Active-site titration, with its relative insensitivity to precise reaction conditions, and giving absolute

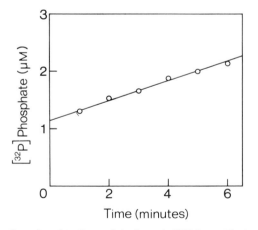

FIG. 4.12. Active-site titration of isoleucyl-tRNA synthetase using only $10\ \mu g$ (100 picomoles) of enzyme ($0\cdot1$ ml of $1\ \mu M$ enzyme). The reaction sequence is:

$$E + Ile + [\gamma^{32}P]ATP \xrightarrow[\substack{\searrow \\ ^{32}PP}]{fast} E.Ile \sim AMP \xrightarrow[H_2O]{slow} E + Ile + AMP.$$

(A. R. Fersht and M. M. Kaethner, *Biochemistry* **15,** 818 (1976)) values for the concentrations of enzyme solutions, has been the most important factor in providing highly reproducible data and enabling the comparison of rate constants from steady-state and pre-steady-state kinetics. (The rate constants of pre-steady-state kinetics generally involve exponential processes that do not depend on the concentration of enzyme, whilst steady-state rates are generally directly proportional to the concentration.)

Part 2 Magnitude of rate constants for enzymic processes

A. Upper limits on rate constants[12]

1. Association and dissociation

A simple way of analysing the rate constants of chemical reactions is the *collision theory* of reaction kinetics. The rate constant for a bimolecular reaction is considered to be composed of the product of three terms: the frequency of collisions, Z; a steric factor, p, to allow for the fraction of molecules that are in the correct orientation; and an activation energy term to allow for the fraction of molecules that are sufficiently thermally activated to react. That is

$$k_2 = Zp \exp(-E_A/RT). \tag{4.80}$$

The maximum value for the bimolecular rate constant occurs when the activation energy E_A is zero and the steric factor is unity. The rate is then said to be *diffusion controlled* and is equal to the encounter frequency of the molecules. Assuming that the reacting molecules are uncharged spheres of radius r_a and r_b, the encounter frequency may be calculated to be

$$Z = (2RT/3000\eta)(r_a + r_b)^2/r_a r_b \tag{4.81}$$

(where η is the viscosity). For two molecules of the same radius in water at 25° this is equal to an encounter frequency of $7 \times 10^9 \text{ s}^{-1} \text{M}^{-1}$. It should be noted that two large molecules collide at exactly the same rate as two small ones. This is because the increase of target area exactly compensates for the slower diffusion of the larger molecules. However, the rate of encounter of a small molecule with a large molecule is higher than this value because of the combination of the large target area of one and the high mobility of the other. More sophisticated calculations, allowing for the possibility of favourable electrostatic interactions at one extreme, and an unfavourable geometry for a small molecule hitting a particular target area of a larger one at the other extreme, give a range of 10^9–$10^{11} \text{ s}^{-1} \text{M}^{-1}$ for the encounter frequency. A similar treatment gives a range of 10^9–10^{12} s^{-1} for the upper limit on the *dissociation* rate constants of bimolecular complexes. Many of the second-order rate constants that do not involve the proton or hydroxide ion are found experimentally to be about 10^9 s^{-1} M^{-1}.

2. Chemical processes

The upper limit on the rate constant of any unimolecular or intramolecular reaction is the frequency of a molecular vibration, about 10^{12}–10^{13} s^{-1}.

3. Proton transfers

Favourable proton transfers between electronegative atoms such as O, N, or S are extremely fast. The bimolecular rate constants are generally diffusion controlled, being 10^{10}–10^{11} s^{-1} M^{-1} (Table 4.1). For example, the rate constant for the transfer of a proton from H_3O^+ to imidazole, a favourable transfer since imidazole is a stronger base than H_2O, is $1 \cdot 5 \times 10^{10}$ s^{-1} M^{-1}. The rate constant for the reverse reaction, the transfer of a proton from the imidazolium ion to water may be calculated from the difference in their pK_as using the following equations:

$$\frac{[B][H^+]}{[BH^+]} = K_a, \qquad \frac{[A^-][H^+]}{[HA]} = K_a' \qquad (4.82)$$

$$\frac{[B][HA]}{[BH^+][A^-]} = K_A/K_A' \qquad (4.83)$$

$$[B]+[HA] \underset{k_{-1}}{\overset{k_1}{\rightleftharpoons}} [BH^+]+[A^-] \qquad (4.84)$$

so that

$$k_{-1}/k_1 = K_a/K_a'. \qquad (4.85)$$

The rate constant for the transfer of a proton from the imidazolium ion (p$K_a = 6 \cdot 95$) to water ([H_2O] = 55 M, p$K_a = -1 \cdot 74$) is calculated from (4.85) to be $1 \cdot 7 \times 10^3$ s^{-1}.

Proton transfers from carbon acids and to carbon bases are generally much slower. This is because the lower electronegativity of carbon requires that the negative charge on a carbon base be stabilized by electron delocalization. The consequent reorganization of structure and solvent may slow down the overall transfer rate.

It was once thought that the rate of equilibration of the catalytic acid and basic groups on an enzyme with the solvent limited the rates of acid- and base-catalysed reactions to turnover numbers of 10^3 s^{-1} or less. This is because the rate constants for the transfer of a proton from the imidazolium ion to water and from water to imidazole are about 2×10^3 s^{-1}. However, protons are transferred between imidazole or imidazolium and buffer species in solution with rate constants many times

TABLE 4.1. *Proton transfer rate constants* $(25°, \text{ s}^{-1} \text{ M}^{-1})$

H^+ and	k	OH^- and	k
OH^-	1.4×10^{11}		
Inorganic acid anions	$10^{10}-10^{11}$	Inorganic acids	$\sim 10^{10}$
Carboxylates	$\sim 5 \times 10^{10}$	Carboxylic acids	$\sim 10^{10}$
Phenolates	$\sim 5 \times 10^{10}$	Phenols	$\sim 10^{10}$
Enolates	$\sim 5 \times 10^{10}$	Enols	$\sim 10^{10}$
Amines	10^{10}	Ammonium ions	$\sim 3 \times 10^{10}$
Carbanions	$< 1-10^{10}$	Carbon acids	$< 1-10^{9}$
		Phosphoric acids	$10^{8}-10^{10}$

M. Eigen, *Nobel Symposium* **5**, 245 (1967)

faster than this. For example, the rate constants with ATP, which has a pK_a similar to imidazole, are about $10^9 \text{ s}^{-1} \text{ M}^{-1}$ and there is about 2 mM ATP in the cell. Similarly, there are several other metabolites each at millimolar concentrations that have suitable acidic and basic groups so that catalytic groups on an enzyme should be able to equilibrate with the solvent at 10^7-10^8 s^{-1} or faster. Enzyme turnover numbers are usually considerably lower than this, in the range $10-10^3 \text{ s}^{-1}$, although carbonic anhydrase and catalase have turnover numbers of 10^6 and $4 \times 10^7 \text{ s}^{-1}$ respectively. The problem of carbonic anhydrase is discussed in Chapter 12F.

TABLE 4.2. *Proton transfer rates involving imidazole* $(pK_a = 6.95)$

Donor (DH$^+$)	pK_a	$k(DH^+ \rightarrow Im)$ $\text{s}^{-1} \text{ M}^{-1}$	$k(ImH^+ \rightarrow D)$ $\text{s}^{-1} \text{ M}^{-1}$
H_3O^+	-1.74	1.5×10^{10}	31
H_2O	15.74	45	2.3×10^{10}
CH_3CO_2H	4.76	1.2×10^9	7.7×10^6
$HATP^{3-}$	6.7	2×10^9	1×10^9
p-Nitrophenol	7.14	4.5×10^8	7.0×10^8
$HP_2O_7^{3-}$	8.45	1.1×10^8	3.6×10^9
Phenol	9.95	1×10^7	1×10^{10}
CO_3^{2-}	10.33	1.9×10^7	2×10^{10}
Glucose	12.3	1.6×10^5	2×10^{10}

$25°$, ionic strength $= 0$. H_2O rates calculated on $[H_2O] = 55 \text{ M}$
M. Eigen and G. G. Hammes, *Adv. Enzymol.* **25**, 1 (1963)

B. Enzymic rate constants and rate-determining processes

1. Association of enzymes and substrates

Calculations suggest that the diffusion-controlled encounter frequency of an enzyme and substrate should be about $10^9 \, s^{-1} \, M^{-1}$. The observed values in Table 4.3 tend to fall in the range $10^6-10^8 \, s^{-1} \, M^{-1}$. The faster ones are close to diffusion controlled, but the slower ones are significantly lower than the limit. This may be partly due to desolvation requirements in some cases, or, more likely in others, to a two-step process that appears as a single step. For example, at low concentrations the binding of NAG_2 to lysozyme appears to occur at about $5 \times 10^6 \, s^{-1} \, M^{-1}$. But extending the measurements to higher concentrations shows that the binding is a two-step process with an association rate constant of $4 \times 10^7 \, s^{-1} \, M^{-1}$ at pH 4·4 and 31° (see Section C6 and Fig. 4.9).[10,13]

$$E + NAG_2 \underset{1\cdot2\times10^5 \, s^{-1}}{\overset{4\times10^7 \, s^{-1} \, M^{-1}}{\rightleftharpoons}} E.NAG_2 \underset{1\cdot3\times10^3 \, s^{-1}}{\overset{1\cdot7\times10^4 \, s^{-1}}{\rightleftharpoons}} E'.NAG_2 \qquad (4.58)$$

2. Association can be rate determining for k_{cat}/K_M

It is seen in Table 4.4 that for some efficient enzymes k_{cat}/K_M may be as high as $3 \times 10^8 \, s^{-1} \, M^{-1}$. In these cases, the rate-determining step for this parameter, which is the apparent second-order rate constant for the reaction of free enzyme with free substrate, is close to the diffusion-controlled encounter of the enzyme and substrate. Briggs–Haldane kinetics holds for these enzymes (see Chapter 3B).

3. Dissociation of enzyme–substrate and enzyme–product complexes

Dissociation rate constants are much lower than the diffusion-controlled limit since the forces responsible for the binding must be overcome in this step. In some cases, enzyme–substrate dissociation is slower than the subsequent chemical steps and this gives rise to Briggs–Haldane kinetics.

4. Enzyme–product dissociation can be rate determining for k_{cat}

Product dissociation is sometimes rate determining at saturating substrate concentrations with some dehydrogenases. Examples of this are the dissociation of NADH from glyceraldehyde-3-phosphate dehydrogenase at high pH,[14] NADH from horse liver alcohol dehydrogenase at low salt[15,16] and NADPH from glutamate dehydrogenase.[17]

5. Conformational changes

There are many documented cases of substrate-induced conformational changes with rate constants in the range $10-10^4 \, s^{-1}$, and also instances

TABLE 4.3. *Association and dissociation rate constants for enzyme–substrate interactions*

Enzyme	Substrate	k_1 $(s^{-1}\,M^{-1})$	k_{-1} (s^{-1})	Ref.
(a) Small ligands				
Catalase	H_2O_2	5×10^6		1
Catalase-H_2O_2	H_2O_2	$1 \cdot 5 \times 10^7$		1
Chymotrypsin	Proflavin	$1 \cdot 2 \times 10^8$	$8 \cdot 3 \times 10^3$	2
	Acetyl-L-tryptophan p-nitrophenyl ester	6×10^7	6×10^4	3
	Furylacryloyl-L-tryptophanamide	$6 \cdot 2 \times 10^6$		4
	Trifluorylacetyl-D-tryptophan	$1 \cdot 5 \times 10^7$		5
Creatine kinase	ADP	$2 \cdot 2 \times 10^7$	$1 \cdot 8 \times 10^4$	6
	MgADP	$5 \cdot 3 \times 10^6$	$5 \cdot 1 \times 10^3$	
Glyceraldehyde-3-phosphate dehydrogenase	NAD	$1 \cdot 9 \times 10^7$	1×10^3	7
		$1 \cdot 4 \times 10^6$	$2 \cdot 1 \times 10^2$	
Lactate dehydrogenase (rabbit muscle)	NADH	$\sim 10^9$	$\sim 10^4$	8
Lactate dehydrogenase (pig heart)	NADH	$5 \cdot 5 \times 10^7$	39	9
	Oxamate	$8 \cdot 1 \times 10^6$	17	9
Liver alcohol dehydrogenase	NADH	$2 \cdot 5 \times 10^7$	9	10
Lysozyme	$(NaG)_2$	4×10^7	1×10^5	11, 12
Malate dehydrogenase	NADH	5×10^8	50	13
Pyruvate carboxylase-Mn^{2+}	Pyruvate	$4 \cdot 5 \times 10^6$	$2 \cdot 1 \times 10^4$	14
Ribonuclease	Uridine 3'-phosphate	$7 \cdot 8 \times 10^7$	$1 \cdot 1 \times 10^4$	15
	Uridine 2', 3'-cyclic phosphate	1×10^7	2×10^4	16
Tyrosyl-tRNA synthetase	Tyrosine	$2 \cdot 4 \times 10^6$	24	17
(b) Protein–nucleic acids				
Phenylalanyl-tRNA synthetase	tRNA[Phe]	$1 \cdot 6 \times 10^8$	27	18
Seryl-tRNA synthetase	tRNA[Ser]	$2 \cdot 1 \times 10^8$	11	19
Tyrosyl-tRNA synthetase	tRNA[Tyr]	$2 \cdot 2 \times 10^8$	$1 \cdot 5$	20
		$1 \cdot 4 \times 10^8$	53	
(c) Protein–protein				
Trypsin	Basic pancreatic	$1 \cdot 1 \times 10^6$	$6 \cdot 6 \times 10^{-8}$	21
Anhydrotrypsin	trypsin inhibitor	$7 \cdot 7 \times 10^5$	$8 \cdot 5 \times 10^{-8}$	21

TABLE 4.3 (*contd.*)

Enzyme	Substrate	k_1 $(s^{-1}\,M^{-1})$	k_{-1} (s^{-1})	Ref.
Trypsin	Pancreatic	$6 \cdot 8 \times 10^6$	$2 \cdot 2 \times 10^{-4}$	21
Anhydrotrypsin	secretory trypsin inhibitor	4×10^6	$1 \cdot 4 \times 10^{-3}$	21
Insulin	Insulin	$1 \cdot 2 \times 10^8$	$1 \cdot 5 \times 10^4$	22
β-Lactoglobulin	β-Lactoglobulin	$4 \cdot 7 \times 10^4$	$2 \cdot 1$	22
α-Chymotrypsin	α-Chymotrypsin	$3 \cdot 7 \times 10^3$	$0 \cdot 68$	22

1　B. Chance in *Currents in biochemical research* (ed. D. E. Green), John Wiley, p. 308 (1956).
2　U. Quast, J. Engel, H. Heumann, G. Krause, and E. Steffen, *Biochemistry* **13,** 2512 (1974).
3　M. Renard and A. R. Fersht, *Biochemistry* **12,** 4713 (1973).
4　G. P. Hess, J. McConn, E. Ku, and G. McConkey, *Phil. Trans. R. Soc.* **B257,** 89 (1970).
5　S. H. Smallcombe, B. Ault, and J. H. Richards, *J. Am. chem. Soc.* **94,** 4585 (1972).
6　G. G. Hammes and J. K. Hurst, *Biochemistry* **8,** 1083 (1069).
7　K. Kirschner, M. Eigen, R. Bittman, and B. Voigt, *Proc. natn. Acad. Sci. U.S.A.* **56,** 1661 (1966).
8　G. H. Czerlinski and G. Schreck, *J. biol. Chem.* **239,** 913 (1964).
9　H. D'A Heck, *J. biol. Chem.* **244,** 4375 (1969).
10　J. D. Shore and H. Gutfreund, *Biochemistry* **9,** 4655 (1970).
11　E. Holler, J. A. Rupley, and G. P. Hess, *Biochem. biophys. Res. Commun.* **37,** 423 (1969).
12　J. H. Baldo, S. E. Halford, S. L. Patt, and B. D. Sykes, *Biochemistry,* **14,** 1893 (1975).
13　G. Czerlinski and G. Schreck, *Biochemistry* **3,** 89 (1963).
14　A. S. Mildvan and M. C. Scrutton, *Biochemistry* **6,** 2978 (1967).
15　C. G. Hammes and F. G. Walz, Jr. *J. Am. chem. Soc.* **91,** 7179 (1969).
16　E. J. del Rosario and C. G. Hammes, *J. Am. chem. Soc.* **92,** 1750 (1970).
17　A. R. Fersht, R. S. Mulvey, and G. L. E. Koch, *Biochemistry* **14,** 13 (1975).
18　G. Krauss, R. Römer, D. Riesner, and G. Maass, *FEBS Letts,* **30,** 6 (1973).
19　A. Pingoud, D. Riesner, D. Boehme, and G. Maass, *FEBS Letts,* **30,** 1 (1973).
20　A. Pingoud, D. Boehme, D. Riesner, R. Kownatski, and G. Maass, *Eur. J. Biochem.* **56,** 617 (1975).
21　J.-P. Vincent, M. Peron-Renner, J. Pudles, and M. Lazdunski, *Biochemistry,* **13,** 4205 (1974).
22　R. Koren and G. G. Hammes, *Biochemistry* **15,** 1165 (1976).

where discrepancies in rate constants indicate rate-determining protein isomerizations.[17] Isomerizations are often associated with slow steps, for example the dissociation of NADH from some dehydrogenases involves a concomitant conformational change. However, there are few, if any, direct demonstrations that a conformational change is, by itself, rate determining.

It should be noted that the rate-determining step of a reaction changes with substrate concentration since the rate is proportional to k_{cat} at saturating concentrations of substrate, and is proportional to k_{cat}/K_M at low concentrations. When a step is said to be rate determining without stating the reaction conditions, it usually refers to k_{cat}.

TABLE 4.4. *Enzymes where k_{cat}/K_M is close to the diffusion-controlled association rate*

Enzyme	Substrate	k_{cat} (s^{-1})	K_M (M)	k_{cat}/K_M (s^{-1} M^{-1})	Ref.
Acetylcholin-esterase	Acetylcholine	$1\cdot4\times10^4$	9×10^{-5}	$1\cdot6\times10^8$	1
Carbonic	CO_2	1×10^6	$1\cdot2\times10^{-2}$	$8\cdot3\times10^7$	2
anhydrase	HCO_3^-	4×10^5	$2\cdot6\times10^{-2}$	$1\cdot5\times10^7$	3
Catalase	H_2O_2	4×10^7	$1\cdot1$	4×10^7	4
Crotonase	Crotonyl-CoA	$5\cdot7\times10^3$	2×10^{-5}	$2\cdot8\times10^8$	5
Fumarase	Fumarate	$8\cdot0\times10^2$	5×10^{-6}	$1\cdot6\times10^8$	6
	Malate	$9\cdot0\times10^2$	$2\cdot5\times10^{-5}$	$3\cdot6\times10^7$	6
Triosephosphate isomerase	Glyceraldehyde-3-phosphate	$4\cdot3\times10^3$	$4\cdot7\times10^{-4}$	$2\cdot4\times10^{8a}$	7

[a] The observed value is $9\cdot1\times10^6$ s^{-1} M^{-1}. The tabulated value is calculated on the basis of only 3.8% of the substrate being reactive since 96·2% is hydrated under the conditions of the experiment.

1 T. I. Rosenberry, *Adv. Enzymol.* **43**, 103 (1975).
2 J. C. Kernohan, *Biochim. biophys. Acta* **81**, 346 (1964).
3 J. C. Kernohan, *Biochim. biophys. Acta* **96,** 304 (1965).
4 Y. Ogura, *Archs Biochem. Biophys.* **57,** 288 (1955).
5 R. M. Waterson and R. L. Hill, *Fedn Proc.* **30,** 1114 (1971).
6 J. W. Teipel, G. M. Hass, and R. L. Hill, *J. biol. Chem.* **243,** 5684 (1968).
7 S. J. Putman, A. F. W. Coulson, I. R. T. Farley, B. Riddleston, and J. R. Knowles, *Biochem. J.* **129,** 301 (1972).

References

1 H. Hartridge and F. J. W. Roughton, *Proc. R. Soc.* **A104,** 376 (1923).
2 F. J. W. Roughton, *Proc. R. Soc.* **B115,** 475 (1934).
3 B. Chance, *J. Franklin Inst.* **229,** 455, 613, 637 (1094).
4 Q. Gibson, *J. Physiol.* **117,** 49P (1952).
5 A. R. Fersht and R. Jakes, *Biochemistry* **14,** 3350 (1975).
6 G. Czerlinski and M. Eigen, *Z. Electrochem.* **63,** 652 (1959).
7 A. S. Mildvan and M. C. Scrutton, *Biochemistry* **6,** 2987 (1967).
8 B. D. Sykes, *J. Am. chem. Soc.* **91,** 949 (1969).
9 S. H. Smallcombe, B. Ault, and J. H. Richards, *J. Am. chem. Soc.* **94,** 4585 (1972).
10 J. H. Baldo, S. E. Halford, S. L. Patt, and B. D. Sykes, *Biochemistry* **14,** 1893 (1975).

11   A. R. Fersht, J. S. Ashford, C. J. Bruton, R. Jakes, G. L. E. Koch, and B. S. Hartley, *Biochemistry* **14,** 1 (1975).

12   M. Eigen and G. G. Hammes, *Adv. Enzymol.* **25,** 1 (1963).

13   E. Holler, J. A. Rupley, and G. P. Hess, *Biochem. biophys. Res. Commun.* **37,** 423 (1969).

14   D. R. Trentham, *Biochem. J.* **122,** 71 (1971).

15   H. Theorell and B. Chance, *Acta chem. scand.* **5,** 1127 (1951).

16   J. D. Shore and H. Gutfreund, *Biochemistry* **9,** 4655 (1970).

17   A. di Franco, *Eur. J. Biochem.* **45,** 407 (1974).

Further reading

M. Eigen and G. G. Hammes, 'Elementary steps in enzyme reactions', *Adv. Enzymol.* **25,** 1 (1963).

G. G. Hammes and P. R. Schimmel, 'Rapid reactions and transient states', *The Enzymes* **2,** 67 (1970).

H. Gutfreund, *Enzymes: physical principles*, Wiley-Interscience (1970).

S. Claesson (ed.), *Nobel symposium 5 (Fast reactions and primary processes in chemical kinetics)*, Wiley–Interscience (1967).

B. Chance, R. H. Eisenhardt, Q. H. Gibson, and K. Karl. Lonberg-Holm (eds.), *Rapid mixing and sampling techniques in biochemistry*, Academic Press (1964).

G. Careri, P. Fasella, and E. Gratton, 'Statistical time events in proteins: a physical assessment', *CRC Critical Reviews in Biochemistry*, **141** (1975).

K. Kustin (ed.) *Methods in Enzymology XVI*, Academic Press (1969).

Chapter 5

The pH dependence of enzyme catalysis

The activities of many enzymes vary with pH in the same way that simple acids and bases ionize. This is not surprising since we saw in Chapter 1 that the active sites generally contain important acidic or basic groups. It is to be expected that if only one protonic form of the acid or base is catalytically active, the catalysis should somehow depend on the concentration of the active form. In this chapter we shall see that k_{cat}, K_M, and k_{cat}/K_M, are affected in different ways by the ionizations of the enzyme and enzyme–substrate complex.

A. Ionization of simple acids and bases: the basic equations

It is usual to discuss the ionization of a base B in terms of its conjugate acid BH^+ in order to use the same set of equations for both acids and bases. The ionization constant K_a is defined by:

$$K_a = [B][H^+]/[BH^+].$$ (5.1)

Or for an acid HA and its conjugate base A^-:

$$K_a = [A^-][H^+]/[HA].$$ (5.2)

The pK_a is defined by:

$$pK_a = -\log K_a.$$ (5.3)

Eqns (5.1) and (5.3) (or (5.2)) may be rearranged to give the Henderson–Hasselbalch equation:

$$pH = pK_a + \log([B]/[BH^+]).$$ (5.4)

It is readily seen from this equation that the pK_a of an acid or a base is the pH of half neutralization when the concentrations of B and BH^+ are equal.

The variation of the concentrations of HA and A^- with the proton

TABLE 5.1. $pK_a s$ *of ionizing groups*[a]

Group	Model compounds (small peptides)	pK_a Usual range in proteins
Amino acid α-CO_2H	3·6 ⎫	
Asp (CO_2H)	4·0 ⎬	2–5·5
Glu (CO_2H)	4·5 ⎭	
His (Imidazole)	6·4	5–8
Amino acid α-NH_2	7·8	
Lys (ε-NH_2)	10·4	
Arg (Guanidine)	~12	
Tyr (OH)	9·7	9–12
Cys (SH)	9·1	
Phosphates	1·3, 6·5	

[a]Data mainly from C. Tanford, *Adv. Prot. Chem.* **17**, 69 (1962); C. Tanford and R. Roxby, *Biochemistry* **11**, 2192 (1972)

concentration is found from rearranging eqn (5.2) as

$$[HA] = [A_0][H^+]/([K_a + [H^+]) \tag{5.5}$$

and

$$[A^-] = [A_0]K_a/(K_a + [H^+]) \tag{5.6}$$

where

$$[A_0] = [HA] + [A^-].$$

Suppose there is some quantity L (an absorption coefficient, a rate constant etc.) such that the property of a solution (the absorbance, reaction rate etc.) is the product of this quantity and a concentration. If the value of L for the molecule HA is L_{HA}, and for the molecule A^- is L_{A^-}, then the observed value of the property at a particular pH, $L_H[A_0]$, is given by $L_H[A_0] = L_{HA}[HA] + L_{A^-}[A^-]$. That is:

$$L_H[A_0] = L_{HA}[A_0][H^+]/(K_a + [H^+]) + L_{A^-}[A_0]K_a/(K_a + [H^+]) \tag{5.7}$$

so that

$$L_H = \frac{L_{HA}[H^+] + L_{A^-}K_a}{K_a + [H^+]}. \tag{5.8}$$

A *doubly ionizing* system, such as

$$H_2A \underset{}{\overset{K_1}{\rightleftharpoons}} HA^- \underset{}{\overset{K_2}{\rightleftharpoons}} A^{2-}$$
$$\quad + \qquad + \tag{5.9}$$
$$\quad H^+ \qquad 2H^+$$

may be analysed to give:

$$L_H = \frac{[H^+]^2 L_{H_2A} + [H^+]K_1 L_{HA^-} + K_1 K_2 L_{A^{2-}}}{K_1 K_2 + [H^+]K_1 + [H^+]^2}. \tag{5.10}$$

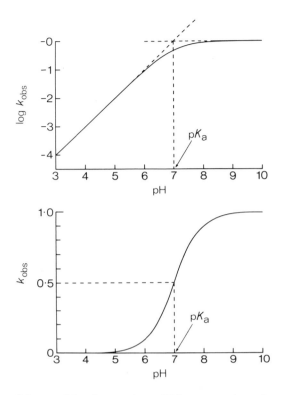

FIG. 5.1. Plots of k_{obs} and log k_{obs} against pH for a reaction whose rate depends on the basic form of an acid of pK_a 7

1. Extraction of pK_as by inspection of equations

The crucial point in eqns (5.8) and (5.10) is that the points of inflection in the plots of L_H against pH, i.e. the pK_as, are determined by the *denominator* of the fraction. The numerator determines the amplitudes of the functions. The apparent pK_as of a complex kinetic equation may be found by rearranging the equation to the form of eqn (5.8) for a singly ionizing system or that of eqn (5.10) for a doubly ionizing system and then comparing the denominators. This may be illustrated by considering the pH dependence of $1/L_H$ instead of L_H in eqn (5.8), i.e.

$$1/L_H = (K_a + [H^+])/(L_{HA}[H^+] + L_{A^-}K_a). \qquad (5.11)$$

Rearranging (5.11) to be of the same form as eqn (5.8) gives

$$1/L_H = (K_a/L_{HA} + [H^+]/L_{HA})/(K_aL_{A^-}/L_{HA} + [H^+]). \qquad (5.12)$$

We can say directly from the examination of the denominator of eqn (5.12) that the plot of $1/L_H$ against pH will have an apparent ionization

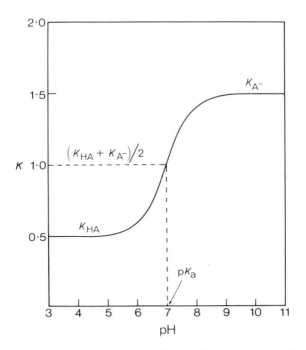

FIG. 5.2. Plot of the pH dependence of an arbitrary constant K which has the value of K_{HA} for the acid form and K_{A^-} for the basic form of an acid of pK_a 7

constant, K_{app}, given by

$$K_{app} = K_a(L_{A^-}/L_{HA}). \tag{5.13}$$

B. Effect of ionizations of groups on enzymes on kinetics

Although enzymes contain a multitude of ionizing groups, it is usually found that plots of rate against pH take the forms of simple single or double ionization curves. This is because the only ionizations that are of importance are those of groups at the active site which are directly involved in catalysis or those elsewhere that are responsible for maintaining the active conformation of the enzyme.

We shall now analyse some of the simple examples beginning with the Michaelis–Menten mechanism. Some simplifying assumptions will be made that may break down in some circumstances but often hold in practice. There are four assumptions.

(a) The groups act as perfectly titrating acids or bases (in practice, generally a fair approximation).
(b) Only one ionic form of the enzyme is active (usually true).

(c) All intermediates are in protonic equilibrium, i.e. proton transfers are faster than chemical steps (generally true—see Chapter 4).

(d) The rate-determining step does not change with pH (this may break down with interesting consequences).

1. Simple theory: the Michaelis–Menten mechanism

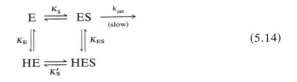

$$(5.14)$$

In (5.14) the ionization constant of the free enzyme is K_E, that of the enzyme–substrate complex K_{ES}, and the dissociation constants of HES and ES are K_S' and K_S respectively.

All four equilibrium constants cannot vary independently due to the cyclic nature of the equilibria. Once three are fixed, the fourth is defined by:

$$K_E K_S' = K_{ES} K_S. \qquad (5.15)$$

Eqn (5.15) may be derived by multiplying the various dissociation and ionization constants or simply by inspection, since the process HES→ ES→E must give the same energy change as for HES→HE→E.

The two important conclusions from eqn (5.15) are:

(a) if $K_E = K_{ES}$, then $K_S = K_S'$ and there is no pH dependence of binding;

(b) if $K_E \neq K_{ES}$, i.e. the pK_a is perturbed on binding,

$$K_S = K_S'(K_E/K_{ES}) \qquad (5.16)$$

and the binding of S must of necessity be pH dependent.

2. pH Dependence of k_{cat}, k_{cat}/K_M, K_M, and $1/K_M$[1,2]

The pH dependence of v is obtained by expressing the concentration of ES in terms of E_0 to give, after the necessary algebra,

$$v_H = \frac{k_{cat}[E_0][S]}{K_S + [S](1 + [H^+]/K_{ES}) + K_S[H^+]/K_E}. \qquad (5.17)$$

The pH dependence of k_{cat} is given from eqn (5.17) when [S] is much greater than K_S, i.e.

$$(V_{max})_H = [E_0](k_{cat})_H = \frac{k_{cat}[E_0]K_{ES}}{K_{ES} + [H^+]}. \qquad (5.18)$$

Comparison of the denominator of eqn (5.18) with that of (5.8) shows that the pH dependence of V_{max} or k_{cat} follows the ionization constant of the enzyme–substrate complex, K_{ES}.

The apparent value of K_M at each pH may be found by rearranging eqn (5.17) to the form of the basic Michaelis–Menten equation (eqn (3.1)).

$$(K_M)_H = \frac{K_S K_{ES} + [H^+] K_S K_{ES}/K_E}{K_{ES} + [H^+]}. \qquad (5.19)$$

K_M also follows the ionization of the enzyme–substrate complex.

The pH dependence of k_{cat}/K_M is given by the variation of v at low values of [S] (or alternatively by the ratios of eqns (5.18) and (5.19)). Simplifying eqn (5.17) by putting [S] close to zero and rearranging gives:

$$(k_{cat}/K_M)_H = \frac{(k_{cat}/K_S) K_E}{K_E + [H^+]}. \qquad (5.20)$$

k_{cat}/K_M (and v for [S] much less than K_M) follows the ionization of the free enzyme.

3. Simple rule for the prediction and assignment of pK_as

The above results are part of the general rule that 'the plot of the equilibrium constant K or the rate constant k for the process X→Y as a function of pH follows the ionization constants of X'.

Let us now apply this rule to scheme (5.14), bearing in mind that the substrate may also have ionizing groups.

a. pH dependence of k_{cat}
The process concerned is

$$ES \xrightarrow{k_{cat}} EP. \qquad (5.21)$$

k_{cat} follows the pK_a of the enzyme–substrate complex.

b. pH dependence of K_M
The process concerned is

$$ES \xrightarrow{K_M} E + S. \qquad (5.22)$$

The pH dependence of K_M follows the ionization of the enzyme–substrate complex.

c. pH dependence of $1/K_M$†

The process concerned is

$$E + S \xrightarrow{1/K_M} ES. \qquad (5.23)$$

The pH dependence of $1/K_M$ follows the ionizations in the free enzyme and the free substrate.

d. pH dependence of k_{cat}/K_M

The process concerned is

$$E + S \xrightarrow{k_{cat}/K_M} EP. \qquad (5.24)$$

The pH dependence of k_{cat}/K_M follows the ionizations in the free enzyme and free substrate.

C. Modifications and breakdown of the simple theory

The simple theory adumbrated above has to be modified to account for the pH dependence of the catalytic parameters in mechanisms more complicated than the basic Michaelis–Menten.

1. Modifications due to additional intermediates

a. Intermediates on the reaction pathway[3]

$$E + S \xrightleftharpoons{K_S} ES \xrightarrow{k_2} EA \xrightarrow{k_3} EP \qquad (5.25)$$

The presence of additional intermediates does not affect the pH dependence of k_{cat}/K_M or $1/K_M$ since they represent changes from the free enzyme and free substrate only. The pH dependence of these still gives the pK_as of the free enzyme and free substrate. But the pH dependence of k_{cat} and K_M now concerns changes from the intermediate as well as the enzyme–substrate complex. If the intermediate EA in eqn (5.25) is the major enzyme-bound species, the pH dependence of k_{cat} and K_M will give the pK_as of this. If both ES and EA accumulate, the pH profiles give pK_as that are the weighted mean of those of ES and EA.[4]

b. Non-productive binding modes[5]

When a substrate binds in a non-productive mode as well as in the productive mode, the pH dependence of k_{cat} and K_M may give an apparent pK_a for the catalytically important group at the active site which

---

† Some find it confusing that plotting the inverse of a function, e.g. $1/K_M$ instead of K_M, gives a different pK_a. The mathematical reason for this was given in eqns (5.11) to (5.13). The pH dependence of K_M contains the information for determining the pK_as of E, S, and ES but they are manifested in different ways in different plots. An alternative procedure for analysing a plot of $\log K_M$ versus pH has been given by Dixon.[1]

is far from its real value in the productive complex. This happens if the ratio of productive to non-productive binding changes on ionization of the group. Suppose the activity of the enzyme is dependent on a group being in the basic form at high pH. If, say, the substrate is bound with more being in the productive mode at low pH than at high pH, then as the pH decreases through the pK_a of the catalytic group, the decrease in rate as the group becomes protonated will be partially compensated for by an increase in productive binding. This has the effect of lowering the apparent pK_a controlling the pH dependence of k_{cat} from the value in the productive complex. The pH dependence of K_M is affected in an identical manner.

The algebraic solution of such a situation is as follows.

$$
\begin{array}{ccc}
\text{HE} & \underset{}{\overset{K_a}{\rightleftharpoons}} & \text{E} \\
{\scriptstyle K'_S}\big\Updownarrow & & \big\Updownarrow{\scriptstyle K_S} \\
\text{HES} & \underset{}{\overset{K'_a}{\rightleftharpoons}} \text{ES} & \xrightarrow[\text{slow}]{k_2} \text{EP} \\
{\scriptstyle K'}\big\Updownarrow & & \big\Updownarrow{\scriptstyle K} \\
\text{HES}' & \underset{}{\overset{K''_a}{\rightleftharpoons}} \text{ES}'
\end{array}
\qquad (5.26)
$$

In (5.26) HES and ES are the productively-bound complexes, and K' and K the equilibrium constants between these and the non-productively bound complexes HES' and ES' as defined in the scheme. Solving the rate equation by the usual means gives

$$
k_{cat} = \frac{k_2}{K(1+[H^+]/K''_a)+1+[H^+]/K'_a}
\qquad (5.27)
$$

$$
K_M = \frac{K_S(1+[H^+]/K_a)}{K(1+[H^+]/K''_a)+1+[H^+]/K'_a}
\qquad (5.28)
$$

Rearranging eqns (5.27) and (5.28) to the form of eqn (5.8) and inspection of the denominator shows that the observed pK_a is given by

$$
pK_{obs} = pK'_a - \log\frac{(1+K)}{(1+K')}
\qquad (5.29)
$$

Non-productive binding modes do not affect the pH dependence of k_{cat}/K_M or $1/K_M$ for the reasons discussed in the previous case.

2. Breakdown of the simple rules: Briggs–Haldane kinetics and change of rate-determining step with pH: kinetic pK_as[4,6–8]

$$
E + S \underset{k_{-1}}{\overset{k_1}{\rightleftharpoons}} ES \xrightarrow{k_2} EP
\qquad (5.30)
$$

In the extreme case of Briggs–Haldane kinetics, the rate constant for the chemical step is larger than that for the dissociation of the enzyme–substrate complex. In this case, k_{cat}/K_M is equal to k_1, the association of the enzyme and substrate. Suppose again that the catalytic activity of the enzyme depends on a group being in its basic form. At high pH k_2 is faster than k_{-1}. But as the pH is lowered k_2 decreases as the base becomes protonated until it is slower than k_{-1}. This leads to an apparent pK_a in the pH profile of k_{cat}/K_M that is lower than the pK_a of the important base (see Fig. 5.3).[7] It may be shown that if the ionization constant of the

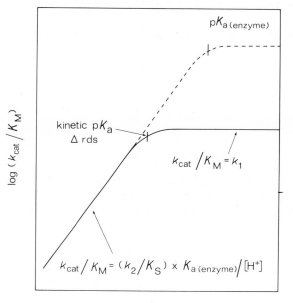

Fig. 5.3. Illustration of the 'kinetic pK_a' caused by a change in rate-determining step with pH. At low pH, k_{cat}/K_M falls off as $[k_2/(k_{-1}/k_1)]K_a/(K_a+[H^+])$ (which simplifies as $K_S = k_{-1}/k_1$, and $[H^+] \gg K_a$ at low pH). At high pH k_{cat}/K_M levels off at k_1, the second-order rate constant for the association of enzyme and substrate

group on the enzyme is K_a, the apparent ionization constant is given by

$$K_{app} = K_a(k_2+k_{-1})/k_{-1}. \qquad (5.31)$$

K_{app} is termed a *kinetic pK_a* since it does not represent a real ionization but is composed of the ratios of rate constants which are not for proton transfers. Kinetic pK_as occur whenever there is a change of rate-determining step with pH.

3. An experimental distinction between kinetic and equilibrium pK_as[7,8]

pK_as that are composed of the combination of equilibrium constants and the individual pK_as of titrating groups, such as that for the non-productive binding scheme (eqn (5.29)), titrate as real pK_as. If one were to add acid or base to the enzyme–substrate complexes in scheme (5.26), the catalytic group would titrate according to the pK_a given by eqn (5.29). Similarly, if one were to measure the fraction of the base in the ionized form by some spectroscopic technique, it would be found to ionize with the pK_a of eqn (5.29). This does not happen with kinetic pK_as. In the case of the change of rate-determining step with pH for k_{cat}/K_M in the Briggs–Haldane mechanism, although the pK_a from the kinetics will be that given by eqn (5.31), direct measurement of the titration of the catalytic group will give its true ionization constant K_a.

4. Microscopic and macroscopic pK_as

When an acid exists in different forms, such as HES and HES' in (5.26), the pK_a in each of the forms is termed a *microscopic* pK_a. The pK_a with which the system is found to titrate, e.g. pK_{obs} in eqn (5.29), is called variously a *macroscopic*, an *apparent*, or a *group* pK_a. To all intents and purposes it is a real pK_a, unlike a kinetic pK_a.

D. Influence of surface charge on pK_as of groups in enzymes

The surface of an enzyme contains many polar groups. Chymotrypsinogen, for example, has on its surface 4 arginine and 14 lysine residues, which are positively charged, and 7 aspartate and 5 glutamate residues, which are negatively charged. These provide an ionic atmosphere, or electrostatic field, which may stabilize or destabilize buried, or partly buried, ionic groups. Calculations have been made by Linderstrøm–Lang which show that the perturbation of the pK_a of a buried group is a complex function of the size and shape of the protein and also the ionic strength of the solution.[9] The magnitude of these effects is nicely illustrated by some studies in which the surface carboxylates or ammonium ions are chemically modified.

The active site of chymotrypsin bears a negative charge at high pH and a net zero charge at low pH. It is expected that the basic form of the active site at high pH will be stabilized by a positively charged surface and destabilized by a negatively charged one. It is seen in Table 5.2 that this is borne out experimentally. The conversion of the 14 positively charged lysines to negatively charged carboxylates on reaction of chymotrypsin with succinic anhydride causes the pK_a controlling the pH dependence of k_{cat} for the hydrolysis of acetyltryptophan methyl ester to increase from 7·0 to 8·0. Conversely, the conversion of 13 negatively charged carboxyls

TABLE 5.2. *Influence of surface charge on* pK_as

Enzyme	Modification	pK_a	k_{cat} (s^{-1})
Chymotrypsin[a]	—	7·0[b]	47
Succinyl-chymotrypsin[a]	Lys (—NH_3^+) → —$NHOCCH_2CH_2CO_2^-$	8·0[b]	74
Ethylenediamine-chymotrypsin	Asp, Glu (—CO_2^-) → —$CONH(CH_2)_2NH_3^+$	6·1[b]	50
Trypsin[c]	—	7·0[d]	0·7
Acetyl-trypsin[c]	Lys (—NH_3^+) → —$NHCOCH_3$	7·2[d]	1·2

[a] δ-chymotrypsin, 25° ionic strength 0·1, P. Valenzuela and M. L. Bender, *Biochim. biophys. Acta* **256,** 538 (1971)
[b] pK_a for k_{cat} for the hydrolysis of acetyl-L-tryptophan methyl ester
[c] W. E. Spomer and J. F. Wootton, *Biochim. biophys. Acta* **235,** 164 (1971)
[d] pK_a for k_{cat}/K_M for the hydrolysis of benzoyl-L-arginine amide

to positively charged amines lowers the pK_a to 6·1. Similarly, the acylation of the surface amino groups of trypsin by acetic anhydride causes an increase of 0·2 units in the pK_a of the free enzyme.

These effects are mimicked by changes of pH. Below pH 5 the carboxylates become protonated, and above pH 9 the ammonium groups of lysine start to lose their positive charge as they deprotonate. This causes perturbations in the titration curves of the enzyme at extremes of pH. These effects are most marked at low ionic strengths of 0·2 M or less, but may be depressed by high ionic strength to become negligible at 1 M. Fortunately, many titration curves are measured between pH 5 and pH 9 where few surface groups titrate, and excellent results are obtained at ionic strengths of 0·1 M.[8,10]

As well as perturbing the pK_a of the catalytic acid or base, the large change of surface charge may alter the rate constants for the chemical steps. Change in surface charge is analogous to the change of ionic strength in non-enzyme ionic reactions, and the two effects are analogous to secondary and primary salt effects in physical organic chemistry.

E. Graphical representation of data

Suppose one of the kinetic parameters, such as k_{cat}, depends on the enzyme being in the acidic form. Then the rate will depend upon the ionization constant, K_a, and the pH according to

$$(k_{cat})_H = k_{cat}[H^+]/(K_a + [H^+]) \qquad (5.32)$$

where $(k_{cat})_H$ is the observed value at the particular $[H^+]$ (eqn (5.8)). Eqn

(5.32) is of identical form to the Michaelis–Menten equation and so K_a may be found by plots analogous to the Lineweaver–Burk or Eadie plots (eqns (3.28) and (3.29)). For example,

$$(k_{cat})_H = k_{cat} - K_a(k_{cat})_H/[H^+] \qquad (5.33)$$

and so k_{cat} and K_a may be obtained by plotting $(k_{cat})_H$ against $(k_{cat})_H/[H^+]$. Similarly, if the rate depends on the enzyme being in the basic form it may be shown that

$$(k_{cat})_H = k_{cat} - (k_{cat})_H[H^+]/K_a \qquad (5.34)$$

Here $(k_{cat})_H$ must be plotted against $(k_{cat})_H[H^+]$.

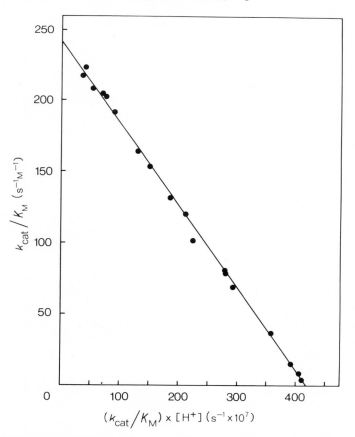

FIG. 5.4. Illustration of the use of eqn (5.34) for determining the pK_a of the active site of an enzyme. k_{cat}/K_M rather than k_{cat} is used in this example (the hydrolysis of acetyl-L-tyrosine p-acetylanilide by α-chymotrypsin at 25°, ref. 8)

For more complicated examples where the parameter does not fall to zero at either high or low pH, as is often found in plots of association or dissociation constants against pH, the procedure is modified by plotting the difference between the observed value at the particular pH and one of the limiting values at the extremes of pH.

F. Illustrative examples and experimental evidence

The best-studied enzyme in this context is chymotrypsin. Apart from being well characterized in both its structure and catalytic mechanism, it has the advantage of a very broad specificity. Substrates may be chosen to obey the simple Michaelis–Menten mechanism, to accumulate intermediates, to show non-productive binding, and to exhibit Briggs–Haldane kinetics with a change of rate-determining step with pH.

The pH dependence of k_{cat}/K_M for the hydrolysis of substrates follows a bell-shaped curve with a maximum at pH 7·8 and pK_as of 6·8 and 8·8 for α-chymotrypsin, and a maximum at pH 7·9 and pK_as of 6·8 and 9·1 for δ-chymotrypsin. The pK_a of 6·8 represents the ionization of the catalytically important base at the active site, whilst the high-pH ionization is due to the α-amino group of Ile-16 which holds the enzyme in a catalytically active conformation. This conformationally important ionization does not affect k_{cat}, which usually gives a sigmoid curve of pK_a 6–7, but causes an increase of K_M at high pH.

1. pK_a of the active site of chymotrypsin

a. The free enzyme

The theory predicts that, unless there is a change of rate-determining step with pH, the pH dependence of k_{cat}/K_M for all non-ionizing substrates should give the same pK_a, that for the free enzyme. Apart from one exception, this is found. At 25° and ionic strength 0·1 M, the pK_a of the active site is $6\cdot80\pm0\cdot03$. The most accurate data available fit very precisely the theoretical ionization curves between pH 5 and 8, after allowance has been made for the fraction of the enzyme in the inactive conformation. The relationship holds for amides where no intermediate accumulates and the Michaelis–Menten mechanism holds, and also for esters where the acylenzyme accumulates (Table 5.3 and Figs 5.5 and 5.6.)

b. Acetyl-L-tryptophan p-nitrophenyl ester: Briggs–Haldane kinetics with change of rate-determining step with pH

The one exception is the pH dependence of k_{cat}/K_M for the hydrolysis of this ester. This gives an apparent pK_a of 6·50 for the free enzyme. The reason for the low value is that described in Section C2. At high pH, the association of enzyme and substrate is partially rate-determining for

TABLE 5.3. pK_a *of active site of chymotrypsin*

Substrate	pK_a k_{cat}/K_M	k_{cat}	Ref.
Acetyl-L-phenylalanine alaninamide[a]	6·80	6·6	1
Formyl-L-phenylalanine semicarbazide[a]	6·84	6·32	1
Acetyl-L-tyrosine p-acetylanilide[a]	6·77	—	1
p-Nitrophenylacetate[a]	6·85	—	2
Acetyl-L-phenylalanine ethyl ester[a]	6·8	6·85	3
Acetyl-L-tyrosine p-acetylanilide[b]	6·83	—	4
Acetyl-L-tryptophan p-nitrophenyl ester[b]	6·50	6·9	4

[a] α-chymotrypsin, 25°, ionic strength 0·1
[b] δ-chymotrypsin, 25°, ionic strength 0·95

1   A. R. Fersht and M. Renard, *Biochemistry* **13,** 1416 (1974).
2   M. L. Bender, G. E. Clement, F. J. Kezdy, and H. d'A. Heck, *J. Am. Chem. Soc.* **86,** 3680 (1964).
3   B. R. Hammond and H. Gutfreund, *Biochem. J.* **61,** 187 (1955).
4   M. Renard and A. R. Fersht, *Biochemistry* **12,** 4713 (1973).

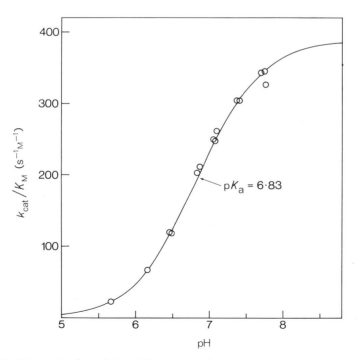

FIG. 5.5. Determination of the pK_a of the active site of δ-chymotrypsin from the pH dependence of k_{cat}/K_M for the hydrolysis of N-acetyl-L-tyrosine p-acetylanilide (ref. 7)

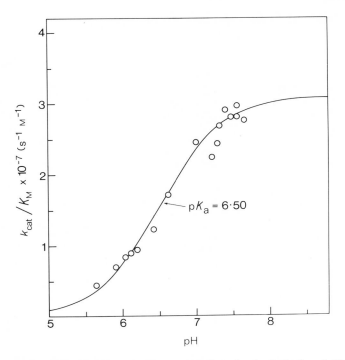

FIG. 5.6. The anomalously low pK_a found for the hydrolysis of N-acetyl-L-tryptophan p-nitrophenyl ester by δ-chymotrypsin (ref. 7)

k_{cat}/K_M. The limiting value of k_{cat}/K_M at high pH is $3 \times 10^7 \, s^{-1} \, M^{-1}$, a value close to that for the diffusion-controlled encounter of enzyme and substrate. However, at low pH, the chemical steps slow down and become rate-determining as the enzyme becomes protonated and less active. The rate constant for the association step may be calculated from eqn (5.31) to be $6 \times 10^7 \, s^{-1} \, M^{-1}$. The data are consistent with scheme (5.35) and a pK_a of 6·8 for the active site.[7]

$$\text{Ac-TrpONPh} + \text{CT} \underset{6 \times 10^4 \, s^{-1}}{\overset{6 \times 10^7 \, s^{-1} \, M^{-1}}{\rightleftharpoons}} \text{Ac-TrpONPh.CT}$$

$$\downarrow \; 7 \times 10^4 \, s^{-1}$$

$$\text{Ac-Trp-CT} \qquad\qquad (5.35)$$

$$\downarrow \; 65 \, s^{-1}$$

$$\text{Ac-Trp} + \text{CT}$$

This is a rare example where steady-state kinetics may be analysed to give rate constants for several specific steps on the reaction pathway.

c. *The enzyme–substrate complex*

The pK_a of the enzyme–substrate complex is not constant like that of the free enzyme since the binding of substrates perturbs the pK_a of the active site. The pK_a values of k_{cat} for the hydrolysis of amides range from 6–7. In consequence, the value of K_M increases at low pH according to eqn (5.16).

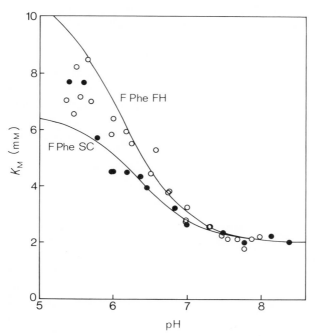

Fig. 5.7. Increase in K_M at low pH for the chymotrypsin-catalysed hydrolysis of N-formyl-L-phenylalanine semicarbazide (FPheSC) and formylhydrazide (FPheFH) linked to the pK_as of the enzyme substrate complexes being lower than the pK_a of the free enzyme (ref. 8)

The precise determination of the pK_a for the enzyme–substrate complex from the pH dependence of K_M is difficult for two reasons. First, the variation of K_M is relatively small and highly accurate data are required. Secondly, two plateau regions must be determined, one at low pH as well as the one at high pH. This means that measurements are required over a wider range of pH than for the determination of k_{cat} or k_{cat}/K_M, and are more susceptible to perturbations from the ionizations of other groups.

d. *The ionization at high pH*[11]

The pK_a at high pH is due to a substrate independent conformational change of the enzyme which may be monitored directly by physical techniques such as optical rotation and fluorescence yield. Kinetic measurements give the same pK_a as found by these methods.

G. Direct titration of groups in enzymes

Several methods have been developed for the direct titration of some ionizing side chains in proteins. One of the major problems is to identify the desired group amongst all the similar ones.

1. The effect of D_2O on pH/pD and pK_as

Many of the spectroscopic methods use D_2O as the solvent rather than H_2O in order to separate the required signals from those from the solvent. It has been found experimentally that the glass electrode gives a lower reading in D_2O by 0.4 units:

$$pH = pD + 0.4. \tag{5.36}$$

In mixtures of H_2O and D_2O the glass electrode reading is related to pH by [12]

$$pH = pD + 0.3139\alpha + 0.0854\alpha^2 \tag{5.37}$$

where α is the atom fraction of deuterium, $[D]/([D]+[H])$. However, pK_as are higher in D_2O and D_2O/H_2O mixtures than in H_2O. The increase in pK_a sometimes balances the decreased reading on the glass electrode so that the measured pK_a in D_2O is often assumed to be the true pK_a in water. The effect of solvent on ionization constants is somewhat variable, though, and the simplification can lead to errors of a few tenths of a pH unit.

2. Methods

a. Nuclear magnetic resonance: histidine residues [13]
NMR has been particularly useful in determining the pK_as of histidines in proteins. The signals from the C(2) and C(4) protons are further downfield than the bulk of the resonances of the protein and may be resolved in D_2O solutions. The chemical shift changes on protonation and so the histidines may be readily titrated. Where there is more than one histidine there is the difficulty of assigning the individual pK_as. In the case of ribonuclease, chemical modification, selective deuteration, and a knowledge of the crystal structure has enabled the assignment of the pK_as of all four histidines. [14,15]

The resonances of protons in hydrogen bonds may be shifted downfield to such an extent that they may be observed in H_2O solutions. The proton between Asp-102 and His-57 in chymotrypsinogen, chymotrypsin, and other serine proteases has been located and its resonance found to titrate with a pK_a of $7 \cdot 5^{16}$ (although the pK_a is for the dissociation of the proton on the other nitrogen of the imidazole ring).

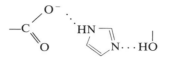

One of the serine proteases, α-lytic protease, which is thought to have the same 'charge relay system' at the active site, has only one histidine. The chemical shift of the ^{13}C carbon of this histidine enriched in the C(2) position is found to titrate with a pK_a of 6·7. But, unlike that in model compounds, the coupling constant between this carbon and its proton does not change with pH. It appears that Asp-102 is the group that titrates with the pK_a of 6·7, and not the histidine as always assumed.[17] This conclusion is supported by an infrared titration of the carboxyl (see below).[18]

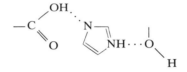

^{13}C NMR may be used to measure the pK_as of lysine and aspartate residues. Proton NMR has been used to titrate the ionization of tyrosine residues.[19]

b. Infrared spectroscopy[18,20,21]
Carboxyl groups absorb at about 1710 cm^{-1} and carboxylates at about 1570 cm^{-1}. Infrared difference spectra in D_2O solutions have been used to measure the pK_as of abnormal carboxyl groups in α-lactoglobulin (7·5), lysozyme (2·0, 6·5), and trypsin (Asp-102, 7).

c. Ultraviolet difference spectroscopy[22,23]
Difference spectra have been frequently used to measure the ionization of the phenolic hydroxyl of tyrosines. The sulphydryls of cysteines and the imidazoles of histidines are also amenable to difference spectroscopy.

d. Fluorescence[22,24]
This is useful when the ionization of a group perturbs the spectrum of a neighbouring tryptophan, the major fluorescent species in proteins, or causes a conformational change which perturbs the fluorescence of the protein as a whole. Tyrosines may be titrated in the absence of tryptophans (which fluoresce more strongly).

e. Difference titration[25]

The direct titration of a protein with acid or base usually gives an uninterpretable ionization curve because of the overlapping titrations of the many groups. However, one or two ionizations may be isolated by comparing the titration curve with that of the enzyme after a particular group has been blocked. Asp-52 of lysozyme may be specifically esterified with triethoxonium fluoroborate. The difference titration between this modified form and the native protein gives not only the pK_a of Asp-52 but also the effect of this ionization on the other important acid at the active site, Glu-35.

f. Denaturation difference titration[26]

It is not practically possible to determine by titration whether or not a buried group in a protein, such as Asp-102 in chymotrypsin and chymotrypsinogen, is ionized. Accurate titration is possible only between pH 3 and 11 because of the high background concentrations of hydroxyl ions and protons outside this range. For this reason it is not possible to detect a missing ionization in the overall titration curve if the protein has any other abnormally titrating groups. Chymotrypsin, for example, has three abnormally low titrating carboxyls which are still ionized below pH 3. However, these groups titrate normally when the protein is denatured and the number of carboxyls titrating in the denatured proteins is easily determined. It was shown that Asp-102 is ionized in the pH forms of chymotrypsin and chymotrypsinogen by measuring the proton uptake on denaturation, and adding this to the number of carboxyls known to be ionized in the denatured proteins. In order to improve the accuracy, the majority of the surface carboxyls were converted to amides to lower the number of ionizing groups.

g. Chemical modification

The rate of inhibition of enzymes by irreversible inhibitors has been used in the same way as normal kinetics to give the pK_as of the free enzyme and the enzyme–inhibitor complex. The pK_as of other residues have been measured from the pH dependence of their reaction with chemical reagents. For example, since the basic forms of amines react with acetic anhydride[27] or dinitrofluorobenzene,[28–30] their extent of ionization is given from the relative rates of reaction as a function of pH. This has been used to measure the pK_as of amino groups in proteins by modifying them with radioactive reagents, digesting the protein, separating the peptides, and measuring their specific radioactivities. In this way the pK_as of several groups may be determined simultaneously as has been done for elastase and chymotrypsin.

A useful variation of this technique is to measure the rate of exchange of tritium from tritiated water with the C(2) proton of the imidazole ring of histidines.[18,31,32] The rate constant for the exchange depends on the state of ionization of the imidazole, being faster for the unprotonated form. This procedure is a useful adjunct to NMR experiments which also measure the pK_as of histidine residues.

These experiments are tedious to perform, but the assignments of pK_a are unambiguous.

H. Effect of temperature, polarity of solvent, and ionic strength on pK_as of groups on enzymes and in solution

Ions are stabilized by polar solvents. The electrostatic dipoles of the solvent directly interact with the electrical charges of the ions, and the dielectric constant decreases the tendency of the ions to reassociate. The ionization of a neutral acid, as in eqn (5.38), is depressed by the addition of a solvent of low polarity to an aqueous solution (Table 5.4).

$$HA = A^- + H^+ \qquad (5.38)$$

On the other hand, the ionization of a cationic acid is insensitive to solvent polarity since there is no change of charge in the equilibrium.

$$BH^+ = B + H^+ \qquad (5.39)$$

TABLE 5.4. *Effect of organic solvents on* pK_as *at* 25°

| Weight % Dioxan | pK_a | | | | Glycine | |
	Acetic acid	Tris[a]	Benzoylarginine	—CO_2H	—NH_3^+
0	4·76	8·0	3·34	2·35	9·78
20	5·29	8·0		2·63	9·29
45	6·31	8·0		3·11	8·49
50		8·0	4·59		
70	8·34	8·0	4·60	3·96	7·42

[a] $(HOCH_2)_3CNH_2$
H. S. Harned and B. B. Owen, *The physical chemistry of electrolytic solutions*, Reinhold, New York, 755–6 (1958); T. Inagami and J. M. Sturtevant, *Biochim. Biophys. Acta* **38,** 64 (1960)

The procedure of using the effect of solvent polarity on the pK_a of a group on an enzyme to tell whether it is due to the ionization of a cationic or neutral acid is unreliable. A partly buried acid is shielded from the full effects of the solvent and the electrostatic interactions with the protein may be more important. For example, the pK_a of the acylenzyme of benzoylarginine and trypsin is almost invariant in 0–50% dioxane/water and increases only slightly in 88% dioxane/water.[33] This would appear to be consistent with a cationic acid such as imidazole ionizing, since the pK_a of acetic acid under these conditions increases by about 6 units. Yet NMR and infrared experiments suggest that the carboxyl of Asp-102 is ionizing.[17,18]

Similarly, although increasing ionic strength decreases the pK_as of carboxylic acids in solution, the pK_as of Asp-52 and Glu-35 in lysozyme are increased by increasing ionic strength due to an effect on the surface charge of the protein.[25]

Imidazole groups in solution have enthalpies of ionization of about 30 kJ/mol (7 kcal/mol) whilst carboxylic acids have negligible enthalpies of ionization. But the changes in the enthalpy of the solvating water molecules also make important contributions to these values so the solution values cannot be extrapolated to partly buried groups in proteins.

I. Highly perturbed pK_as in enzymes

There are many examples of amine bases and carboxylic acids in proteins which titrate with anomalously high or low pK_as (Table 5.5). The reasons

TABLE 5.5 *Some highly perturbed* pK_as *of groups in proteins*

Enzyme	Residue	pK_a
Lysozyme	Glu-35	6·5
Lysozyme-glycolchitin complex	Glu-35	~8·2
α-Lytic protease	Asp	6·7
Acetoacetate decarboxylase	Lys (ϵ-NH$_2$)	5·9
Chymotrypsin	Ile-16 (α-NH$_2$)	10·0
α-Lactoglobulin	CO$_2$H	7·5

(Refs 25, 25, 17, 6, 11, and 20 respectively).

for this are quite straightforward and depend on the microenvironment. If a carboxyl group, such as the aspartate in the charge relay system of α-lytic protease or chymotrypsin, is in a region of relatively low polarity, its pK_a will be raised since the anionic form is destabilized. Alternatively, if the carboxylate ion forms a salt bridge with an ammonium ion, it will be stabilized by the positive charge and be more acidic. Conversely, if an amino group is buried in a non-polar region, like the lysine in acetoacetate decarboxylase, protonation is inhibited and the pK_a is lowered. An ammonium ion in a salt bridge, such as Ile-16 in chymotrypsin, is stabilized by the negative charge on the carboxylate ion. Deprotonation is inhibited and the pK_a is raised.

References

1 M. Dixon, *Biochem. J.* **55,** 161 (1953).
2 R. A. Alberty and V. Massey, *Biochim. biophys. Acta* **13,** 347 (1954).
3 M. L. Bender, G. E. Clement, F. J. Kézdy, and H. d'A Heck, *J. Am. chem. Soc.* **86,** 3680 (1964).
4 A. R. Fersht and Y. Requena, *J. Am. Chem. Soc.* **93,** 7079 (1971).
5 J. Fastrez and A. R. Fersht, *Biochemistry* **12,** 1067 (1973).
6 D. E. Schmidt Jr. and F. H. Westheimer, *Biochemistry* **10,** 1249 (1971).
7 M. Renard and A. R. Fersht, *Biochemistry* **12,** 4713 (1973).
8 A. R. Fersht and M. Renard, *Biochemistry* **13,** 1416 (1974).
9 J. T. Edsall and J. Wyman, *Biophysical chemistry,* Academic Press, New York, p. 510 (1958).
10 F. J. Kézdy, G. E. Clement, and M. L. Bender, *J. Am. chem. Soc.* **86,** 3690 (1964).
11 A. R. Fersht, *J. molec. Biol.* **64,** 497 (1972) and references therein.
12 L. Pentz and E. R. Thornton, *J. Am. chem. Soc.* **89,** 6931 (1967).
13 J. L. Markley, *Accts chem. Res.* **8,** 70 (1975).
14 J. L. Markley, *Biochemistry* **14,** 3546 (1975).
15 S. M. Dudkin, M. Ya. Karpeiskii, V. G. Sakharovskii, and G. I. Yakovlev, *Dokl. Akad. Nauk SSSR* **221,** 740 (1975).
16 G. Robillard and R. G. Shulman, *J. molec. Biol.* **86,** 519 (1974).
17 M. W. Hunkapiller, S. H. Smallcombe, D. R. Whitaker, and J. H. Richards, *Biochemistry* **12,** 4732 (1973).
18 R. E. Koeppe II and R. M. Stroud, *Biochemistry* **15,** 3450 (1976).
19 S. Karplus, G. H. Snyder, and B. D. Sykes, *Biochemistry* **12,** 1323 (1973).
20 H. Susi, T. Zell, and S. N. Timasheff, *Archs. Biochem. Biophys.* **85,** 437 (1959).
21 S. N. Timasheff and J. A. Rupley, *Archs. Biochem. Biophys.* **150,** 318, (1972).
22 S. N. Timasheff, *The Enzymes* **2,** 371 (1970).
23 M. J. Gorbunoff, *Biochemistry* **10,** 250 (1971).
24 R. W. Cowgill, *Biochim. biophys. Acta* **94,** 81 (1965).
25 S. M. Parsons and M. A. Raftery, *Biochemistry* **11,** 1623, 1630, 1633 (1972).
26 A. R. Fersht and J. Sperling, *J. molec. Biol.* **74,** 137 (1973).
27 H. Kaplan, K. J. Stephenson, and B. S. Hartley, *Biochem. J.* **124,** 289 (1971).
28 A. L. Murdock, K. L. Grist, and C. H. W. Hirs, *Archs. Biochem. Biophys.* **114,** 375 (1966).
29 R. J. Hill and R. W. Davis, *J. biol. Chem.* **242,** 2005 (1967).
30 W. H. Cruickshank and H. Kaplan, *Biochem. biophys. Res. Commun.* **46,** 2134 (1972).
31 H. Matsuo, M. Ohe, F. Sakiyama, and K. Narita, *J. Biochem., Tokyo* **72,** 1057 (1972).
32 M. Ohe, H. Matsuo, F. Sakiyama, and K. Narita, *J. Biochem., Tokyo* **75,** 1197 (1974).
33 T. Inagami and J. M. Sturtevant, *Biochim. Biophys. Acta* **38,** 64 (1960).

Chapter 6

Practical kinetics

An essential element in any kinetic study is the availability of convenient assays for measuring either the rate of formation of products or the rate of depletion of substrates. There exists a variety of methods for doing this, from the classical procedures of manometry, viscometry, and polarimetry, to modern techniques such as NMR and ESR, but the most commonly employed procedures are spectrophotometry, spectrofluorimetry, automatic titration, and the use of radioactively labelled substrates. In this chapter we shall discuss the more common methods with the emphasis on their practical uses and limitations rather than give a general theoretical survey of all the techniques available.

A. Kinetic methods

1. Spectrophotometry

Many substances absorb light in the ultraviolet or visible regions of the spectrum. If the intensity of the light shining on to a solution of the compound is I_0 and that transmitted through is I, the absorbance A of the solution is defined by

$$A = \log(I_0/I) \tag{6.1}$$

The absorbance usually follows *Beer's Law*,

$$A = \epsilon cl \tag{6.2}$$

where ϵ is the *absorption* or *extinction coefficient* of the compound, c is its concentration (usually in units of molarity), and l is the pathlength (cm) of the light through the solution. A compound may be assayed by measuring A if ϵ is known.

Spectrophotometry is particularly useful with naturally occurring chromophores. For example, the rates of many dehydrogenases may be measured from the rate of appearance of NADH at 340 nm ($\epsilon = 6\cdot23 \times 10^3 \, \text{M}^{-1} \, \text{cm}^{-1}$) since NAD does not absorb at this wavelength. Otherwise artificial substrates may be used, such as p-nitrophenyl esters with esterases since the p-nitrophenolate ion absorbs at 400 nm ($\epsilon = 1 \cdot 8 \times 10^4 \, \mathrm{M}^{-1} \, \mathrm{cm}^{-1}$).

The sensitivity of the method depends on the extinction coefficient involved; for $\epsilon = 10^4$ the lower limit of detectability is about 0·5 nanomoles using a conventional spectrophotometer requiring at least 0·5 ml of solution and absorbance of 0·01.

Some possible errors

The usual source of error is the breakdown of Beer's law. This may happen in several ways: the chromophore aggregates or forms micelles at high concentrations with a change in absorption coefficient; at high background absorbances so much light of the correct wavelength is absorbed that light of other wavelengths leaking through the monochromator becomes significant ('stray light'); the bandwidth of the monochromator is set so wide that wavelengths other than those absorbed by the chromophore are transmitted to the sample; the solutions are turbid and scatter light.

2. Spectrofluorimetry

Some compounds absorb light and then re-emit it at a longer wavelength. This is known as fluorescence. The efficiency of the process is termed the quantum yield q, which is equal to the number of quanta emitted per number absorbed (always less than one). Natural fluorophores include NADH, which absorbs at 340 nm and emits at about 460 nm. The major fluorophore in proteins is tryptophan, absorbing at 275–295 nm and re-emitting at 330–340 nm. Tyrosine fluoresces weakly in the same region. Many synthetic substrates for esterases, phosphatases, sulphatases, and glycosidases are based on 4-methylumbelliferone, a highly fluorescent derivative of phenol.

Fluorimetry is theoretically about 100 times more sensitive than spectrophotometry for the detection of low concentrations. Fluorescence is measured at right-angles to the exciting beam against a dark background. A drift of 5% in the excitation intensity leads to only a 5% change in the signal. Absorbance measurements involve detecting a small change in the transmitted light. Any fluctuation in intensity is magnified; a 5% change is equivalent to an absorbance of 0·02.

Although the intensity of the fluoresced light is proportional to the intensity of the exciting beam, it is not possible to compensate indefinitely for decreasing concentration by increasing the excitation intensity. This is because the solvent scatters light—Rayleigh scattering at the same wavelength as the excitation, and Raman scattering at a longer wavelength—and this eventually swamps the emitted light. Also, the compound that is being excited will be destroyed by photolysis at sufficiently high intensities.

Some properties of fluorescence

Fluorescence occurs by a photon being absorbed by a compound to give an excited state of lifetime about 10 ns which decays by re-emission of a photon. Decay may also take place by a collision with another molecule, such as an iodide ion, or a transfer of energy to another group in the molecule. The fluorescence is then said to be *quenched*. The quenching of the fluorescence of tryptophans of a protein on the binding of ligands is a useful way of measuring the extent of binding.

Fluorescence may also be *enhanced*. Sometimes a compound has a low quantum yield in aqueous solution but a higher one in non-polar media. The dyes toluidinyl- and anilinyl-naphthalene sulphonic acids fluoresce very weakly in water but strongly when bound in the hydrophobic pockets of proteins. Interestingly enough, if they are bound next to a tryptophan residue, they may be exited by light absorbed by the tryptophan at 275–295 nm whose energy is transferred to them. Tryptophan and NADH fluoresce relatively weakly in water and their fluorescence may be enhanced in the non-polar regions of proteins.

When a small molecule is excited by light which is plane polarized, the fluoresced light is not polarized because the small molecule rapidly tumbles in solution: its rotational relaxation time is short compared with the 10 ns lifetime of the excited state. However, large molecules tumble slowly; a rule of thumb is that the rotational relaxation time in nanoseconds is approximately equal to the molecular weight in thousands (chymotrypsin with a molecular weight of 25 000 has a rotational relaxation time of about 20 ns). Fluoresced light from residues which are part of, or bound to, the protein will be partly plane polarized if the residues rotate at only the same rate as the protein. This is a useful guide to the mobility of the residues being studied.

Some possible errors

Errors usually arise through some form of quenching such as the inadvertent addition of something such as potassium iodide, or more commonly through *concentration quenching*. This occurs when the fluorophore or ligand absorbs significantly through a Beer's law effect and reduces the intensity of the exciting or emitted light. For example, if the solution has an absorbance of A, then the average intensity of the exciting light in the cell is given from eqn (6.1):

$$I_{ave} = I_0 \times 10^{-\frac{1}{2}A} \tag{6.3}$$

The true fluorescence may be calculated from the above equation, from more sophisticated correction formulae, or from standard curves.[1]

3. Automated spectrophotometric and spectrofluorimetric procedures

The spectroscopic assays are simplest and most accurate when the products have a different spectrum from the reagents and may be observed directly. Assays in which aliquots of the reaction mixture are developed with a reagent to give a characteristic colour, such as amino acids with ninhydrin, may be automated by using a proportionating pump which mixes the reagents in the desired ratios and pumps them through a flow cell in a spectrophotometer or fluorimeter (see Fig. 6.1). Apart from being less tedious for routine work, these procedures are much more reproducible and accurate than the conventional approach.

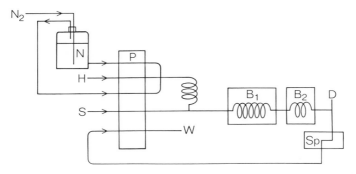

Fig. 6.1. Flow diagram for the automated ninhydrin assay for the production of an amino acid during a reaction. Samples of the reaction mixture are continuously drawn off and the amino acid assayed as follows. The peristaltic pump (P) mixes ninhydrin (N) with hydrazine sulphate (H) and nitrogen. (Nitrogen bubbles form in the flow tube and separate the liquid into successive segments that do not mix with each other). The hydrazine and ninhydrin mix thoroughly in the first coil to give the active assay solution. They then meet the reaction mixture (S) that is incubating and mix with this in the second coil. The colour develops during the 20 minute travel time at 95° through the coil in B_1. After cooling to 25° in coil B_2, the bubbles are removed by the 'debubbler' (D) and the absorbance is monitored by passing through a flow cell in the spectrophotometer (Sp). The amount of each solution mixed depends on the bore of the tubing used. As well as pumping all the solutions into the spectrophotometer, the peristaltic pump sucks out the solution and pumps it to waste: the bore of the sucking tube is sufficiently smaller than the sum of the tubes pumping in so that some of the solution and all of the nitrogen bubbles are forced out through the debubbler (D). (A. R. Fersht and M. Renard, *Biochemistry* **13**, 1416 (1974), modified from J. Lenard, S. L. Johnson, R. W. Hyman, and G. P. Hess, *Anal. Biochem.* **11**, 30 (1965))

4. Coupled assays

Some reactions that do not give chromophoric or fluorescent products may be coupled with another enzymic reaction that does. Many of these are based on the formation or disappearance of NADH. The formation of pyruvate may be linked with the conversion of NADH to NAD by lactate

dehydrogenase (6.4); the formation of ATP may be coupled to the above
by the use of pyruvate kinase etc. as shown below.

$$\text{Pyruvate} + \text{NADH} \xrightleftharpoons{\text{Lactate dehydrogenase}} \text{Lactate} + \text{NAD} \qquad (6.4)$$

For example, phosphofructokinase may be assayed by the following
scheme.

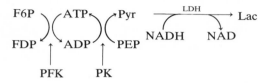

(F6P = fructose-6-phosphate,   FDP = fructose-1,6-diphosphate,
Pyr = pyruvate, PEP = phosphoenolpyruvate, PFK = phosphofructokinase,
PK = pyruvate kinase, LDH = lactate dehydrogenase)

Another assay for phosphofructokinase involves converting the fructose
diphosphate to dihydroxyacetone phosphate and glyceraldehyde-3-
phosphate with aldolase, equilibrating the triose phosphates with triose
phosphate isomerase, and then measuring the production of NADH on
the oxidation of the glyceraldehyde phosphate by glyceraldehyde-3-
phosphate dehydrogenase.

5. Automatic titration of acid or base

A hydrolytic reaction which releases acid may be followed by titration
with base. This is best done automatically by use of the pH-stat. A glass
electrode registers the pH of the solution which is kept constant by the
automatic addition of base from a syringe controlled by an electronic
circuit. Reaction volumes down to 1 ml may be used and the limit of
detectability is about 50 nmol (5–10 μl of 5×10^{-3}–1×10^{-2} M base). The
usual source of error with this apparatus is the buffering effect of
dissolved CO_2.

6. Radioactive procedures

The most sensitive assay methods available involve the use of radioactively labelled substrates and reaction volumes of 20–100 μl.
   Radioactivity is measured in Curies (Ci): 1 Ci is $2 \cdot 22 \times 10^{12}$ decomposi-
tions per minute. In practice, radioactive decay is measured in terms of
counts per minute using a scintillation counter with the common isotopes
^{3}H, ^{14}C, ^{35}S, and ^{32}P. These isotopes emit β radiation (electrons) when
they decay. The radiation may be monitored by using a scintillant (see
Appendix) which converts the radiation into light quanta which are
registered as 'counts' by a photomultiplier. The low energy emission of

TABLE 6.1. *Some common radioactive isotopes*

Isotope	Half Life	Specific activity of 100% isotopic abundance (Ci/Mol)	Type of emission	Energy of emission (max)(MeV)
^{14}C	5730 yr	62·4	β	0·156
^{3}H	12·35 yr	$2·9 \times 10^4$	β	0·0186
^{35}S	87·4 days	$1·49 \times 10^6$	β	0·167
^{32}P	14·3 days	$9·13 \times 10^6$	β	1·709
^{125}I	60 days	$2·18 \times 10^6$	γ	
^{131}I	8·06 days	$1·62 \times 10^7$	β, γ	0·247–0·806
^{75}Se	120 days	$1·09 \times 10^6$	γ	

^{3}H is counted with an efficiency of between 15 and 40% and the higher energy emissions from ^{14}C, ^{35}S, and ^{32}P at about 80%. Because the energy of the ^{3}H emission is so different from the others, it may be counted in their presence in a 'double labelling' experiment by monitoring different regions of the energy spectrum separately (Table 6.1).

The energy of the emission from ^{32}P is so high that it may be monitored in the absence of scintillant due to the Čerenkov effect. On passing through water or a polythene scintillation vial the electrons move so rapidly that they spontaneously emit photons which may be detected at about 40% efficiency using the counter at an open window setting. This has been used to count ^{32}P and ^{14}C in each other's presence by first monitoring the Čerenkov radiation and then counting again after the addition of scintillant.

The sensitivity of detection depends on the specific activity of the compound. Some examples are given in Table 6.2. As low as 10^{-17} moles may be detected, some 6 or 7 orders of magnitude below the lower limit of spectrophotometry.

TABLE 6.2. *Some radioactively labelled substrates available from the Radiochemical Centre, Amersham*

Substrate	Specific Activity (Ci/Mol)	Limit of detectability[a] (picomoles)
[^{14}C]Methionine (90% enriched)	280	0·1
[Me-^{3}H]Methionine	8×10^3	0·01
[^{35}S]Methionine	3×10^5	1×10^{-4}
[$\gamma^{32}P$]ATP	$1·5 \times 10^6$	2×10^{-5}

[a] Based on 50 counts per minute as the minimum measurable.

Some possible errors

(a) *Quenching.*   The emission from tritium is so weak that it is readily absorbed by a filter paper or chromatographic strip on which it is adsorbed, or by precipitates. Fewer problems occur with ^{14}C but the position of the energy maximum may be lowered on rare occasions. Coloured materials or charcoal may absorb the light emitted from the scintillant.

(b) ^{3}H *transfer.*   Tritium is sometimes transferred from substrates to solvents or proteins. For this reason, and more importantly the above, the use of ^{14}C is preferable to ^{3}H, although the higher specific activities and relative cheapness of ^{3}H-labelled compounds often more than compensate for the disadvantages.

(c) *Miscellaneous.*   The addition of water to scintillants often lowers the efficiency of counting so it should always be added in constant amounts. Sometimes the addition of base causes artefacts. Strip lighting can cause some scintillants to phosphoresce and the effect can last several minutes.

Separation of reaction products

Clearly, in order to assay a reaction it is necessary to separate the products from the starting materials to measure the radioactivity. Chromatography and high-voltage electrophoresis are very useful since the supporting paper may be cut into strips and the *relative* activities of the various regions accurately measured. Much more convenient, though, is the use of a filter pad to selectively adsorb or trap the desired compound. For example, labelled ATP may be adsorbed on to charcoal and collected on a glass fibre disk, or adsorbed on to a disk impregnated with charcoal, and the remaining reagents washed away. Filter disks of diethylammoniumethyl-cellulose have been used to selectively adsorb anionic reagents, and disks of carboxymethyl-cellulose to trap cations. Proteins, and any strongly bound ligand, often adsorb to nitrocellulose filters. Covalently labelled proteins or polynucleotides may be precipitated with acid and collected on glass fibre filters if a heavy precipitate is formed, or on nitrocellulose if the precipitate is light. These disks do not lower the efficiency of counting of ^{14}C or ^{32}P.

B. Plotting kinetic data

1. Exponentials

a. Single

$$A \xrightarrow{\ k\ } B \qquad\qquad (6.5)$$

In the first-order reaction of eqn (6.5) the concentration of B at time t,

$[B_t]$, is related to the final concentration of B, $[B_\infty]$, (the endpoint) by

$$[B_t] = [B_\infty](1 - \exp(-kt)) \tag{6.6}$$

(cf., eqn (4.5)).

k is usually obtained from the following semilogarithmic plot of $[B_\infty] - [B_t]$ against t:

$$\ln([B_\infty] - [B_t]) = \ln[B_\infty] - kt. \tag{6.7}$$

It is important to determine the endpoint $[B_\infty]$ accurately since errors in this cause serious errors in the derived rate constant. Least-squares methods are often used to fit the data directly to the theoretical equation. It is better to use the observed endpoint in this method rather than treat it as a variable parameter since the deviations caused by small changes in the endpoint are usually insignificant compared with the 'noise' in the data.

b. The Guggenheim method

In cases where the endpoint cannot be determined, the Guggenheim method may be used.[2] In this, the differences between pairs of readings at t and $t + \Delta t$ (where Δt is a constant time which must be at least 2 or 3 times the half-time) are plotted against t in the semilogarithmic plot since it may be shown that

$$\ln([B_{t+\Delta t}] - [B_t]) = \text{Constant} - kt. \tag{6.8}$$

c. Consecutive exponentials

A series of exponentials of the form (cf. eqn (4.29))

$$[B] = X\{\exp(-k_1 t) - \exp(-k_2 t)\} \tag{6.9}$$

are relatively straightforward to solve if one of the rate constants is more than 5 or 10 times faster than the other. The slower process is plotted as a simple semilogarithmic plot by using the data from the tail end of the curve after the first process has died out (that is after 5 to 10 half-lives of the first process have occurred). The data may then be fed back into eqn (6.9) to give the faster rate constant. This is often done graphically as shown in Fig. 6.2. If the rate constants are not separated by this factor, it is simpler to try to change their ratio by a change in the reaction conditions than to use least-squares methods.

2. Second-order reactions

$$A + B \xrightarrow{k_2} C \tag{6.10}$$

Second-order kinetics are best dealt with by converting them to pseudo-first-order by using one of the reagents in large excess over the

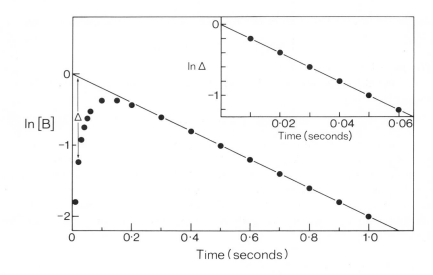

Fig. 6.2. Graphical method of analysing the kinetics of a reaction obeying eqn (6.9). The logarithm of [B] is plotted against time. The rate constant for the slower process is obtained from the slope of the linear region after the faster process has died out. The rate constant for the faster process is obtained plotting the logarithm of Δ (the difference between the value of [B] at a particular time and the value of [B] extrapolated back from the linear portion of the plot as in the figure) against time for the earlier points. The rate constants for this example are 20 and $2\ \mathrm{s}^{-1}$

other. If $[A] \gg [B]$, its concentration changes little during the reaction and

$$d[C]/dt = -d[B]/dt = k_2[A_0][B]. \tag{6.11}$$

The disappearance of B and the appearance of C follow exponential first-order kinetics with a pseudo-first-order rate constant of $k_2[A_0]$. Plotting a series of such reactions with varying $[A_0]$ gives k_2.

The analytical solution for the reaction when $[A_0]$, the initial concentration of A, is similar to $[B_0]$, the initial concentration of B, is

$$(\ln[A_0][B]/[B_0][A])/([B_0]-[A_0]) = k_2 t, \tag{6.12}$$

an equation that is tedious to plot. However, this is greatly simplified when $[A_0]=[B_0]$. The analytical solution is now

$$1/([A_0]-[C]) - 1/[A_0] = k_2 t. \tag{6.13}$$

This simple equation holds to a good approximation even when the initial concentrations are not exactly equal. (In this case the average of the concentrations of A_0 and B_0 should be used in eqn (6.13) instead of $[A_0]$).

3. Michaelis–Menten kinetics

Although it has become fashionable to fit data directly to the integrated form of the Michelis–Menten equation using a computer, the most satisfactory method of determining k_{cat}, k_{cat}/K_M, and K_M is to use the classical approach of the measurement of initial rates (i.e. the first 5% or less of the reaction) and a plot such as that of Eadie and Hofstee (eqn (3.29)).[3-5] A good range of substrate concentrations are the following multiples of the K_M: 8; 4; 2; 1; 0·5; 0·25; 0·125.

C. Determination of enzyme–ligand dissociation constants

1. Kinetics

It was seen in Chapter 3 that the K_M for an enzymic reaction is not always equal to the dissociation constant of the enzyme–substrate complex but may be lower or higher depending on whether or not intermediates accumulate or Briggs–Haldane kinetics hold. Enzyme–substrate dissociation constants cannot be derived from steady-state kinetics unless mechanistic assumptions are made or there is corroborative evidence. Pre-steady-state kinetics are more powerful since the chemical steps may often be separated from those for binding.

The *dissociation constants of competitive inhibitors* are readily determined from inhibition studies using eqn (3.32). This equation holds whether or not the K_M for the substrate is a true dissociation constant. The inhibition must first be shown to be competitive by determining the apparent K_M for the substrate at different concentrations of the inhibitor, and the K_I calculated from the apparent K_M and eqn (3.32). Significant changes in K_M are obtained only at relatively high values of the inhibitor concentration.

2. Equilibrium dialysis

This is a method of directly measuring the concentrations of free and enzyme–bound ligand. A solution of the enzyme (and ligand) is separated from a solution of the ligand by a semipermeable membrane across which only the small ligand may equilibrate. After equilibration, a sample from the chamber containing protein gives the sum of the concentrations of free and bound ligand, whilst a sample from the other chamber gives the concentration of free ligand. Measurements may be made using chamber volumes of only $20\ \mu l$ and sampling triplicate aliquots of $5\ \mu l$, using radioactively labelled ligands. However, since the apparatus requires at least 1 to 2 hours equilibration with even the most porous membranes, the method cannot be used with unstable ligands or enzymes. A *non-equilibrium dialysis* technique has been used to make rapid measurements.[6] In this, the *rate* of diffusion of the ligand across the membrane

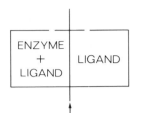

SEMIPERMEABLE MEMBRANE

FIG. 6.3. The principle of equilibrium dialysis

from the side containing enzyme is measured as a function of concentration. Binding slows down the rate.

3. Equilibrium gel-filtration[7]

Certain gels, such as the commercial products Sephadex and Biogel, are made with pores that are large enough to be occupied by small ligands but not by proteins. If a chromatography column is packed with one of these gels and a solution of protein and ligand applied, the protein travels faster through the column than the ligand since the ligand has to pass through the volume of solution surrounding the beads of gel and in the pores, whilst the protein has only to travel through the surrounding water. In equilibrium gel-filtration, such a column is equilibrated with a solution of the ligand, and a sample of the enzyme in the equilibrating buffer is applied to the column. As the protein travels through the column it drags

FIG. 6.4. Microcells for equilibrium dialysis constructed to the design of P. T. Englund, J. A. Huberman, T. M. Jovin, and A. Kornberg, *J. biol. Chem.* **244,** 3038 (1969). A sheet of dialysis membrane is placed between the two Perspex (Plexiglass) chambers and the cells are clamped together. The unit shown contains two sets of cells: the smaller Perspex half-cells fit either end of the central block which contains a half-cell at each end

FIG. 6.5. The microcells of Fig. 6.4 mounted on a rotating drum, ready for insertion into the thermostatted housing

any bound ligand with it at the same flow rate. The result is that a peak of ligand travels through the column in the position of the protein whilst a trough follows at the position normally occupied by a small ligand. The area under the peak should be the same as that in the trough and is equal to the amount of bound ligand.

One advantage of this method is that the enzyme is in contact with any particular substrate molecule for a short time only and so can be used in cases where the enzyme slowly hydrolyses the ligand. Another advantage is that gels are available that are able to distinguish between the size of one polymer and another so that, for example, the binding of a tRNA (molecular weight = 25 000) to an aminoacyl-tRNA synthetase (molecular weight = 100 000) may be measured.[8]

The method may be scaled down. A convenient size for the binding of small ligands is a 1 ml tuberculin syringe packed with Sephadex G-25. A sample volume of 100 μl is applied and individual drops (about 35–40 μl) collected in siliconized tubes. The volume of each drop is measured by drawing it into a syringe. However, where the gel has to distinguish between a protein and a large ligand, much larger columns have to be used (a typical example being 0.7×18 cm for the case of tRNA/tRNA synthetase).

Some possible errors
The criteria for an accurate experiment are that (a) the base line must be constant, (b) there must be a region of base line between the peak and the trough to show that the ligand and protein have sufficiently different mobilities, and (c) the area of the peak must equal that of the trough. (Since it is very difficult to make up the solution of the enzyme and the ligand exactly the same as that of the ligand in the equilibrating buffer, it is advisable to assay an aliquot of the enzyme solution before addition to the column in order to make any necessary correction to the area of the trough.)

Artefacts can occur due to the specific retardation of a ligand. ATP, ADP, and AMP, for example, bind to Sephadex and are retarded. The faster mobility of phosphate and pyrophosphate produced in reactions in

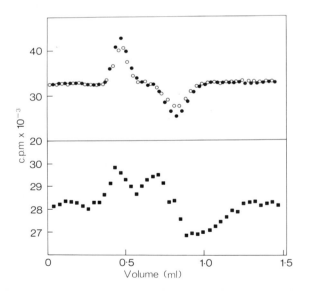

FIG. 6.6. Experimental curves for equilibrium gel-filtration. Upper curves, the 1 ml tuberculin syringe was incubated in 109 μM [^{14}C]Valine (O) or 109 μM [^{14}C]Valine and 4 mM ATP (●). 100 μl of a solution of 26 μM valyl-tRNA synthetase was added in the same solution. Stoichiometries of 0.8 and 1.1 respectively are found for the binding of the amino acid. Note the return to baseline between the peak and trough—the mark of a good equilibrium gel-filtration experiment. Lower curve, an artefactual double peak obtained on binding [γ^{32}P]ATP and valine to the enzyme. Some of the labelled ATP hydrolyses to [^{32}P]orthophosphate which travels faster down the column than the [γ^{32}P]ATP

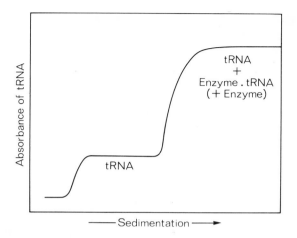

FIG. 6.7. Sedimentation of a mixture of tRNA and aminoacyl-tRNA synthetase. The complex of tRNA and enzyme sediments faster than the free tRNA. The free enzyme and its complex with tRNA migrate together since they are in rapid equilibrium, the tRNA exchanging between the two

which $[\gamma^{32}P]ATP$ is hydrolysed causes additional peaks and troughs (see Fig. 6.6).

4. Ultracentrifugation

The binding of a small polymer to a larger, such as tRNA to an aminoacyl-tRNA synthetase, cannot be determined by equilibrium dialysis, and also requires relatively large volumes for equilibrium gel-filtration. In this case, the binding may be measured on a 100–200 μl scale using the analytical ultracentrifuge. The cell is filled with a mixture of, say, the tRNA and aminoacyl-tRNA synthetase, and the absorbances of the bound tRNA and the free tRNA directly measured by the ultraviolet optics during sedimentation. The higher molecular weight complex of the enzyme and the tRNA sediments faster than the free tRNA, and there is a sharp moving boundary of absorbance as the complex moves down the cell (see Fig. 6.7).[9]

5. Filter assays[10]

Many proteins, along with any slowly dissociating bound ligands, are adsorbed on nitrocellulose filters, whilst the free ligands are not retained. In special cases, this provides a very economical procedure for assaying binding. Although this has provided much useful data on the binding of nucleic acids to proteins, there are several possible errors. Binding is often less than 100% efficient and may be variable from batch to batch of filters. The filters also saturate at fairly low concentrations of protein.

6. Spectroscopic methods

Apart from NMR experiments where the concentrations of free and bound ligand may often be individually measured, spectroscopic methods do not usually give a direct measurement of the number of bound ligand molecules. On binding of the ligand there is just a change in spectroscopic signal which is related to the fraction of the protein that is binding. But, without additional evidence, the number of ligand molecules binding is not known. Typical examples of these spectroscopic methods are the binding of a ligand to an enzyme causing a change in the fluorescence of the protein, or the protein causing a change in the fluorescence or the optical spectrum of the ligand. The usual procedure is to add increasing amounts of the ligand to a relatively dilute solution of the protein and to plot the change in the spectroscopic signal by one of the procedures given in the following section. A variation of the procedure for ligands that are not chromophores is to measure their competitive inhibition of the binding of chromophoric ligands.

7. Titration procedures

If the dissociation constant of the complex is sufficiently low, it may be possible to determine the number of equivalents of ligand that are required to give the maximum spectral change that occurs when all the binding sites are occupied. For example, increasing amounts of the ligand are added to a solution of the protein whose concentration is at least ten times higher than the dissociation constant. The results are plotted as in Fig. 6.8 to give the stoichiometry.

Another procedure is that of Job.[11]

D. Plotting binding data

1. Single binding site

The binding of a ligand to a single site on a protein is described by the following equations:

$$EL \underset{K_S}{\rightleftharpoons} E + L \tag{6.14}$$

$$K_S = [E][L]/[EL], \tag{6.15}$$

where [E] and [L] are the concentrations of the unbound enzyme and ligand.

In terms of the total enzyme concentration $[E_0]$,

$$[EL] = [E_0][L]/([L] + K_S). \tag{6.16}$$

Eqn (6.16) is of the same form as the Michaelis–Menten equation and may be manipulated in the same way. A good way of plotting the data is

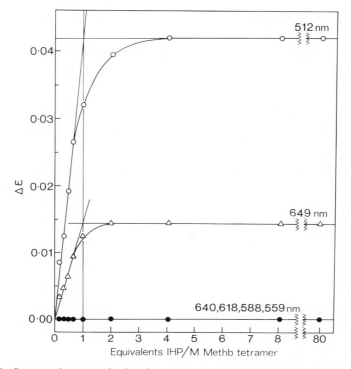

FIG. 6.8. Spectrophotometric titration of the binding of inositolhexaphosphate to methaemoglobin. The complex has an increased absorbance at 512 and 649 nm and no increases at 640, 618, 588 and 559 Å. The concentration of methaemoglobin, (20 μM) is about 14 times higher than the dissociation constant of 1·4 μM for the complex. The intersection of the slope of the increase in absorbance with the maximum value gives the stoichiometry (1 in this case). Note that this simple procedure cannot be used if the protein is not initially present at such high concentration relative to the dissociation constant, since it assumes that all the added ligand is bound to the protein for the early additions.

to use the equivalent of the Eadie plot:

$$[EL] = [E_0] - K_S([EL]/[L]).$$ (6.17)

A plot of $[EL]$ against $[EL]/[L]$ gives K_S.

Eqn (6.17) cannot be used directly with spectroscopic data since $[EL]$ is not known. However, since the concentration of EL is usually directly proportional to the change of the spectroscopic signal being observed,

$$\Delta F = \Delta F_{max} - K_S(\Delta F/[L])$$ (6.18)

where ΔF is the change in spectroscopic signal on adding $[L]$ to the protein solution. A plot of ΔF against $\Delta F/[L]$ gives K_S and ΔF_{max}, the change in signal when all the protein is converted into complex.

2. Multiple binding sites

a. Identical

If there are n identical non-interacting sites on the protein, eqn (6.17) may be modified to the *Scatchard Plot,*

$$\nu = n - K_S(\nu/[L]) \tag{6.19}$$

where ν is the number of moles of ligand bound per mole of protein. The stoichiometry n and K_S are obtained from the plot of ν against $\nu/[L]$.

b. Non-identical

If there are two classes of sites, one weak and the other strong, the Scatchard plot will be biphasic and composed of the sum of two different Scatchard plots. The determination of the values of K_S from such plots is only satisfactory when they differ by at least a factor of ten.

Appendix: Two convenient scintillants

1. BBOT

Although this water-miscible scintillant does not have the capacity of Bray's solution,[12] it is simpler to make up and does not suffer from the effects of strip lighting. It can take up to about 1% water. Two litres of scintillant contain 9 g BBOT (2,5-*bis*-[5′tert-butylbenzoxazolyl-(2′)]thiophene), 1·5 l toluene and 0·5 l methoxyethanol.

2. PPO/POPOP

Nitrocellulose disks may be suspended in this water-immiscible scintillant. 12·5 g PPO (2,5-diphenyloxazole) and 0·75 g POPOP (1,4-*bis*[2(5-phenyloxazolyl)]benzene) are dissolved in 2·5 l toluene.

References

1   M. Ehrenberg, E. Cronvall, and R. Rigler, *FEBS Letts.* **18,** 199 (1971).
2   E. A. Guggenheim, *Phil. Mag.* **2,** 538 (1926).
3   I. A. Nimmo and G. L. Atkins, *Biochem. J.* **141,** 913 (1974).
4   R. R. Jennings and C. Niemann, *J. Am. chem. Soc.* **77,** 5432 (1955).
5   A. Cornish-Bowden, *Biochem. J.* **149,** 305 (1975).
6   S. P. Colowick and F. C. Womack, *J. biol. Chem.* **244,** 774 (1969).
7   J. P. Hummel and W. J. Dreyer, *Biochim. biophys. Acta* **63,** 530 (1962).
8   R. M. Waterson, S. J. Clarke, F. Kalousek, and W. H. Konigsberg, *J. biol. Chem.* **248,** 181 (1973).
9   G. Krauss, A. Pingoud, D. Boehme, D. Riesner, F. Peters, and G. Maass, *Eur. J. Biochem.* **55,** 517 (1975).
10 M. Yarus and P. Berg, *Analys. Biochem.* **35,** 450 (1970).
11 P. Job, *Annls. Chim.* (*Ser. 10*) **9,** 113 (1928).
12 G. A. Bray, *Analyt. Biochem.* **1,** 279 (1960).

Chapter 7

Some examples of the use of kinetics in the solution of enzyme mechanisms

The mechanism of an enzymic reaction is ultimately defined when all the intermediates, complexes, and conformational states of the enzyme are characterized and the rate constants for their interconversion determined. The task of the kineticist in this elucidation is to detect the number and sequence of these intermediates and processes, define their approximate nature, that is whether covalent intermediates are formed or conformational changes occur, measure the rate constants, and, from studying pH dependence, search for the participation of acidic and basic groups. The chemist seeks to identify the chemical nature of the intermediates by what chemical paths they form and decay and what types of catalysis are involved. These results can then be combined with those from X-ray diffraction and calculations from theoretical chemists to give a complete description of the mechanism.

We shall now discuss some of the techniques that have been used to detect intermediates and delineate reaction pathways using some well-known enzymes as examples.

A. Pre-steady-state versus steady-state kinetics

It is often said that kinetics can never prove mechanisms but only rule out alternatives. While this is certainly true of steady-state kinetics in which the only measurements made are those of the rate of appearance of products or disappearance of reagents, it is not true of pre-steady-state kinetics. If the intermediates on a reaction pathway are directly observed and their rates of formation and decay measured, kinetics can prove a particular mechanism. This is the basic strength of pre-steady-state kinetics; the enzyme is used in *substrate* quantities and the events on the enzyme directly observed. There may, of course, be intermediates that remain undetected because they do not accumulate or give rise to spectral signals or are simply beyond the time scale of the measurements, but an overall reaction pathway may be proved in a scientifically acceptable manner by pre-steady-state kinetics. The basic weakness of steady-state kinetics is that the evidence obtained is always ambiguous. No direct information is obtained about the number of intermediates and so the minimum number is always assumed. This is not to say that pre-steady-state kinetics should be performed to the exclusion of steady-state kinetics, but rather that a combination of the two approaches be used. Once pre-steady-state kinetics have given information about the intermediates on the pathway, steady-state kinetics become much more powerful.

There are certain practical advantages of pre-steady-state kinetics. Essentially very simple processes may be measured, such as the stoichiometry of a burst process, the rate constant for the transfer of an enzyme-bound intermediate to a second substrate, or a ligand-induced conformational change. Also, the first-order processes that are usually measured are independent of enzyme concentration, unlike the rate constants of steady-state measurements. Although very high concentrations of enzymes are required for the rapid reaction measurements, these are usually close to those that occur *in vivo*. Furthermore, these are usually similar to those used for making direct measurements on the physical state of the protein so that data may be obtained for the state of aggregation etc. under the reaction conditions.

Pre-steady-state kinetics involve direct measurements, and direct measurements are always preferable, especially considering the tendency of enzymes to 'misbehave' (see Section B4).

1. Detection of intermediates: what is 'proof'?

Much of the following involves determining the chemical pathways of enzymic reactions. This usually requires the detection of the chemical intermediates involved since this gives *positive* evidence. We shall consider that an intermediate is 'proved' to be on a reaction pathway if the following criteria are satisfied.

(a) The intermediate is isolated and characterized.
(b) The intermediate is formed sufficiently rapidly to be on the reaction pathway.
(c) The intermediate reacts sufficiently rapidly to be on the reaction pathway.

These criteria require that pre-steady-state kinetics are used at some stage in order to measure the relevant formation and decomposition rate constants of the intermediate. But, the rapid reaction measurements are not sufficient by themselves since the rate constants must be shown to be consistent with the activity of the enzyme under steady-state conditions. Hence the power, and the necessity, of combining the two approaches.

It must always be borne in mind that an intermediate that has been isolated could be the result of a rearrangement of another intermediate and is not itself on the reaction pathway. This is why criteria (b) and (c) are needed. Conversely, it is possible that a genuine intermediate is isolated but that during the experimental work-up the enzyme takes up a different conformation that is of low activity. In this case, criteria (b) and (c) will not hold even though the chemical nature of the intermediate is correct.

B. Chymotrypsin: detection of intermediates by stopped-flow spectrophotometry, steady-state kinetics, and product partitioning

The currently accepted mechanism for the hydrolysis of amides and esters catalysed by this archetypal serine protease involves the initial formation of a Michaelis complex followed by the acylation of Ser-195 to give an acylenzyme (Chapter 1). Much of the kinetic work with the enzyme has been directed to detecting the acylenzyme. This gives a good illustration of the methods that are available using pre-steady-state and steady-state kinetics. The acylenzyme accumulates in the hydrolysis of activated or

$$E + RCO—X \xrightleftharpoons{K_S} RCO—X.E \xrightarrow{k_2} \underset{\substack{+ \\ XH}}{RCO—E} \xrightarrow{k_3} RCO_2H + E \quad (7.1)$$

specific ester substrates $(k_2 > k_3)$ so that the detection is relatively straightforward. Accumulation does not occur with the physiologically relevant peptides $(k_2 < k_3)$ and detection is difficult.

1. Detection of acylenzme in the hydrolysis of esters by pre-steady-state kinetics

a. 'Burst' experiments
In 1954, Hartley and Kilby[1] examined the reaction of substrate quantities of chymotrypsin with excess *p*-nitrophenyl acetate or *p*-nitrophenyl ethyl carbonate. They noted that the release of *p*-nitrophenol did not extrapolate back to zero but involved an initial 'burst', equal in magnitude to the concentration of enzyme (Fig. 4.10). They postulated that the ester initially rapidly acylated the enzyme in a mole to mole ratio and the subsequent turnover of the substrate involved the relatively slow hydrolysis of the acylenzyme as the rate-determining step. This was subsequently verified by the stopped-flow experiments described below.

Such burst experiments have subsequently been performed on many other enzymes. However, bursts may be due to effects other than the accumulation of intermediates, and artefacts can occur. Some examples are the following.

(a) The enzyme is converted to a less active conformational state on combination with the first mole of substrate.

(b) The dissociation of the product is rate determining.

(c) There is severe product inhibition.

It is not a trivial matter to eliminate possibilities (a) and (b). But this has been done for chymotrypsin in the following series of stopped-flow experiments.

b. Stopped-Flow methods

(i) *Chromophoric leaving group.*[2-4] The original work on *p*-nitrophenyl acetate has been extended by synthesizing *p*-nitrophenyl esters of specific acyl groups, such as acetyl-L-phenylalanine, -tyrosine, and -tryptophan. The rate of acylation of the enzyme is determined from the rate of appearance of the nitrophenol or nitrophenolate ion which absorbs at a different wavelength from the parent ester.

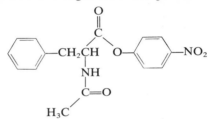

Acetyl-L-phenylalanine *p*-nitrophenyl ester

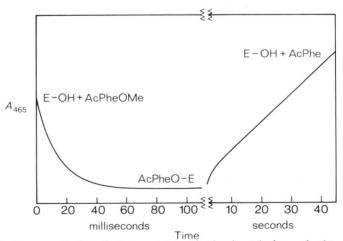

FIG. 7.1. The proflavin displacement method. A solution of chymotrypsin (10 μM), proflavin (50 μM), and AcPheOMe (2mM) was mixed at pH 6 and 25° in a stopped-flow spectrophotometer. The substrate binding step is too fast to be observed. The rapid exponential decrease in absorbance is caused by the displacement of proflavin from the enzyme on formation of the acylenzyme. The slow increase in absorbance is due to the depletion of the substrate and the consequent decrease in the steady-state concentration of the acylenzyme. (A. Himoe, K. G. Brandt, R. J. DeSa, and G. P. Hess, *J. biol. Chem.* **244**, 3483 (1969))

On mixing the ester with the enzyme there is a rapid exponential phase followed by a linear increase in the absorbance due to nitrophenol. The rate constant for acylation and the dissociation constant of the enzyme–substrate complex may be calculated from the concentration dependence of the rate constant for the exponential phases (eqn (4.44)). The rate constant of the linear portion gives the deacylation rate. Unfortunately, nitrophenyl esters are often so reactive that the acylation rate is too fast for stopped-flow measurement.

(ii) *Chromophoric acyl group.*[4,5] The spectrum of the furylacryloyl group depends on the polarity of the surrounding medium and also on the nature of the moiety to which it is attached. The spectrum of furylacryloyl-L-tyrosine ethyl ester changes slightly on binding to chymotrypsin. There are also further changes on forming the acylenzyme and on the subsequent hydrolysis. The rate constants for acylation, deacylation, and the dissociation constant of the Michaelis complex may be measured by the appropriate experiments.

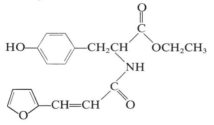

Furylacryloyl-L-tyrosine ethyl ester

On mixing the ester with the enzyme there is an initial change in absorbance which is due to the formation of the Michaelis complex. The rate constant for this is beyond the time scale of stopped flow, but the magnitude of the change can be used to calculate the dissociation constant. The absorbance then changes exponentially as the acylenzyme accumulates. There are then further changes in the spectrum of the furylacryloyl group as the ester is gradually hydrolysed to the free acid.

(iii) *Chromophoric inhibitor displacement.*[6,7] The spectrum of the dye proflavin changes significantly with solvent polarity. It is a competitive inhibitor of chymotrypsin, trypsin, and thrombin, and undergoes a large

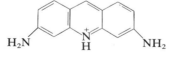

Proflavin (3,6-diaminoacridine)

increase in absorbance at 465 nm ($\Delta\epsilon \sim 2 \times 10^4\,\mathrm{M^{-1}\,cm}$) on binding (see Fig. 7.1).

On mixing an ester such as acetyl-L-phenylalanine ethyl ester with a solution of chymotrypsin and proflavin, the following events occur. There is a rapid displacement of some of the proflavin from the active site as the substrate combines with the enzyme, leading to a decrease in A_{465}. Then, as the acylenzyme is formed, the binding equilibrium between the ester and the dye is displaced, leading to the displacement of all the proflavin. The absorbance remains constant until the ester is depleted and the acylenzyme disappears. The dissociation constant of the enzyme–substrate complex may be calculated from the magnitude of the initial displacement, whilst the rate constant for acylation may be obtained from the exponential second phase.

The use of the proflavin displacement method is far more convenient than the furylacryloyl group since no special substrates have to be synthesized and one readily available compound can be used with all substrates. In general, it is better not to use modified substrates since, apart from the chemical inconvenience of synthesis, they are always open to criticism on the grounds that they could be artefactual.

(iv) *Deacylation.* It is possible to synthesize certain non-specific acylenzymes and store them at low pH.[8-11] On restoring to high pH, they are found to deacylate at the rate expected from the steady-state kinetics. This approach has been extended to cover specific acylenzymes. On incubating acyl-L-tryptophan derivatives at pH 3–4 the acylenzyme accumulates. The solution may then be 'pH-jumped' by mixing with a concentrated high-pH buffer in the stopped-flow spectrophotometer.[12,13] The deacylation rate has been measured by the proflavin displacement method or by using furylacryloyl compounds. Alternatively, the acylenzymes have been made *in situ* by mixing a nitrophenyl ester substrate with excess enzyme in the stopped-flow spectrophotometer.[14]

These experiments prove the intermediacy of the acylenzyme in the hydrolysis of esters. The intermediate can be isolated at low pH and its deacylation rate constant measured and found to be the same as that from steady-state kinetics. The experiments on the acylation reaction show that the intermediate is rapidly formed, and with a $1:1$ stoichiometry. Of course, none of these experiments show that acylation is on Ser-195. But the ultimate proof of this has come from the X-ray diffraction experiments on indolylacryloyl-chymotrypsin, where the enzyme is seen to be acylated on this residue (Chapter 1).[15] The classical identification of reaction at Ser-195 was from irreversibly inhibiting the enzyme with phosphate esters, such as diisopropyl fluorophosphate, and isolating the peptide containing the phosphate after partly hydrolysing the protein.[16,17]

2. Detection of acylenzyme in the hydrolysis of esters by steady-state kinetics

In the last section it was seen that stopped-flow kinetics can detect intermediates that *accumulate*. Detection of these intermediates by steady-state kinetics is of necessity indirect and relies on inference. Proof ultimately depends on relating the results to the direct observations of the pre-steady-state kinetics. But steady-state kinetics can also detect intermediates that do not accumulate and, by extrapolation from the cases where accumulation occurs, *prove* their existence and nature.

Detection of intermediates by steady-state kinetics depends on:

(a) the accumulation of an intermediate that is able to react with an acceptor whose concentration may be varied, or preferably with several different acceptors;

(b) a common intermediate E—R being generated by a series of different substrates all containing the structure R. This intermediate must be able to react with different acceptors.

The hydrolysis of esters (and amides) by chymotrypsin satisfies these criteria. The hydrolysis of, say, acetyl-L-tryptophan p-nitrophenyl ester forms an acylenzyme that reacts with various amines, such as hydroxylamine, alaninamide, hydrazine etc., and also with alcohols such as methanol, to give the hydroxamic acid, dipeptide, hydrazide, and methyl ester of acetyl-L-tryptophan respectively. The same acylenzyme is generated in the hydrolysis of the phenyl, methyl, ethyl etc. esters of the amino acid (and also during the hydrolysis of amides).

The kinetic consequences of the common intermediate can be used to diagnose its presence.

a. Rate-determining breakdown of a common intermediate implies common V_{max} or k_{cat}

If several different substrates generate the same intermediate and if its breakdown is rate determining, then they should all hydrolyse with the same k_{cat}. This has been found for many series of ester substrates of chymotrypsin (Table 7.1) since the original study of Gutfreund and Hammond in 1959.[18] For weakly activated esters, the value of k_{cat} decreases below that of k_3 since k_2 becomes partly rate determining (eqn (7.1)). With amides k_{cat} is very low and k_2 is completely rate determining.

However, the occurrence of a common k_{cat} in the reaction of a series of substrates is not sufficient evidence for the accumulation of a common covalent intermediate whose breakdown is rate determining. k_{cat} is constant for the hydrolysis of a wide range of phosphate esters by alkaline phosphatase.[19,20] This was once interpreted as evidence for the rate-determining hydrolysis of a phosphorylenzyme. But it now seems likely that, at alkaline pH, dephosphorylation is rapid[21] and there is rate-determining dissociation of inorganic phosphate from the enzyme–product complex (E.P$_i$).[22,23] The common value of k_{cat} is caused by a common intermediate, but it is a non-covalent one.

b. Partitioning of the intermediate between competing acceptors.

If an intermediate is generated that may react with different acceptors, two procedures may be used for its detection.

TABLE 7.1. *Comparison of values of* k_{cat} *for the hydrolysis of substrates by* α-*chymotrypsin at pH 7.0 and 25°*

Derivative	k_{cat} (s^{-1})	k_M (mM)	Rate-determining step	Ref.
	N-acetyl-L-tryptophan derivatives			
Amide	0·026	7·3	Acylation	1
Ethyl ester	27	0·1	Deacylation	1
Methyl ester	28	0·1	Deacylation	1
p-Nitrophenyl ester	30	0·002	Deacylation	1
	N-acetyl-L-phenylalanine derivatives			
Amide	0·039	37	Acylation	1
Ethyl ester	63	0·09	Deacylation	1
Methyl ester	58	0·15	Deacylation	1
p-Nitrophenyl ester	77	0·02	Deacylation	1
	N-benzoylglycine derivatives			
Ethyl ester	0·1	2·3	Mainly acylation	2
Methyl ester	0·14	2·4	Mainly acylation	2
Isopropyl ester	0·05	2·3	Acylation	2
Isobutyl ester	0·17	2·4	Mainly acylation	2
Choline ester	0·43	1·2	Deacylation	2
4-Pyridinemethyl ester	0·51	0·092	Deacylation	2
p-Methoxyphenyl ester	0·61	0·1	Deacylation	3
Phenyl ester	0·54	0·14	Deacylation	3
p-Nitrophenyl ester	0·54	0·03	Deacylation	3

1. B. Zerner, R. P. M. Bond, and M. L. Bender, *J. Am. chem. Soc.* **86,** 3674 (1964)
2. R. M. Epand and I. B. Wilson, *J. biol. Chem.* **238,** 1718 (1963)
3. A. Williams, *Biochemistry* **9,** 3383 (1970). (The data are corrected to pH 7·0 from pH 6·91 assuming a pK_a of 6·8)

The first involves determining product ratios. For example, the hydrolysis of a series of esters of hippuric acid by chymotrypsin in solutions containing hydroxylamine leads to the formation of the free hippuric acid and hippurylhydroxamic acid in a constant ratio (Table 7.2). The non-enzymic hydrolysis under the same conditions leads to variable product ratios.[24,25]

$$E + RCOOR' \rightarrow RCO—E \underset{H_2O}{\overset{NH_2OH}{\rightleftarrows}} \begin{array}{l} RCONHOH \\ \\ RCO_2H \end{array} \qquad (7.2)$$

This is good evidence for a common intermediate.

TABLE 7.2. *Product ratios in the hydrolysis of N-benzoylglycine esters by α-chymotrypsin in 0·1 M hydroxylamine*

| Ester | Hydroxyaminolysis/Hydrolysis | |
	Enzymic pH 6·6–6·8	Non-enzymic pH 12
Methyl	0·37	0·99
Isopropyl	0·38	0·29
Homocholine	0·37	1·73
4-Pyridinemethyl	0·37	3·03

R. M. Epand and I. B. Wilson, *J. biol. Chem.* **238**, 1718 (1963); **240**, 1104 (1965)

The second procedure involves measuring the rates of formation of the products for a particular substrate at varying acceptor concentrations. This gives information on the rate-determining step of the reaction as well as detecting the intermediate. Suppose that the rate-determining step in the reaction is the formation of the acylenzyme. Then, since the acceptor reacts with the acylenzyme after the rate-determining step, it cannot increase the rate of destruction of the ester. The overall formation rates will be as in Fig. 7.2. If the rate-determining step is the hydrolysis of the acylenzyme, the acceptor increases the rate of its reaction and hence increases the overall reaction rate. The product formation rates will be as in Fig. 7.3.

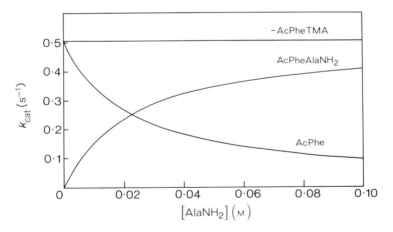

FIG. 7.2. The chymotrypsin-catalysed hydrolysis and transacylation reactions of acetylphenylalanine *p*-trimethylammoniumanilide (AcPheTMA) in the presence of various concentrations of AlaNH$_2$. The values of k_{cat} for the depletion of AcPheTMA and the production of AcPhe and AcPheAlaNH$_2$ are calculated from eqn (7.3) using $k_2 = 0·504$ s^{-1}, $k_3'[H_2O] = 144$ s^{-1}, and $k_4 = 6340$ s^{-1} M^{-1} (ref. 14)

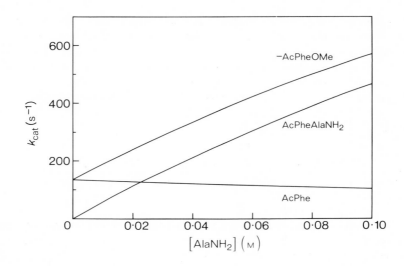

FIG. 7.3. The chymotrypsin-catalysed hydrolysis and transacylation reactions of AcPheOMe in the presence of various concentrations of AlaNH$_2$. The values of k_{cat} for the depletion of AcPheOMe and the production of AcPhe and Ac-PheAlaNH$_2$ are calculated from eqn (7.3) using $k_2 = 2200$ s^{-1}, $k_3'[H_2O] = 144$ s^{-1} and $k_4 = 6340$ s^{-1} M^{-1} (ref. 14)

The steady-state kinetics for partition may be calculated from:

$$E + RCOOR' \underset{}{\overset{K_S}{\rightleftharpoons}} E.RCOOR' \xrightarrow{k_2} RCO\!-\!E \begin{array}{l} \xrightarrow{k_3'[H_2O]} RCO_2H \\ \searrow_{k_4[N]} \\ RCO\text{-}N \end{array} \qquad (7.3)$$

$$\downarrow R'OH$$

The following expressions for k_{cat} and K_M may be derived by the usual procedures for the reactions of chymotrypsin.[26] The kinetics are simplified in this example because N does not bind to the enzyme,

$$K_M = K_S \left(\frac{k_3'[H_2O] + k_4[N]}{k_2 + k_3'[H_2O] + k_4[N]} \right). \qquad (7.4)$$

For the formation of RCO_2H,

$$k_{cat} = \frac{k_2 k_3'[H_2O]}{k_2 + k_3'[H_2O] + k_4[N]}; \qquad (7.5)$$

for the formation of $RCO\text{-}N$,

$$k_{cat} = \frac{k_2 k_4[N]}{k_2 + k_3'[H_2O] + k_4[N]}. \qquad (7.6)$$

Note that k_{cat}/K_M for the rate of disappearance of RCO_2R' is obtained by dividing the sum of the two values of k_{cat} from (7.5) and (7.6) by the K_M and is equal to k_2/K_S. It may be recalled from Chapter 3F that this is always true for such a series of sequential reactions following an equilibrium binding step. This provides another criterion for the detection of an intermediate. If the reaction of the nucleophiles involves the direct attack on the Michaelis complex, as in (7.7), k_{cat}/K_M will be a function of the concentration of N. For example, if

$$E + RCOOR' \underset{}{\overset{K_S}{\rightleftharpoons}} E.RCOOR' \quad \begin{array}{c} \xrightarrow{k_2[H_2O]} RCO_2H \\ \xrightarrow[k_3[N]]{} RCO\text{-}N \end{array} \qquad (7.7)$$

then, for the disappearance of ester,

$$k_{cat}/K_M = (k_2[H_2O] + k_3[N])/K_S. \qquad (7.8)$$

(Complications will, of course, arise if the nucleophiles bind to the enzyme and both compete for a single site.)

The speeding up of the deacylation rate by added nucleophiles has been used to give the rate constant for deacylation (k_2) for substrates where k_3 is normally rate determining. 1,4-butanediol is a sufficiently good nucleophile that moderate concentrations cause the deacylation rate to become faster than k_2.[27] Values of some rate constants obtained by this method are listed in Table 7.3.

Partition experiments are seen to provide a very powerful approach for the detection of intermediates.

3. Detection of acylenzyme in the hydrolysis of amides and peptides[14]

The acylenzyme does not accumulate in the hydrolysis of amides, so detection is indirect and difficult. Fortunately, use can be made of the direct detection in ester substrates to provide a rigorous proof of the acylenzyme with amides.

The acylenzyme mechanism was proved for derivatives of acetyl-L-phenylalanine (AcPhe) as follows.

(a) The hydrolysis of amides in the presence of acceptor nucleophiles gives the same product ratios as found for the hydrolysis of the methyl ester (AcPhe-OMe) under the same conditions (Table 7.4). Furthermore, these product ratios are the same as those expected from direct rate measurements of the attack of the nucleophiles on AcPhe-chymotrypsin, generated *in situ* in the stopped-flow spectrophotometer (Table 7.5).

(b) Under conditions where over 94% of the amide that is reacting forms AcPhe-nucleophile, there is no significant increase in the rate of disappearance of the amide. This is consistent with attack by the nucleophile after the rate-determining step, that is after the formation of an

TABLE 7.3. *Kinetic constants for the hydrolysis of N-acyl-L-amino acid esters by α-chymotrypsin at 25°, pH 7.8, and ionic strength 0.1 determined by partitioning experiments*

Acyl	Amino acid	Ester	k_{cat} (s^{-1})	K_M (mM)	k_2 (s^{-1})	k_3 (s^{-1})	K_S (mM)
Acetyl	Gly	OCH_3	0·109	862	0·49	0·14	3380
Acetyl	Gly	OC_2H_5	0·051	445	0·094	0·11	823
Benzoyl	Gly	OCH_3	0·31	4·24	0·42	1·17	5·78
Acetyl	But	OCH_3	1·41	66·7	8·81	1·68	417
Benzoyl	But	OCH_3	0·32	1·41	0·41	1·52	1·79
Benzoyl	Ala	OC_2H_5	0·069	5·97	0·069	0.6	5·97
Acetyl	Norval	OCH_3	5·08	14·3	35·6	5·93	100
Benzoyl	Norval	OCH_3	2·45	0·85	4·16	5·93	1·45
Acetyl	Val	OCH_3	0.173	87·7	0·98	0·21	500
Acetyl	Val	OC_2H_5	0·152	110	0·55	0·21	398
Acetyl	Val	$i\text{-}OC_3H_7$	0·096	177	0·178	0·21	327
Chloroacetyl	Val	OCH_3	0·127	43	0·32	0·21	108·8
Benzoyl	Val	OCH_3	0·064	4·17	0·09	0·22	5·84
Acetyl	Norleu	OCH_3	16·1	5·37	103	19·1	34·4
Acetyl	Phe	OCH_3	97·1	0·93	796	111	7·63
Acetyl	Phe	OC_2H_5	68·6	1·85	265	92·7	7·14
Acetylala (L)	Phe	OCH_3	57·3	0·296	176	85	0·909
Benzoyl	Phe	OCH_3	30·7	0·0349	45·8	91·6	0·0524
Acetyl	Tyr	OC_2H_5	192	0·663	5000	200	17·2
Benzoyl	Tyr	OCH_3	90·9	0·018	364	121	0·072
Benzoyl	Tyr	OC_2H_5	85·9	0·022	249	131	0·0638
Acetylleu (L)	Tyr	OCH_3	65·7	0·0192	158	113	0·0461
Furoyl	Tyr	OCH_3	50	0·417	66·7	200	0·56

I. V. Berezin, N. F. Kazanskaya, and A. A. Klyosov, *FEBS Letts.* **15,** 121 (1971)

TABLE 7.4. *Product ratios in the hydrolysis of substrates by δ-chymotrypsin*[a]

Substrate		Transacylation/Hydrolysis (M^{-1})		
	Acceptor =	$AlaNH_2$	$GlyNH_2$	H_2NNH_2
AcPhe-OMe		43	13	2·2
AcPhe—NH— (ring) —N^+Me_3		45	11	1·8
AcPhe-AlaNH₂		43	9	—

[a] 25°, pH 9.3
(J. Fastrez and A. R. Fersht, *Biochemistry* **12,** 2025 (1973))

TABLE 7.5. *Rate constants for the attack of nucleophiles on AcPhe-δ-chymotrypsin*[a]

	$k\,(\text{s}^{-1}\,\text{M}^{-1})$	
Nucleophile	Direct kinetic measurement	Calculated from product ratios[b]
AlaNH$_2$	4.8×10^3	6.2×10^3
GlyNH$_2$	1.5×10^3	1.6×10^3
H$_2$NNH$_2$	3.3×10^2	2.8×10^2
(H$_2$O	$142\,\text{s}^{-1}$)	

[a] 25°, pH 9.3. [b] From Table 7.4
(J. Fastrez and A. R. Fersht, *Biochemistry* **12**, 2025 (1973))

intermediate, and at least 94% of the reaction going through this intermediate.

(c) The final *proof* of the acylenzyme route comes from calculating the rate constant for the hydrolytic reaction from the rate constant for the reverse reaction (the synthesis of the substrate by the acylenzyme route) and the *Haldane equation* (Chapter 3H). It is found that amines will react with the acylenzyme to produce amides and peptides. Hence, by the principle of microscopic reversibility, the reverse reaction (the hydrolysis of peptides by the acylenzyme mechanism) must also occur. The question is whether or not this reaction is rapid enough to account for the observed hydrolysis rate. This was answered by measuring $(k_{cat}/K_M)_S$ for the synthesis of a peptide by the acylenzyme route, K_{eq} for the hydrolysis of the peptide, and then calculating $(k_{cat}/K_M)_H$ for the hydrolytic reaction from the Haldane equation $(K_{eq} = (k_{cat}/K_M)_H/(k_{cat}/K_M)_S)$. The calculated value is close to the experimental value.

4. Validity of partitioning experiments and some possible experimental errors

Neither the occurrence of a constant value of V_{max} nor a constant product ratio is sufficient proof of the presence of an intermediate. It was seen for alkaline phosphatase that a constant value for V_{max} is artefactual, and also there is no *a priori* reason why the attack of acceptors on a Michaelis complex should not also give constant product ratios. In order for partitioning experiments to provide a satisfactory proof of the presence of an intermediate they must be linked with rate measurements. When the rate measurements are restricted to steady-state kinetics the most favourable situation is when the intermediate accumulates. If the kinetics of eqns (7.4)–(7.6) hold, it may be concluded beyond reasonable doubt that an intermediate occurs. The ideal situation is that partitioning experiments should be combined with pre-steady-state studies as described for chymotrypsin and amides.

Errors can arise through the enzymes 'misbehaving'. Chymotrypsin is often treated as a solution of imidazole and serine. But proteins are quite sensitive to their environment; they often bind organic molecules and ions non-specifically to alter their kinetic properties slightly. The first experimental rule is that reactions should be carried out as far as is possible at the same concentration of enzyme. Many proteins aggregate somewhat and this can cause changes in rate constants. The second rule is that product ratios should be determined directly by analysing the products rather than from indirect measurements. For example, the rate of attack of $GlyNH_2$ on the acylenzyme BzTyr-chymotrypsin was once measured from the decrease in k_{cat} for the hydrolysis of BzTyr-$GlyNH_2$ on addition of $GlyNH_2$. (The $GlyNH_2$ inhibits the reaction since it reacts with the acylenzyme to regenerate the BzTyr-$GlyNH_2$.) Unfortunately, it was subsequently found that amines bind to chymotrypsin causing increases of up to 30% in k_{cat}.[14] This increase is of the same order as the expected decreases due to the reversal of the reaction. In general, small changes of rate are not reliable in enzymic reactions in circumstances where they would be in chemical kinetics.

There are circumstances where the simple rules for the partition of intermediates break down. If the acceptor nucleophile reacts with the acylenzyme before the leaving group has diffused away from the enzyme-bound intermediate, the partition ratio could depend on the nature of the leaving group (e.g. due to steric hindrance of attack etc.). Also, the measurement of rate constants for the attack of the nucleophiles on the intermediate could be in slight error due to the non-specific binding effects mentioned above.

C. Further examples of detection of intermediates by partition and kinetic experiments

1. Alkaline phosphatase

There is little doubt that a phosphorylenzyme is formed during the hydrolysis of phosphate esters by this enzyme. The phosphorylenzyme is stable at low pH and may be isolated.[28,29] Partition experiments using Tris buffer as a phosphate acceptor give a constant product ratio (Table 7.6).[30,31] Earlier kinetic experiments have to be evaluated in the light of the recent findings that the enzyme as isolated may contain tightly, although non-covalently, bound phosphate,[32] and that the rate-determining step in the reaction at high pH is the dissociation of the tightly bound phosphate[22,23] (although there is not unanimity on the kinetic points or the stoichiometry of binding).[33,34] As mentioned earlier, it seems that the constant value of V_{max} for the hydrolysis of a wide series of phosphate esters (Table 7.7)[19,20] is the result of a slow, rate-determining

TABLE 7.6. *Relative values of* V_{max} *for the hydrolysis of phosphate esters by alkaline phosphatase*

Phosphate	V_{max}
5'AMP	1
Pyrophosphate	1
3'AMP	0·9
ApAp	0·6
ATP	1·05
dATP	1·05
dGTP	1·05
UDP	1·0
5'UMP	0·85
dCTP	1·05
Ribose-5-phosphate	0·7
β-Glycerol phosphate	0·9
Ethanolamine phosphate	0·7
Glucose-1-phosphate	0·8
Glucose-6-phosphate	0·9
Histidinol phosphate	0·8
p-Nitrophenyl phosphate	1·0

L. A. Heppel, D. R. Harkness, and R. J. Hilmoe, *J. biol. Chem.* **237,** 841 (1962)

TABLE 7.7. *Product ratios in the hydrolysis of phosphate esters and phosphoramidates by alkaline phosphatase*

Phosphate	Transphosphorylation/ Hydrolysis[a]
Phenyl	1·42
Cresyl	1·41
Chlorophenyl	1·38
p-t-Butylphenyl	1·37
p-Nitrophenyl	1·37
o-Methoxy-p-methylphenyl	1·40
α-Naphthyl	1·40
β-Naphthyl	1·40
p-Nitrophenyl	1·2[b]
Phosphoramidates	1·1[b,c]

[a] acceptor = 1 M Tris, pH 8; [b] acceptor = 2 M Tris, pH 8·2; [c] average for several phosphoramidates
(H. Barrett, R. Butler, and I. B. Wilson, *Biochemistry* **8,** 1042 (1969); S. Snyder and I. B. Wilson, *Biochemistry* **11,** 3220 (1972))

dissociation of the enzyme–product complex, $E.P_i$.

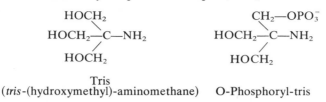

Tris
(*tris*-(hydroxymethyl)-aminomethane)　　O-Phosphoryl-tris

2. Acid phosphatase

This is a straightforward example of the accumulation of a phosphorylenzyme intermediate with rate-determining breakdown. A wide series of esters is hydrolysed with the same V_{max} (Table 7.8);[35] constant

TABLE 7.8. *Hydrolysis of phosphate esters by prostatic acid phosphatase at pH 5 and 37°*

Phosphate ester	K_M (mM)	V_{max}(relative)
β-Glycerol phosphate	1·1	1
2'AMP	0·28	1
Acetyl phosphate	0·17	1
3'AMP	0·068	1
p-Nitrophenyl phosphate	0·034	1

G. S. Kilsheimer and B. Axelrod, *J. biol. Chem.* **227**, 879 (1957)

product ratios are found (Table 7.9); stopped-flow studies using p-nitrophenyl phosphate find a burst of one mole of p-nitrophenolate released per enzyme subunit; and the enzyme is covalently labelled by diisopropyl fluorophosphate.[36]

$$E\text{—}OH + ROPO_3^-H \rightleftharpoons E\text{—}OH.ROPO_3^-H \xrightarrow{\text{fast}} E\text{—}OPO_3^-H \xrightarrow{\text{slow}}$$
$$ROH$$
$$E\text{—}OH + H_2PO_4^- \quad (7.9)$$

TABLE 7.9. *Product ratios in the hydrolysis of phosphate esters by prostatic acid phosphatase*

Phosphate ester	Transphosphorylation/Hydrolysis	
	Acceptor = Ethanol	Acceptor = Ethanolamine
p-Nitrophenyl phosphate	0·29	0·044
Phenyl phosphate	0·26	0·044
3'UMP	0·28	—
3'AMP	0·30	0·046
β-Glycerol phosphate	0·28	0·041

W. Ostrowski and E. A. Barnard, *Biochemistry* **12**, 3893 (1973)

3. β-Galactosidase: retention of configuration implying intermediate

The alcoholysis of β-D-galactosides by β-galactosidase occurs with retention of configuration at the C_1 carbon.[37,38] This rules out the simple

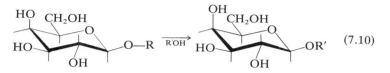

$$\text{(7.10)}$$

bimolecular displacement reactions (S_N2) that often occur in organic chemistry since it is known that these always occur with inversion.[39]

$$X^{\ominus} \quad B \cdots C - Y \rightarrow X - C \cdots B + Y^{\ominus} \qquad \text{(7.11)}$$

Retention of configuration will occur if there are two successive inversions at the carbon or if there is a carbonium ion intermediate that can be attacked by a nucleophile on one side only. Retention of configuration is therefore consistent with a multistep reaction involving an intermediate. By this criterion, lysozyme also involves an intermediate.[40,41]

Evidence for an intermediate from partitioning experiments is presented in Table 7.10. There is constant partitioning between water and methanol.[42,43] Examination of the values of V_{max} suggests that formation

TABLE 7.10. *Product ratios and relative values of* V_{max} *for the hydrolysis of* β*-galactosides by* β*-galactosidase at* 25° *and pH* 7·0–7·5

β-Galactoside	Methanolysis / Hydrolysis M^{-1}	V_{max} (relative)	Rate-determining step
2,4-Dinitrophenyl		1·3	Degalactosylation
3,5-Dinitrophenyl		1·1	Degalactosylation
2,5-Dinitrophenyl		1·1	Degalactosylation
2-Nitrophenyl	1·97	1·0	
3-Nitrophenyl	1·96	0·9	
3-Chlorophenyl	2·08	0·5	Galactosylation
4-Nitrophenyl	1·99	0·2	Galactosylation
Phenyl	1·94	0·1	Galactosylation
4-Methoxyphenyl	2·14	0·1	Galactosylation
4-Chlorophenyl	2·13	0·02	Galactosylation
4-Bromophenyl	2·02	0·02	Galactosylation
Methyl	2·2	0·06	Galactosylation

T. M. Stokes and I. B. Wilson, *Biochemistry* **11**, 1061 (1972); M. L. Sinnott and O. M. Viratelle, *Biochem. J.* **133**, 81 (1973); M. L. Sinnott and I. J. Souchard, *Biochem. J.* **133**, 89 (1973)

of the intermediate is rate determining for the weakly activated substrates since V_{max} is variable, but hydrolysis of the intermediate is rate determining for the highly activated dinitro compounds since the rate levels off. Also consistent with this is the observation that the rate of disappearance of the 2,4-dinitro and 3,5-dinitro substrates is increased by added methanol, but not that of the less reactive substrates (see Figs. 7.2 and 7.3). A likely mechanism for this reaction involves the initial formation of a carbonium ion.[44,45]

D. Aminoacyl-tRNA synthetases: detection of intermediates by quenched flow, steady-state kinetics, and isotope exchange

1. The reaction mechanism

These enzymes catalyse the formation of aminoacyl-tRNA from the free amino acid (AA) and ATP. In the absence of tRNA, the enzymes will, with a few exceptions, activate amino acids to the attack of nucleophiles, and ATP to the attack of pyrophosphate.[46-49] This is done by forming a tightly bound complex with the aminoacyl adenylate, the mixed anhydride of the amino acid, and AMP. (The chemistry of activation is discussed in Chapter 2D2c.)

$$AA + ATP + tRNA \xrightarrow{\text{E}} AA\text{-tRNA} + AMP + PP \qquad (7.12)$$

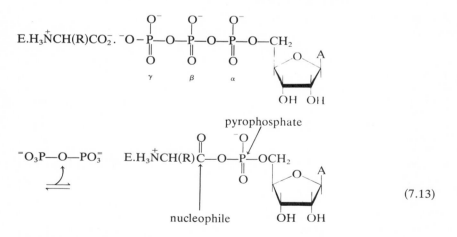

The activation to the attack of pyrophosphate is measured by the pyrophosphate exchange technique. The enzyme, amino acid, and ATP are incubated with [^{32}P]-labelled pyrophosphate so that β,γ-labelled ATP is formed by the continuous recycling of the $E.AA \sim AMP$ complex. The complex is formed as in (7.13) and the reaction is reversed by the attack of labelled pyrophosphate to generate labelled ATP. This process is repeated until the isotopic label is uniformly distributed amongst all the reagents.

The acylation of tRNA proceeds by the attack of a ribose hydroxyl of the terminal adenosine of the tRNA on the carbonyl group as indicated in (7.13) and (7.14).

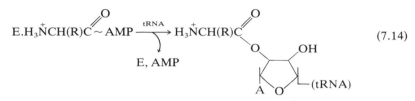

(7.14)

There is no doubt that the enzyme-bound aminoacyl adenylate is formed in the absence of tRNA. It may be isolated by chromatography and the free aminoacyl adenylate obtained by precipitation of the enzyme with acid.[50,51] Furthermore, the isolated complex will transfer its amino acid to tRNA.

The following mechanism is derived logically from these observations.

$$E + ATP + AA \xrightarrow[PP]{} E.AA \sim AMP \xrightarrow{tRNA} AA\text{-}tRNA + AMP + E \quad (7.15)$$

Despite this, it seemed at one stage that not all evidence was consistent with the aminoacyl adenylate pathway. An alternative mechanism is possible; in the presence of tRNA, an aminoacyl adenylate is *not* formed but reaction occurs by the simultaneous reaction of tRNA, amino acid and ATP.[52] As is seen below, this mechanism is not correct, but at the time it was suggested it was a valid possibility. Furthermore, it makes an important point: the finding of a partial reaction in the absence of one of the substrates (for example, the formation of the aminoacyl adenylate from the amino acid and ATP in the absence of tRNA) does not mean that the same reaction occurs in the presence of *all* the substrates. Indeed, there are examples in which such partial reactions have been found to be artefacts.

The aminoacyl adenylate pathway is proved very simply from three quenched-flow experiments using the three criteria for proof: the intermediate is isolated, it is formed fast enough, and reacts fast enough to be on the reaction pathway.[53] The following is found for the isoleucyl-tRNA synthetase (IRS). (a) On mixing the preformed and isolated IRS. $[^{14}C]Ile \sim AMP$ with tRNA in the pulsed quenched-flow apparatus, the first-order rate constant for the transfer of the $[^{14}C]Ile$ to the tRNA is measured to be the same as the k_{cat} for the steady-state aminoacylation of

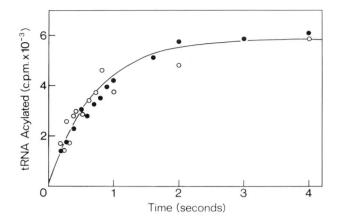

FIG. 7.4. Transfer of $[^{14}C]$Ile from IRS.$[^{14}C]$Ile ~ AMP to tRNA^{Ile} on mixing the complex with excess tRNA in the pulsed quenched-flow-apparatus

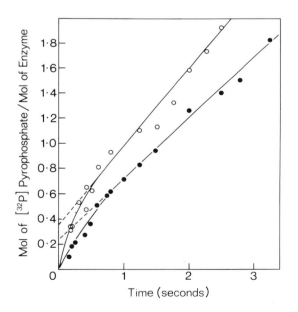

FIG. 7.5. Release of $[^{32}P]$ pyrophosphate on mixing $[^{32}P]$ATP, isoleucine, tRNA, and enzyme in the pulsed quenched-flow apparatus. The extrapolated burst of product formation is below one mole per mole of enzyme because the concentrations of ATP are not saturating (Chapter 4D). Open circles (O) are for $[ATP] = 2 \times K_M$; filled circles are for $[ATP] = 1 \times K_M$

the tRNA under the same reaction conditions. The rate constant for the reaction of the intermediate is thus fast enough to be on the reaction pathway and, furthermore, appears to be the rate-determining step of the reaction. (b) On mixing IRS, isoleucine, tRNA, and $[\gamma^{32}P]ATP$ (i.e. labelled in the terminal phosphate) in the pulsed quenched-flow apparatus, there is a burst of release of labelled pyrophosphate before the steady-state rate of aminoacylation of tRNA. This means that either the aminoacyl adenylate is formed before the aminoacylation of tRNA, thus proving the mechanism, *or* there could be a pathway which involves the formation of aminoacyl-tRNA in a rapid process followed by a subsequent slow step, such as the dissociation of the IRS. Ile-tRNA complex (7.16). The latter reaction, which was thought to occur by many workers,

$$E + AA + ATP + tRNA \xrightarrow{\text{fast}} E.AA\text{-tRNA} \xrightarrow{\text{slow}} E + tRNA \qquad (7.16)$$
$$\searrow$$
$$AMP + PP$$

is disproved by a third quenching experiment. Eqn (7.16) predicts that there should be a burst of charging of tRNA since one mole of enzyme-bound aminoacyl-tRNA is formed rapidly whilst the subsequent turnover is slow. (c) On mixing IRS, $[^{14}C]$Ile, tRNA, and ATP in the quenched-flow apparatus, the initial rate of charging of tRNA extrapolates back

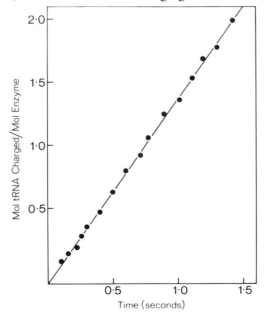

FIG. 7.6. Initial rate of charging of tRNA^{Ile} on mixing saturating concentrations of $[^{14}C]$Ile, ATP, and tRNA with the enzyme

through the origin without any indication of a burst of charging. The burst of pyrophosphate release is due to the formation of the aminoacyl adenylate before the transfer of the amino acid to tRNA (Fig. 7.6).

2. The editing mechanism

In Chapter 11 it is pointed out that during protein biosynthesis the cell distinguishes between certain amino acids with an accuracy far greater than expected from their differences in structure. This could be caused by the specifically catalysed hydrolysis by the aminoacyl-tRNA synthetase of the aminoacyl adenylate complex of the 'wrong' amino acid or the mischarged tRNA.[51] One example of this is the rejection of threonine by the valyl-tRNA synthetase (VRS).[54] This enzyme catalyses the pyrophosphate exchange reaction in the presence of threonine and also forms a stable VRS.Thr ~ AMP complex. In the presence of tRNA and threonine, the VRS acts as an ATP pyrophosphatase, hydrolysing ATP to AMP and pyrophosphate, and does not catalyse the net formation of Thr-tRNA^{Val}. During this reaction there is the intermediate formation of VRS.Thr ~ AMP shown by the occurrence of the pyrophosphate exchange reaction.

Rapid quenching experiments show that the editing mechanism for rejecting threonine is due to the mischarging of tRNA followed by its rapid hydrolysis. The transiently mischarged tRNA may be trapped, isolated, and found to be hydrolysed at the necessary rate. On mixing the VRS.[^{14}C]Thr ~ AMP complex with tRNA^{Val}. in the quenched-flow apparatus, there is the transient formation of [^{14}C]Thr-tRNA^{Val} (Figure 7.7). This may be isolated by rapidly quenching the reaction with phenol and precipitating the mischarged tRNA from the aqueous layer. The mischarged tRNA is hydrolysed by the VRS with a rate constant of $40\,\text{s}^{-1}$ (Figure 7.8). The rate of transfer of the threonine from the VRS.Thr ~ AMP complex may be measured independently from the rate of liberation of AMP (using the VRS.Thr ~ [^{32}AMP] compound, Figure 7.9). The solid curve in Figure 7.8 is calculated from the independently measured formation and hydrolysis rate constants given in (7.17).

$$\text{VRS.Thr} \sim \text{AMP.tRNA}^{Val} \xrightarrow{36\,\text{s}^{-1}} \text{VRS.Thr-tRNA}^{Val} + \text{AMP} \qquad (7.17)$$

$$\downarrow 40\,\text{s}^{-1}$$

$$\text{VRS} + \text{Thr} + \text{tRNA}^{Val}$$

One experimental point worth noting is that the rate constant for the deacylation of the mischarged tRNA can be measured by the pre-steady-state kinetics even in the presence of a large fraction of uncharged tRNA. This is not easily done by steady-state kinetics because of the competitive

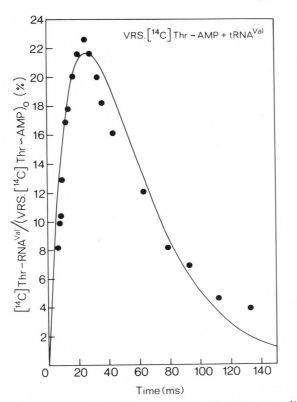

FIG. 7.7. The transient formation of $[^{14}C]$Thr-tRNA^{Val} on mixing VRS.$[^{14}C]$Thr~AMP with excess tRNA in the quenched-flow apparatus

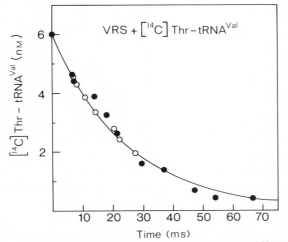

FIG. 7.8. The VRS-catalysed hydrolysis of mischarged $[^{14}C]$Thr-tRNA^{Val}

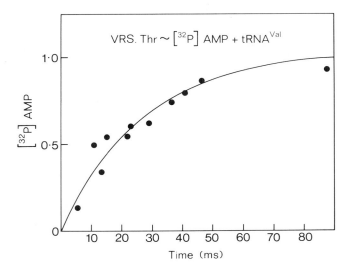

Fig. 7.9. The rate of breakdown of Thr ~ AMP on mixing VRS.Thr ~ [^{32}P]AMP with excess tRNA^{Val}

inhibition by the uncharged material. But in the rapid quenching experiment an excess of enzyme over total tRNA is used.

E. Myosin ATPase: determination of enzyme–substrate association/dissociation rate constants, intermediate conformational states and equilibria by stopped-flow fluorescence, proton release, linked assays, light scattering, and quenched flow

Muscle contraction is achieved by the sliding of two sets of overlapping filaments—the thick filaments containing myosin and the thin filaments containing actin. Cross-bridges, which are part of the myosin molecule, extend between the two filaments and repeatedly become detached and reattached during the cycles of events of contraction. The cross-bridges contain enzymically active sites which catalyse the hydolysis of ATP. In the resting muscle the ATPase activity is very low, but when contraction is stimulated the activity increases by a thousand-fold. The Gibbs energy released on hydrolysis provides the energy for contraction. The problem is: how are the chemical and mechanical energies linked? One hypothesis is that this is done by conformational changes in the protein being linked to the hydrolysis of ATP.

The sites on myosin that contain the ATPase activity and combine with actin are in globular regions known as the S_1 sub-units. These may be cleaved from the rest of the molecule by proteolysis to give the S_1

fragments which are more amenable to kinetic study than the parent molecule.[55] The currently accepted mechanism for the hydolysis of ATP by S_1 is a formidable looking scheme involving 7 steps (eqn (7.18)). Although this appears complex, it has been derived in an instructively logical manner.[56–68]

$$M + ATP \underset{k_{-1}}{\overset{k_1}{\rightleftharpoons}} M.ATP \underset{k_{-2}}{\overset{k_2}{\rightleftharpoons}} M^*.ATP \underset{k_{-3}}{\overset{k_3}{\rightleftharpoons}} M^{**}.ADP.P \underset{k_{-4}}{\overset{k_4}{\rightleftharpoons}} M^*.ADP.P$$

$$\underset{k_{-5}}{\overset{k_5}{\rightleftharpoons}} M^*.ADP + P \underset{k_{-6}}{\overset{k_6}{\rightleftharpoons}} M.ADP \underset{k_{-7}}{\overset{k_7}{\rightleftharpoons}} M + ADP. \qquad (7.18)$$

The conformational states are represented by asterisks, the number of which relate to increasing tryptophan fluorescence. The values of the rate equilibrium constants that have been assigned so far for rabbit skeletal muscle are: $k_1/k_{-1} = 4 \cdot 5 \times 10^3$ M; $k_2 = 400$ s^{-1}; $k_{-2} \leqslant 0 \cdot 0003$ s^{-1}; $k_3/k_{-3} = 9$; $k_3 \geqslant 160$ s^{-1}; $k_4 = 0 \cdot 06$ s^{-1}; $k_5/k_{-5} > 1 \cdot 5 \times 10^{-3}$ M; $k_6 = 1 \cdot 4$ s^{-1}; $k_{-6} = 400$ s^{-1}, and $k_7/k_{-7} = 2 \cdot 7 \times 10^{-4}$ M (all at 21° and pH 8.0).[68] We shall now see how this scheme was derived.

1. The rate-determining step

The turnover number for the hydrolysis of ATP is very low at $0 \cdot 04$ s^{-1}. Quenched-flow experiments using $[\gamma^{32}P]ATP$ show that the rate constant for the cleavage of the ATP during the first turnover of the enzyme is 160 s^{-1}.[60] This means that a step subsequent to the chemical cleavage is rate determining in the steady state. This could be either a slow conformational change or a slow release of products. It was shown to be the former by measuring the dissociation rates and finding that they are relatively fast (see Section 2).[61] The slow step is k_4, a conformational change from a highly fluorescent state to a less fluorescent one. This may be measured by either stopped-flow fluorescence or from proton release. 1 H$^+$/mole is liberated in this step.[64] The rate constant for the proton release is easily measured by stopped-flow spectrophotometry using a chromophoric pH indicator, in this case phenol red.

2. Dissociation rates of ADP and P (phosphate)

The preliminary experiments used to determine the magnitude of the dissociation rates of ADP and P are quite ingenious. The rates of appearance of ADP and P *free in solution* were determined by using the following linked assay systems and the stopped-flow spectrophotometer.[64]

a. ADP release linked to disappearance of NADH via pyruvate kinase and lactate dehydrogenase

$$\text{Phosphoenolpyruvate} \xrightarrow[\text{ATP}]{\text{ADP}} \text{Pyruvate} \xrightarrow[\text{NAD}^+]{\text{NADH}} \text{Lactate} \qquad (7.19)$$

b. P release linked to the formation of NADH via glyceraldehyde-3-phosphate dehydrogenase

$$P + NAD^+ \quad NADH$$

Glyceraldehyde-3-Phosphate $\searrow$ 1,3-Diphosphoglycerate $\rightarrow$ 3-Phosphoglycerate

$$ADP \quad ATP \qquad (7.20)$$

(The glyceraldehyde-3-phosphate is converted to 1,3-diphosphoglycerate which is removed by conversion to 3-phosphoglycerate by phosphoglycerate kinase to prevent product inhibition of the first reaction).

There is no burst of release of P or ADP, so that they either both dissociate slowly at the same rate, or dissociation is after the slow rate-determining step. If the release of say ADP were slow and rate determining, there would be a burst of P release and vice versa. The rate constant for the dissociation of the S_1.ADP complex was found to $1\cdot4\,s^{-1}$ from a 'chase' experiment. *Thio*ITP binds rapidly to free S_1 and undergoes a spectral change. On mixing an excess of *thio* ITP with the S_1.ADP complex in the stopped-flow spectrophotometer, the rate of formation of S_1.*thio*ITP is limited by the dissociation rate of the S_1.ADP and so the latter is simply determined.

3. The two-step processes in the binding of ATP and ADP determined by stopped-flow fluorescence of the protein tryptophans[65]

At low concentrations, ATP binds to S_1 with an apparent second-order rate constant of $1\cdot8\times10^6\,s^{-1}\,M^{-1}$. This is two orders of magnitude below that of a diffusion-controlled reaction and suggests that perhaps binding is a two-step process. Continuing the measurements at higher concentrations reveals that this is so (Fig. 7.10). There is an initial fast step beyond the time scale of stopped-flow measurement followed by an isomerization at $400\,s^{-1}$. k_2 and k_1/k_{-1} may be determined from the concentration dependence of the rate constants as described in Chapter 4. Similar behaviour is found on binding ADP. The second-order rate constant for binding ADP at low concentrations is $1\cdot5\times10^6\,s^{-1}\,M^{-1}$. At high concentrations the rate has a plateau at $400\,s^{-1}$, revealing the isomerization step k_{-6}.

k_{-2} is too low to be measured directly. Values of $3\times10^{-4}\,s^{-1}$ and $2\times10^{-7}\,s^{-1}$ have been estimated for this from experiments on the synthesis of [^{32}P]ATP from ADP and [^{32}P] orthophosphate catalysed by S_1.[66,67]

4. The equilibrium between M*.ATP and M**.ADP.P

This was measured directly by mixing an excess of enzyme with [γ^{32}P]ATP in the quenched-flow apparatus and measuring the concentrations of [γ^{32}P]ATP and [^{32}P] orthophosphate.[63] The equilibrium on the

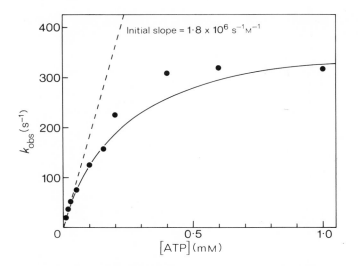

FIG. 7.10. The rate constants for the increase in the fluorescence of S_1 on mixing with various concentrations of ATP in the stopped-flow spectrofluorimeter (ref. 65)

enzyme surface is rapidly achieved and is perturbed only slowly as the $M^{**}.ADP.[^{32}P]P$ isomerizes to $M^{*}.ADP.[^{32}P]P$ at 0.06 s^{-1}.

5. Light scattering measurements

Actin (molecular weight = 42 000) polymerizes to give large aggregates. Each monomer can bind one mole of S_1 (molecular weight 115 000) to form an acto-S_1 complex of some three times greater mass. This is accompanied by a large increase in turbidity since light scattering is approximately related to the mass of the polymer. The ATP-stimulated dissociation of the S_1 from the complex may be measured by the decrease in turbidity using a stopped-flow spectrophotometer. However, in the same way that fluorescence measurements are more sensitive than absorbance since they are measured at right angles to the illuminating beam (Chapter 6), stronger changes in signal are obtained by measuring the decrease in light scattering at right angles using a stopped-flow fluorimeter.[69]

F. Phosphofructokinase: determination of anomeric specificity by stopped flow, quenched flow and locked substrates

Phosphofructokinase catalyses the phosphorylation of fructose-6-phosphate to fructose-1,6-diphosphate. In solution the substrate exists as

$$\text{D-fructose-6-phosphate} + \text{ATP} \longrightarrow \text{D-fructose-1,6-diphosphate} + \text{ADP}$$

$$(7.21)$$

about 80% the β-anomer and 20% the α-anomer. The two forms rapidly equilibrate in solution via the open-chain keto form. The enzyme has

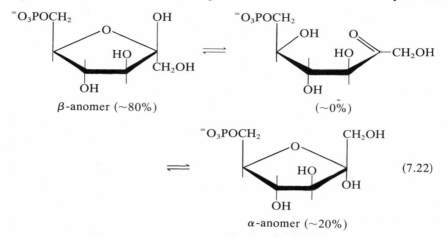

$$\beta\text{-anomer } (\sim 80\%) \qquad (\sim 0\%)$$

(7.22)

$$\alpha\text{-anomer } (\sim 20\%)$$

been shown to be specific for the β form from rapid reaction measurements on a time scale faster than that for the interconversion of the anomers, and also by determining the activity towards model substrates that are locked in either of the configurations. Using sufficient enzyme to phosphorylate all the active anomer of the substrate before the two forms can reequilibrate, it is found that 80% of the substrate reacts rapidly and the remaining 20% at the rate constant for the anomerization. The kinetics were followed by both quenched flow using $[\gamma^{32}P]ATP$,[70] and the coupled spectrophotometric assay of eqn (7.19).[71] The other evidence comes from the steady-state data on the following substrates.[72]

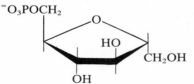

$(-OH, -CH_2OH)$ $V_{max} = 100\%$, $K_M = 0.043$ nM

α,β-D-fructose-6-phosphate

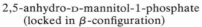

$V_{max} = 87\%$, $K_M = 0.41$ mM

2,5-anhydro-D-mannitol-1-phosphate
(locked in β-configuration)

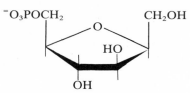

$$V_{max} = 0\%, \qquad K_i = 0\cdot34\ \text{mM}$$

(Competitive inhibitor)

2,5-anhydro-D-glucitol-1-phosphate
(locked in α-configuration)

2,5-Anhydro-D-mannitol-1-phosphate is locked in a configuration which is equivalent to the β-anomer of D-fructose: it lacks the 2-OH group and cannot undergo mutarotation to the equivalent of the α-anomer because the ring cannot open. The glucitol derivative, on the other hand, is locked into the equivalent of the configuration of the α-anomer. It is seen from values of V_{max} and K_M (or K_i) that although both bind, only the β-anomer is bound productively and is phosphorylated.

G. Affinity labels[73]

An affinity label, or active-site-directed irreversible inhibitor, is a chemically reactive compound that is designed to resemble a substrate of an enzyme so it binds specifically to the active site and forms covalent bonds with the protein residues.[74-76] Affinity labels are very useful for identifying catalytically important residues and determining their pK_a-values from the pH dependence of the rate of modification.

$$E + I \underset{}{\overset{K_I}{\rightleftharpoons}} E.I \xrightarrow{k_I} E - I \qquad (7.23)$$

The reaction of an affinity label with an enzyme involves the initial formation of a reversibly bound enzyme–inhibitor complex followed by covalent modification and hence irreversible inhibition. This scheme is analogous to that of the Michaelis–Menten mechanism and should thus show saturation kinetics with increasing inhibitor concentration. The kinetics were solved earlier in eqn (4.71). For the simple case of pre-equilibrium binding followed by a slow chemical step, the solution reduces to

$$-d[E]/dt = \frac{k_I[E][I]}{K_I + [I]} \qquad (7.24)$$

An important consequence of the chemical reaction taking place in the confines of an enzyme–'substrate' complex is that as well as the binding being specific, the rate of the chemical step may be usually rapid because it is favoured entropically over a simple bimolecular reaction in solution in the same way as is a normal enzymic reaction. Thus reagents which are normally only weakly reactive may become very reactive affinity labels.

TABLE 7.11. *High reactivity in affinity labelling*

Reaction	Second-order rate constant $(s^{-1} M^{-1})$
Cbz-L-phenylalanine chloromethylketone + chymotrypsin (alkylation of His-57)	69
Cbz-L-phenylalanine chloromethylketone + acetylhistidine	$4 \cdot 5 \times 10^{-5}$

E. N. Shaw and J. Ruscica, *Archs biochem. Biophys.* **145,** 484 (1971)

The principles of the method are very nicely illustrated by one of the first affinity labelling experiments, the reaction of tosyl-L-phenylalanine chloromethyl ketone (TPCK) with chymotrypsin.[74] TPCK resembles substrates like tosyl-L-phenylalanine methyl ester but it has a chloromethyl

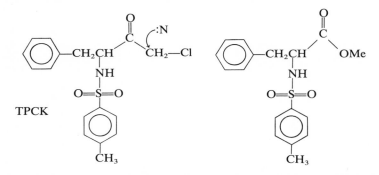

ketone group which is an alkylating reagent. Halomethyl ketones and acids are known to react with thiols and imidazoles. The irreversible inhibition of chymotrypsin by TPCK has the following characteristics.[73,74]

(a)   The rate of inactivation is inhibited by reversible inhibitors or substrates of the enzyme.

(b)   The relative rates of inhibition as a function of pH at low inhibitor concentrations ($[I] \ll K_I$) follow a bell-shaped curve with the same pK_a values as found for the pH dependence of k_{cat}/K_M for the hydrolysis of substrates.

(c)   Using ^{14}C-labelled inhibitor it was found that the inactivated enzyme has 1 mole of inhibitor bound per mole of active sites.

(d)   It was subsequently found that the inhibition follows saturation kinetics with a 'K_M' of about $0 \cdot 3$ mM.[77]

(It was later shown that the site of modification is at His-57).

Characteristics (*a*), (*b*), (*c*), and (*d*) are diagnostic tests for an affinity label that is modifying the group at the active site whose ionization controls activity. Saturation kinetics, (d), show that the label binds to the enzyme—although these may not be observed if the K_I is higher than the solubility of the label. Competitive inhibition by substrates etc. (a), suggests that binding is at the active site. The 1:1 stoichiometry shows that the modification is selective. The pH dependence gives important evidence. It was seen in Chapter 5 that the pH dependence of V_{max}/K_M or k_{cat}/K_M gives the pK_as of the groups in the free enzyme that are involved in catalysis and binding the substrate. Similarly, the pH dependence of k_I/K_I (or the relative rates of reaction at $[I] \ll K_I$ since the rate is proportional to k_I/K_I under this condition) gives the pK_as of the groups in the free enzyme that are involved in binding and reacting with the inhibitor. Thus, identical sets of pK_as should be obtained from the pH dependence of k_{cat}/K_M and k_I/K_I if the groups that are involved in catalysis are also those that react with the affinity label. The pH dependence of k_I gives the pK_as of the enzyme–inhibitor complex just as k_{cat} gives the pK_a of the enzyme–substrate complex (subject to the provisos of Chapter 5).

Enzymes do have some tricks up their sleeves (or rather, in their pockets) to spring on enzymologists. For example, as discussed later in Chapter 12G2, the affinity label 1,2-anhydro-D-mannitol-6-phosphate, labels a glumatic residue in glucose-6-phosphate isomerase.[78] It shows

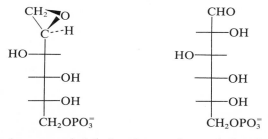

1,2-anhydro-D-mannitol-6-phosphate   glucose-6-phosphate

saturation kinetics, competitive inhibition by substrates, 1:1 stoichiometry, and the pH dependence of k_I gives pK_a-values for the enzyme–inhibitor complex similar to those for enzyme–substrate complexes. However, a crystallographic study has some speculative evidence that the group modified was not the one aimed at but another base that is catalytically important—see Chapter 12G2.[79] If this is correct, the unexpected result is more interesting than the expected one since an important but previously unknown catalytic group has been located.

Another good example of the use of affinity labels is with pepsin and is discussed in eqns (12.29) and (12.30). The enzyme has two catalytically important aspartic residues, one of which is ionized and the other unionized. The ionized carboxylate was trapped with an epoxide, which of course requires the reaction of a nucleophilic group. The unionized carboxyl was trapped with a diazoacetyl derivative of an amino acid ester (7.25). The reaction is catalysed by cupric ions and is presumably due to a

$$\overset{+}{N_2}CHCONHCH(R)CO_2Me \xrightarrow{\quad Cu^{II} \quad} Cu^{II}\overset{\cdot}{C}HCONHCH(R)CO_2Me \quad (7.25)$$

copper-complexed carbene.[80] The electron-deficient carbene with only six electrons in its outer valency shell is known to add across the O—H bonds of unionized carboxyl groups.

Another general approach is the use of *photoaffinity labels.*[81,82] A compound that is stable in the absence of light but is activated by photolysis is reversibly bound to an enzyme and photolysed. The usual reagents are diazo compounds that are photolysed to give highly reactive carbenes, or azides that give highly reactive nitrenes.

$$RC\overset{O}{\underset{CHN_2}{\diagup}} \xrightarrow{h\nu} RC\overset{O}{\underset{CH}{\diagup}} \quad (7.26)$$

$$R—N_3 \xrightarrow{h\nu} R—N\colon \quad (7.27)$$

TABLE 7.12. *Some affinity labels*

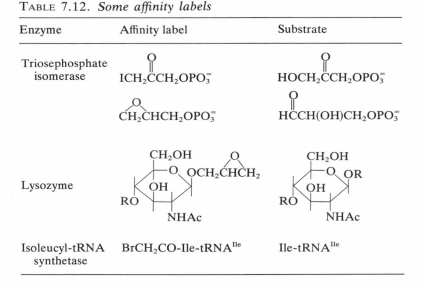

Enzyme	Affinity label	Substrate
Triosephosphate isomerase	$ICH_2\overset{O}{\overset{\|}{C}}CH_2OPO_3^=$	$HOCH_2\overset{O}{\overset{\|}{C}}CH_2OPO_3^=$
	$\overset{O}{\overset{\diagup\diagdown}{CH_2}}CHCH_2OPO_3^=$	$H\overset{O}{\overset{\|}{C}}CH(OH)CH_2OPO_3^=$
Lysozyme		
Isoleucyl-tRNA synthetase	$BrCH_2CO$-Ile-tRNA^{Ile}	Ile-tRNA^{Ile}

F. C. Hartmann, *Biochem. biophys. Res. Commun.* **33**, 888 (1968); S. G. Waley, J. C. Miller, I. A. Rose, and E. L. O'Connell, *Nature, Lond.* **227**, 181 (1970); E. W. Thomas, J. F. McElvy, and N. Sharon, *Nature, Lond.* **222**, 485 (1969); D. V. Santi and W. Marchant, *Biochem. biophys. Res. Commun.* **51**, 370 (1973)

The photoaffinity labels have been useful in mapping out residues at the active sites of enzymes and the binding sites of proteins such as antibodies. The normal affinity labels are more useful for kinetic work since they are selective for the basic and nucleophilic groups that are so prevalent in catalysis and also give information on the pK_as of the groups that are modified.

Many hundreds of affinity labels have now been synthesized, the majority based on halomethyl ketones or epoxides. They are listed each year in the *Specialist Periodical Reports: Amino-Acids, Peptides, and Proteins*, Chemical Society (UK).

References

1   B. S. Hartley and B. A. Kilby, *Biochem. J.* **56,** 288 (1954).
2   H. Gutfreund, *Discuss. Faraday Soc.* **20,** 167 (1955).
3   H. Gutfreund and J. M. Sturtevant, *Biochem. J.* **63,** 656 (1956).
4   A. Himoe, K. G. Brandt, R. J. DeSa, and G. P. Hess, *J. biol. Chem.* **244,** 3483 (1969).
5   T. E. Barman and H. Gutfreund, *Biochem. J.* **101,** 411 (1966).
6   S. A. Bernhard and H. Gutfreund, *Proc. natn. Acad. Sci. U.S.A.* **53,** 1238 (1965).
7   J. McConn, E. Ku, A. Himoe, K. G. Brandt, and G. P. Hess, *J. biol. Chem.* **246,** 2918 (1971).
8   M. L. Bender, G. R. Schonbaum, and B. Zerner, *J. Am. chem. Soc.* **84,** 2540 (1962).
9   M. Caplow and W. P. Jencks, *Biochemistry* **1,** 883 (1962).
10 S. A. Bernhard, S. J. Lau, and H. Noller, *Biochemistry* **4,** 1108 (1965).
11 J. De Jersey, D. T. Keough, J. K. Stoops, and B. Zerner, *Eur. J. Biochem.* **42,** 237 (1974).
12 C. G. Miller and M. L. Bender, *J. Am. chem. Soc.* **90,** 6850 (1968).
13 A. R. Fersht, D. M. Blow, and J. Fastrez, *Biochemistry* **12,** 2035 (1973).
14 J. Fastrez and A. R. Fersht, *Biochemistry* **12,** 2025 (1973).
15 R. Henderson, *J. molec. Biol.* **54,** 341 (1970).
16 E. F. Jansen, M. D. Nutting, and A. K. Balls, *J. biol. Chem.* **179,** 201 (1949).
17 N. K. Schaffer, S. C. May, and W. H. Summeson, *J. biol. Chem.* **202,** 67 (1953).
18 H. Gutfreund and B. R. Hammond, *Biochem. J.* **73,** 526 (1959).
19 A. Garen and C. Levinthal, *Biochem. biophys. Acta.* **38,** 470 (1960).
20 L. A. Heppel, D. R. Harkness, and R. J. Hilmoe, *J. biol. Chem.* **237,** 841 (1962).
21 W. N. Aldridge, T. E. Barman, and H. Gutfreund, *Biochem. J.* **92,** 23C (1964).
22 W. E. Hull and B. D. Sykes, *Biochemistry* **15,** 1535 (1976).
23 W. E. Hull, S. E. Halford, H. Gutfreund, and B. D. Sykes, *Biochemistry* **15,** 1547 (1976).
24 R. M. Epand and I. B. Wilson, *J. biol. Chem.* **238,** 1718 (1963).
25 R. M. Epand and I. B. Wilson, *J. biol. Chem.* **240,** 1104 (1965).

26 M. L. Bender, G. E. Clement, C. R. Gunter, and F. J. Kezdy, *J. Am. chem. Soc.* **86**, 3697 (1964).
27 I. V. Berezin, N. F. Kazanskaya, and A. A. Klyosov, *FEBS Letts.* **15**, 121 (1971).
28 L. Engström, *Biochim, biophys. Acta* **54**, 179 (1961); **56**, 606 (1962).
29 J. H. Schwartz and F. Lipmann, *Proc. natn. Acad. Sci. U.S.A.* **47**, 1996 (1961).
30 H. Barrett, R. Butler, and I. B. Wilson, *Biochemistry* **8**, 1042 (1969).
31 S. L. Snyder and I. B. Wilson, *Biochemistry* **11**, 3220 (1972).
32 W. Bloch and M. J. Schlesinger, *J. biol. Chem.* **248**, 5794 (1973).
33 J. F. Chlebowski, I. M. Armitage, P. P. Tusa, and J. E. Coleman, *J. biol. Chem.* **251**, 1207 (1976).
34 D. Chappelet-Tordo, M. Iwatsubo, and M. Lazdundski, *Biochemistry* **13**, 3754 (1974).
35 G. S. Kilsheimer and B. Axelrod, *J. biol. Chem.* **227**, 879 (1957).
36 W. Ostrowski and E. A. Barnard. *Biochemistry* **12**, 3893 (1973).
37 K. Wallenfels and G. Kurz, *Biochem. Z.* **335**, 559 (1962).
38 K. Wallenfels and O. P. Malhotra, *Adv. Carbohyd. Chem.* **16**, 239 (1961).
39 D. E. Koshland, Jr., in W. D. McElroy and B. Glass (eds.) *Mechanisms of enzyme action*, p. 608, Johns Hopkins Press, Baltimore (1954).
40 J. A. Rupley and V. Gates, *Proc. natn. Acad. Sci. U.S.A.* **57**, 496 (1967).
41 J. J. Pollock, D. M. Chipman, and N. Sharon, *Archs Biochem. Biophys.* **120**, 235 (1967).
42 T. M. Stokes and I. B. Wilson, *Biochemistry* **11**, 1061 (1972).
43 M. L. Sinnott and O. M. Viratelle, *Biochem. J.* **133**, 81 (1973).
44 M. L. Sinnott and I. J. L. Souchard, *Biochem. J.* **133**, 89 (1973).
45 M. L. Sinnott and S. G. Withers, *Biochem. J.* **143**, 751 (1974).
46 M. B. Hoagland, *Biochim. biophys. Acta* **16**, 288 (1955).
47 P. Berg, *J. biol. Chem.* **222**, 1025 (1956).
48 M. B. Hoagland, E. B. Keller, and P. C. Zamecnik, *J. biol. Chem.* **218**, 345 (1956).
49 D. Söll and P. R. Schimmel, *The Enzymes* **10**, 489 (1974).
50 A. Norris and P. Berg, *Proc. natn. Acad. Sci. U.S.A.* **52**, 330 (1964).
51 A. N. Baldwin and P. Berg, *J. biol. Chem.* **241**, 839 (1966).
52 R. B. Loftfield, *Prog. Nucl. Acid Res. (& Mol. Biol.)* **12**, 87 (1972).
53 A. R. Fersht and M. M. Kaethner, *Biochemistry* **15**, 818 (1976).
54 A. R. Fersht and M. M. Kaethner, *Biochemistry* **15**, 3342 (1976).
55 S. Lowey, H. S. Slayter, A. G. Weeds, and H. Baker, *J. mol. Biol.* **42**, 1 (1969).
56 A. Weber and W. Hasselbach, *Biochim. biophys. Acta* **15**, 237 (1954).
57 Y. Tonomura and S. Kitagawa, *Biochim. biophys. Acta* **26**, 15 (1957).
58 R. W. Lymn and E. W. Taylor, *Biochemistry* **9**, 2975 (1970).
59 E. W. Taylor, R. W. Lymn, and G. Moll, *Biochemistry* **9**, 2984 (1970).
60 R. W. Lymn and E. W. Taylor, *Biochemistry* **10**, 4617 (1971).
61 D. R. Trentham, R. G. Bardsley, J. F. Eccleston, and A. G. Weeds, *Biochem. J.* **126**, 635 (1972).
62 C. R. Bagshaw, J. F. Eccleston, D. R. Trentham, D. W. Yates, and R. S. Goody, *Cold Spring Harb. Symp. quant. Biol.* **37**, 127 (1972).

63 C. R. Bagshaw and D. R. Trentham, *Biochem. J.* **133,** 323 (1973).

64 C. R. Bagshaw and D. R. Trentham, *Biochem. J.* **141,** 331 (1974).

65 C. R. Bagshaw, J. F. Eccleston, F. Eckstein, R. S. Goody, H. Gutfreund, and D. R. Trentham, *Biochem. J.* **141,** 351 (1974).

66 H. G. Mannherz, H. Schenck, and R. S. Goody, *Eur. J. Biochem.* **48,** 287 (1974).

67 R. G. Wolcott and P. Boyer, *J. Supramolecular Structure* **3,** 154 (1975).

68 D. R. Trentham, J. F. Eccleston, and C. R. Bagshaw, *Q. Rev. Biophys.* **9,** 2 (1976).

69 R. W. Taylor and A. G. Weeds, *Biochem. J.* **159,** 301 (1976).

70 R. Fishbein, P. A. Benkovic, K. J. Schray, I. J. Siewers, J. J. Steffens, and S. J. Benkovic, *J. biol. Chem.* **249,** 6047 (1974).

71 B. Würster and B. Hess, *FEBS Letts.* **38,** 257 (1974).

72 T. A. W. Koerner Jr., E. S. Younathan, A.-L. E. Ashour, and R. J. Voll, *J. biol. Chem.* **249,** 5749 (1974).

73 E. Shaw, *The Enzymes* **1,** 91 (1970).

74 G. Schoellmann and E. Shaw, *Biochem. biophys. Res. Commum.* **7,** 36 (1962); *Biochemistry* **2,** 252 (1963).

75 B. R. Baker, W. W. Lee, E. Tong, and L. O. Ross, *J. Am. chem. Soc.* **83,** 3713 (1961).

76 L. Wofsy, H. Metzger, and S. J. Singer, *Biochemistry* **1,** 1031 (1961).

77 D. Glick, *Biochemistry* **7,** 3391 (1968).

78 E. L. O'Connell and I. A. Rose, *J. biol. Chem.* **248,** 2225 (1973).

79 P. J. Shaw and H. Muirhead, *FEBS Letts.* **65,** 50 (1976).

80 R. L. Lundblad and W. H. Stein, *J. biol. Chem.* **244,** 154 (1969).

81 A. Singh, E. R. Thornton, and F. H. Westheimer, *J. biol. Chem.* **237,** 3006 (1962).

82 J. R. Knowles, *Acc. chem. Res.* **5,** 155 (1972).

Chapter 8

Cooperative ligand binding and allosteric interactions

A. Positive cooperativity

Many proteins are composed of sub-units and have multiple ligand binding sites. In some cases the ligand binding curves do not follow the Michaelis–Menten equation (Chapter 6D) but instead are *sigmoid*. This is illustrated in Fig. 8.1 where the degree of saturation of haemoglobin with oxygen is plotted against the pressure of oxygen. The sigmoid curve may be compared with the hyperbolic curve that is expected from the Michaelis–Menten equation and is found for the binding to myoglobin. Sigmoid curves are characteristic of the *cooperative* binding of ligands to proteins that have multiple binding sites. Haemoglobin, for example, is composed of four polypeptide chains, each one of which is similar to the single polypeptide chain of myoglobin. The haemoglobin binding curve may be fitted to four successive binding constants, the Adair equation.[1] The surprising feature of this is that the affinity for the fourth oxygen that binds is some hundred to a thousand times higher than for the first oxygen. This increase in affinity with increasing saturation cannot be explained by four non-interacting sites of differing affinities. If this were the case, the high affinity sites would fill first so that the partially ligated molecules would be of lower affinity than the free deoxyhaemoglobin. The increase in affinity with increasing saturation is due instead to the sites *interacting* so that binding at one site causes an increase in affinity at another.

A similar cooperativity of substrate binding is found to occur with some enzymes leading to sigmoid plots of v against [S]. These enzymes are usually the regulatory or control enzymes on metabolic pathways whose activities are subject to feedback inhibition, or activation. The terminology of cooperative interactions stems from the studies on control.[2] The enzymes are termed *allosteric* (Greek, *allos* = other, *stereos* = solid or space) since the allosteric effector (the inhibitor or activator) is generally structurally different from the substrates and binds at its own separate site

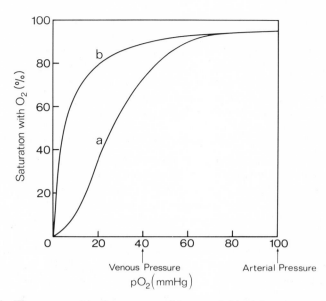

FIG. 8.1. The oxygen binding curves of haemoglobin (a) and myoglobin (b)

TABLE 8.1. *Adair constants for the binding of O_2 to haemoglobin*[a]

2,3-diphosphoglycerate (mM)	K_1	K_2	K_3	K_4
		(mm^{-1} Hg)		
0	0·024	~0·074	~0·086	7·4
2·0	0·01	~0·023	~0·008	11·2

[a] 25°, pH 7·4, and 0·1 M NaCl
I. Tyuma, K. Imai, and K. Shimizu, *Biochemistry* **12**, 1491 (1973)
The Adair equation describes the following:

$$Hb \xrightleftharpoons{K_1, O_2} HbO_2 \xrightleftharpoons{K_2, O_2} Hb(O_2)_2 \xrightleftharpoons{K_3, O_2} Hb(O_2)_3 \xrightleftharpoons{K_4, O_2} Hb(O_2)_4$$

where

$$K_1 = \frac{[HbO_2]}{[Hb][O_2]}, \qquad K_2 = \frac{[Hb(O_2)_2]}{[HbO_2][O_2]} \quad \text{etc.}$$

away from the active site. The term homotropic is sometimes used to denote the interactions amongst the identical substrate molecules, and heterotropic for the interactions of the allosteric effectors with the substrates.

B. Mechanisms of allosteric interactions and cooperativity

The original attempts to explain the mechanism of cooperativity were based on haemoglobin. The best way to understand them is to consider the structures of deoxyhaemoglobin and oxyhaemoglobin. Haemoglobin is composed of two pairs of chains, α and β, arranged in a symmetrical tetrahedral manner. The oxygen binding sites, the haems, are too far apart to interact directly. On oxygenation of deoxyhaemoglobin the overall tetrahedral symmetry is maintained but there are changes in the quaternary and tertiary structures (see Fig. 8.2).[3] It is known from the binding measurements that deoxyhaemoglobin has a low affinity for oxygen whilst oxyhaemoglobin has a high affinity.

1. The Monod–Wyman–Changeux (MWC) concerted mechanism[4]

Monod, Wyman, and Changeux showed that cooperativity can be accounted for in a very simple and elegant manner by assuming that a small fraction of deoxyhaemoglobin exists in the quaternary oxy structure which binds oxygen more strongly. On binding one mole of oxygen, the concentration of the oxy structure is increased since oxygen binds preferentially to it. When a sufficient number of oxygen molecules are bound, the oxy form is sufficiently stabilized to be the major structure in solution, so that subsequent binding is strong. In order to simplify the mathematical equations and the physical concepts they assume that the quaternary structure of the molecule is always symmetrical; a particular partly ligated molecule is either in the oxy state or in the deoxy state and mixed states do not occur. For this reason the MWC model is often described as 'concerted', 'all or none' or 'two state'.

The model is generalized to cover other allosteric proteins using the following assumptions.

(a) The proteins are oligomers.
(b) The protein exists in either of two conformational states, T (= tense), the predominant form when unligated, and R (= relaxed), which are in equilibrium. The two states differ in the energies and numbers of bonds between the sub-units so that the T state is constrained compared with the R state.
(c) The T state has a lower affinity for ligands.
(d) All binding sites in each state are equivalent and have identical binding constants for ligands (the symmetry assumption).

The sigmoid binding curve of any allosteric protein can then be calculated by using just three parameters: L, the allosteric constant, which is equal to the ratio [T]/[R] for the unligated states; and K_T and K_R, the dissociation constants for each site in the T and R states respectively.

Explanation of control through allosteric interactions

The first achievement of the MWC theory was to provide a theoretical curve based on just L, K_R, and K_T which fitted the oxygen binding curve of haemoglobin with high precision. But what was even more impressive was that it provided a very simple explanation for control. Monod,

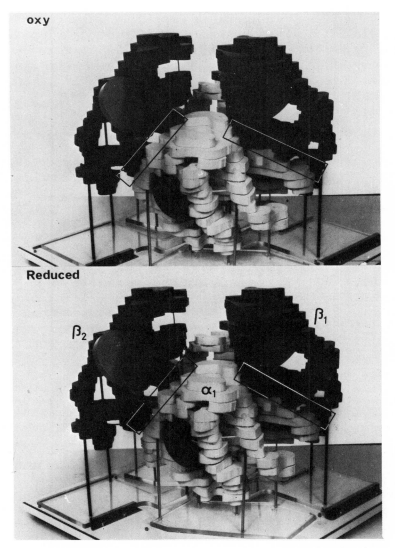

FIG. 8.2. Balsa wood models of oxy- and deoxyhaemoglobin. The haems are represented by disks. Note the increased separation of the β subunits on deoxygenation

Wyman, and Changeux noted that it is a common feature of allosteric enzymes to exhibit cooperativity in v versus [S] plots in the absence of their allosteric activators or inhibitors. They reasoned that if this cooperativity is due to binding and an R–T equilibrium, then control could be explained by the effector altering the R–T ratio by preferentially stabilizing one of the forms. An activator functions by binding to the R state and increasing its concentration. An inhibitor binds preferentially to the T state and so causes the transition to the R state to be more difficult.

In extreme cases, an activator will displace the R–T equilibrium to such an extent that the R state predominates. Cooperativity is then abolished so that Michaelis–Menten kinetics will hold. This has been verified experimentally for several enzymes (Fig. 8.4).

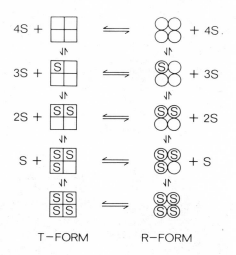

FIG. 8.3. The MWC model for the binding of ligands to a tetrameric protein

It is also predicted that an allosteric inhibitor should bind non-cooperatively to an enzyme that binds its substrates cooperatively since the inhibitor binds to the predominant T state. The converse should be true for activators binding to multiple binding sites in the R state.

The assumption in the MWC model that is most open to criticism is the one of symmetry. By mimimizing the number of intermediate states, the model is only an approximation to reality. But this is also its virtue. The model provides a simple framework in which to rationalize experiments and explain phenomena. The predictions, such as the switch from sigmoid to Michaelis–Menten kinetics in control enzymes at sufficiently high activator concentrations, do not, in any case, depend on the intermediate states. Also, despite the simplicity, the theory accounts for the binding curves of oxygen to a wide series mutant haemoglobins (see later).

The MWC is basically a structural theory. The hypothesis that there are constraints between the sub-units in the T state has provided the basis of much of the structural work by Perutz and others in elucidating the nature and energies of these constraints in haemoglobin.

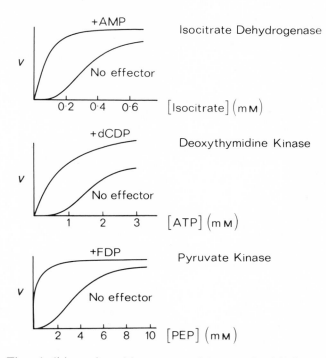

FIG. 8.4. The abolition of positive cooperativity on the binding of allosteric effectors to some enzymes. Note the dramatic increases in activity at low substrate concentrations on the addition of adeosine monophosphate to isocitrate dehydrogenase, deoxycytosine diphosphate to deoxythymidine kinase, and fructose-1,6-diphosphate to pyruvate kinase, showing how the activity may be 'switched on' by an allosteric effector. (J. A. Hathaway and D. E. Atkinson, *J. biol. Chem.* **238,** 2875 (1963); R. Okazaki and A. Kornberg, *J. biol. Chem.* **239,** 275 (1964); R. Haeckel, B. Hess, W. Lauterhorn, and K.-H. Würster, *Hoppe-Seyler's Z. physiol. Chem.* **349,** 699 (1968))

2. The Koshland–Némethy–Filmer (KNF) sequential model[5]

The KNF model avoids the assumption of symmetry but uses another simplifying feature. It assumes that the progress from T to the ligand-bound R state is a sequential process. The conformation of each sub-unit changes in turn as it binds ligand and there is no dramatic switch from one state to another (Fig. 8.5). MWC uses a quaternary structural change, KNF uses a series of tertiary structural changes.

The two assumptions in the KNF model are:

(a) In the absence of ligands the protein exists in one conformation.
(b) Upon binding, the ligand induces a conformational change in the sub-unit to which it is bound. This change may be transmitted to neighbouring vacant sub-units via the sub-unit interfaces.

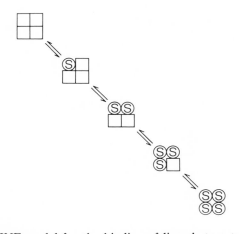

FIG. 8.5. The KNF model for the binding of ligands to a tetrameric protein

This model embodies Koshland's earlier idea of *induced fit*, which postulates that the binding of a substrate to an enzyme may cause conformational changes that align the catalytic groups in their correct orientations.

Using these assumptions it is possible to describe the binding of oxygen by four successive binding constants. This is formally equivalent to the Adair equation—the KNF model may be considered as a molecular interpretation of the Adair equation. In general, the number of constants that is required is equal to the number of binding sites, unlike the MWC model which always uses three.

In sacrificing simplicity, the KNF model is more general and is probably a better description of some proteins than is the MWC model. In turn, the explanation of phenomena is often somewhat more complicated.

3. The general model[6]

Eigen has pointed out that the MWC and KNF models are limiting cases of a general scheme involving all possible combinations (Fig. 8.6). The scheme is more complicated than the 'chessboard' that is illustrated since, for reasons of symmetry, there are 44 possible states for the haemoglobin case. The KNF model moves across the chessboard like a bishop confined to the long diagonal, and the MWC like a rook which is confined to the perimeters. The general model, combining both the MWC and KNF extremes, along with dissociation of the sub-units etc., has been analysed.[7] But the results are too complex for general use and it is far more convenient to interpret experiments in terms of the MWC or KNF simplifications.

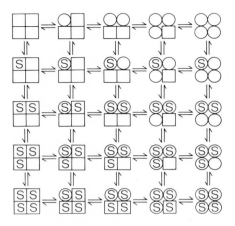

F<small>IG</small>. 8.6. Eigen's general scheme for the binding of ligands to a tetrameric protein. The columns on the extreme left and right represent the MWC simplification. The diagonal from top left to bottom right represents the KNF simplification

C. Negative cooperativity and half-of-the-sites reactivity[8,9]

Some enzymes bind successive ligand molecules with decreasing affinity. As pointed out before, this can be explained by the presence of binding sites of differing affinities so that the stronger are occupied first. But in many cases this is found with oligomeric enzymes composed of identical sub-units. In two cases, the tyrosyl-tRNA synthetase[10] and glyceraldehyde-3-phosphate dehydrogenase[11] from *Bacillus stearother-mophilus*, X-ray diffraction studies on the crystalline enzymes show that the sub-units are arranged symmetrically so that all sites, are initially equivalent. Yet, the dimeric tyrosyl-tRNA synthetase binds only one mole of tyrosine tightly, whilst the binding of the second remains undetected even at millimolar concentrations of tyrosine.[12,13] The four moles of NAD^+ that bind to the rabbit muscle glyceraldehyde-3-phosphate dehydrogenase do so with increasing dissociation constants of $<10^{-11}$, $<10^{-9}$, 3×10^{-7}, and 3×10^{-5} M.[14,15] Similar changes in affinity are found for the bacterial enzyme. This antagonistic binding of molecules is known as negative or anti-cooperativity.

Negative cooperativity cannot be accounted for on the MWC theory; the binding of the first ligand molecule can only stabilize the high affinity state and cannot increase the proportion of the T state. The KNF model accounts for negative cooperativity by the binding of ligand to one site causing a conformational change that is transmitted to a vacant sub-unit (assumption (b) of the KNF model). Negative cooperativity is thus a diagnostic test of the KNF model.

A related phenomenon is half-of-the-sites or half-site reactivity, where

TABLE 8.2. *Some enzymes showing half-of-the-sites reactivity*

Enzyme	Reaction	Number of Sub-units	Ref.
Acetoacetate decarboxylase	Inactivation of active site lysine	12	1
Aldolase	Partial reaction with fructose-6-phosphate	2	2
Aminoacyl-tRNA synthetases (some)	Biphasic formation of aminoacyl adenylate	2	3
Cytidine triphosphate synthetase	Stoichiometry of irreversible inhibition	4	4
Glyceraldehyde-3-phosphate dehydrogenase	Reaction with non-physiological substrates. *But*, full site reactivity with physiological	4	5
Asparatate transcarbamylase	CTP binding Carbamyl phosphate binding	6 (regulatory) 6 (catalytic)	5
Glutamine synthetase	Irreversible inhibition	8	6

1 D. E. Schmidt, Jr. and F. H. Westheimer, *Biochemistry* **10**, 1249 (1971) and references therein.
2 O. Tsolas and B. L. Horecker, *Archs biochem. Biophys.* **173**, 577 (1976).
3 R. S. Mulvey and A. R. Fersht, *Biochemistry* **15**, 243 (1976).
4 A. Levitzki, W. B. Stallcup, and D. E. Koshland, *Biochemistry* **10**, 3371 (1971).
5 F. Seydoux, O. P. Malhotra, and S. A. Bernhard, *Crit. Revs Biochem.* 227 (1974).
6 S. S. Tate and A. Meister, *Proc. natn. Acad. Sci. U.S.A.* **68**, 781 (1971).

an enzyme containing $2n$ sites reacts (rapidly) at only n of them. This can only be detected by pre-steady-state kinetics. The tyrosyl-tRNA synthetase is a good example of this, forming one mole of enzyme-bound tyrosyl adenylate with a rate constant of $18\,s^{-1}$ but the second site reacting 10^4 times more slowly.[12,16]

Half-of-the-sites reactivity is inconsistent with the simple MWC theory since the sites lose their equivalence and symmetry is lost.

A cautionary note must be mentioned at this point. The diagnosis of half-of-the-sites reactivity depends on knowing the exact concentration of binding sites on the enzyme. This depends on determining the precise concentration of protein and its purity. Similarly, an inhomogeneous preparation of enzyme containing molecules of differing affinity for the ligand will give a binding curve similar to that of negative cooperativity. Failure to check these points has led to spurious reports about sub-unit interactions. A further artefact that may be misinterpreted as half-of-the-sites reactivity occurs in the reactions of lactate dehydrogenase and is discussed in Chapter 12—an enzyme-bound intermediate does not appear to fully accumulate because of an unfavourable equilibrium constant.

D. Quantitative analysis of cooperativity

1. The Hill equation—a measure of cooperativity[17]

Consider a case of completely cooperative binding, an enzyme containing n binding sites and all n being occupied simultaneously with a dissociation constant K (eqn (8.1)).

$$E + nS \rightleftharpoons ES_n \tag{8.1}$$

and

$$K = \frac{[E][S]^n}{[ES_n]} \tag{8.2}$$

The degree of saturation Y is given by

$$Y = \frac{ES_n}{[E_0]} \tag{8.3}$$

and

$$1 - Y = \frac{[E]}{[E_0]}. \tag{8.4}$$

Eqns (8.2) and (8.4) may be manipulated to give.

$$\log[Y/(1-Y)] = n \log[S] - \log K. \tag{8.5}$$

A similar equation, called the Hill plot (eqn (8.6)), is found to describe

$$\log[Y/(1-Y)] = h \log[S] - \log K. \tag{8.6}$$

satisfactorily the binding of ligands to allosteric proteins in the region of 50% saturation (10 to 90%). Outside this region the experimental curve deviates from the straight line. The value of h found from the slope of eqn (8.6) in the region of 50% saturation is known as the *Hill constant*. It is a measure of cooperativity. The higher h is, the higher the cooperativity. At the upper limit h is equal to the number of binding sites. If $h = 1$, there is no cooperativity; if $h > 1$, there is positive cooperativity; if $h < 1$, there is negative cooperativity.

The Hill equation may be extended to kinetic measurements by replacing Y by v as in

$$\log v/(V_{max} - v) = h \log[S] - \log K. \tag{8.7}$$

2. The MWC binding curve[4]

Let the dissociation constant of the ligand from the R state be K_R and that from the T state be K_T. Then c is defined by

$$c = K_R/K_T. \tag{8.8}$$

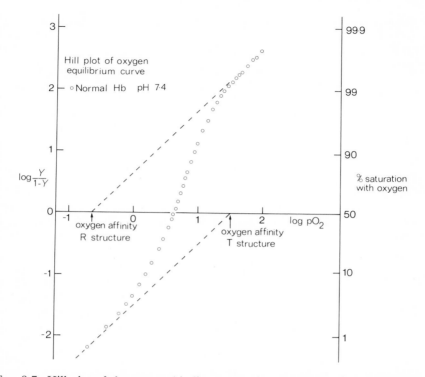

Fig. 8.7. Hill plot of the oxygen binding curve of haemoglobin. (J. V. Kilmartin, K. Imai and R. T. Jones, in *Erythrocyte structure and function*. Alan R. Liss Inc. New York, p. 21 (1975))

The R state with x ligand molecules bound is termed R_x and the equivalent T state T_x. The allosteric constant L is then defined by

$$L = [T_0]/[R_0]. \tag{8.9}$$

The fraction of the protein that is bound with ligand, Y, is then calculated from the mass balance equations. It must be remembered to 'statistically correct' the binding constants. For example, the dissociation constant for the first O_2 binding to haemoglobin is $K_T/4$ since there are 4 sites to which the ligand may bind but only one site for dissociation when it is bound. Similarly, the dissociation constant for the second molecule that binds is given by $2K_T/3$ since there are 3 sites to which it can bind, but once bound there are two bound sites that can dissociate.

Y for a protein containing n sites is given by

$$Y = \frac{([R_1] + 2[R_2] + \cdots + n[R_n]) + ([T_1] + 2[T_2] + \cdots + n[T_n])}{n([R_0] + [R_1] + \cdots + [R_n] + [T_0] + [T_1] + \cdots + [T_n])}.$$

$$\tag{8.10}$$

Solving eqn (8.10) in terms of L and c, and using for convenience

$$\alpha = [S]/K_R \tag{8.11}$$

gives

$$Y = \frac{Lc\alpha(1+c\alpha)^{n+1}+\alpha(1+\alpha)^{n-1}}{L(1+c\alpha)^{n}+(1+\alpha)^{n}}. \tag{8.12}$$

According to the MWC theory the saturation curve for any oligomeric protein composed of n protomers is defined by only three unknown parameters and the concentration of ligand (L, K_R, K_T, and [S], with the latter three disguised as c and α in eqn (8.12)).

Some values of L and c obtained from the computer fitting of eqn (8.12) to the binding curves of some proteins are listed in Table 8.3. (Note: this does not imply that the structural changes follow the MWC model since the KNF model also predicts the same binding curve—see below.)

a. Dependence of the Hill constant on L and c[18]

The value of h may be calculated from a computer analysis of eqn (8.12). It is found that a plot of h against L at a constant value of c gives a bell-shaped curve (Fig. 8.8). h is equal to 1 for L being either much greater or lower than c and is a maximum when

$$L = c^{-n/2} \tag{8.13}$$

(where n is the number of binding sites). The reason for this behaviour is that when L is low there is initially sufficient of the R state to give good binding, and when L is very high, the concentration of the R state is too small to contribute significantly to binding.

TABLE 8.3. *Allosteric constants for some proteins*

Protein	Ligand	Number of binding sites (n)	Hill constant (h)	L	c	Ref.
Haemoglobin	O_2	4	2·8	3×10^5	0·01	1
Pyruvate kinase (yeast)	Phosphoenol pyruvate	4	2·8	9×10^{3a}	0·01[a]	2
Glyceraldehyde-3-phosphate dehydrogenase (yeast)	NAD^+	4	2·3	60	0·04	3

[a] Estimated by the author

References: 1 S. J. Edelstein, *Nature, Lond.* **230,** 224 (1971). 2.   R. Haeckel, B. Hess, W. Lauterhorn, and K.-H. Würster, *Hoppe-Seyler's Z. physiol. Chem.* **349,** 699 (1968); H. Bischofberger, B. Hess, and P. Röschlau, *ibid.* **352,** 1139 (1971). 3.   K. Kirschner, E. Gallego, I. Schuster, and D. Goodall, *J. molec. Biol.* **58,** 29 (1971).

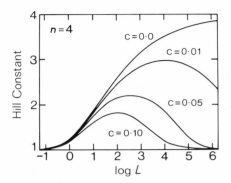

FIG. 8.8. Variation of the Hill constant with L for a tetrameric protein. (M. M. Rubin and J.-P. Changeux, *J. molec. Biol.* **21**, 265 (1966))

The bell-shaped curve is of particular interest in the analysis of the effects of structural changes in a protein which affect L. The Hill constants of a wide series of modified and mutant haemoglobins fit such a curve (Fig. 8.9).[19]

3. The KNF binding curve

The MWC model gives a simple equation for Y due to the assumption of only two dissociation constants. The KNF model requires a different dissociation constant for every intermediate state. There is no simple general equation for Y. As many variables are required as there are binding sites.

4. Diagnostic tests for cooperativity and MWC versus KNF mechanisms

Cooperativity is determined from the value of h from the Hill plot or the characteristic deviations in the straightforward saturation curves or

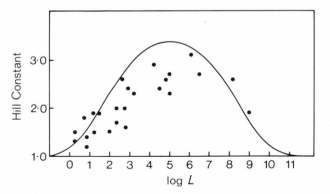

FIG. 8.9. Variation of the Hill constant with L for the binding of oxygen to various mutant haemoglobins. (J. M. Baldwin, *Prog. Biophys. Molec. Biol.* **29**, 3 (1975))

Scatchard plots (Fig. 8.10). The finding of negative cooperativity excludes the simple MWC theory but positive cooperativity is consistent with both models. Analysis of the shape of the binding curve is also ambiguous since both KNF and MWC predict similar shapes. In theory, measurement of the *rates* of ligand binding can distinguish between the two models.[6] The MWC model, for example, predicts fewer relaxation times since fewer states are involved. This approach has been applied with

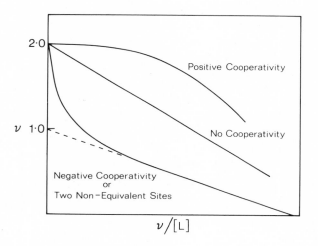

FIG. 8.10. Plots of stoichiometry ν against $\nu/[L]$ for the binding of ligand (L) to a dimeric protein

success to the glyceraldehyde-3-phosphate dehydrogenase from yeast where it has been shown that the kinetics of NAD^+ binding are consistent with the positive cooperativity being due to a MWC model.[20] On the other hand, ligand binding studies on haemoglobin suggest the importance of intermediate states.

The choice of model used often depends on the experiments involved. Workers in the area of, say, the effects of structural changes on the oxygen affinity and Hill constant for haemoglobin prefer the MWC model because it is essentially a *structural* theory. It provides a simple framework for the prediction and interpretation of experiments. Application of the theory to the Hill constant and other *equilibrium* measurements gives very acceptable results. Kineticists prefer the KNF model since the kinetic measurements are more sensitive to the presence of intermediates. There are more variables in the KNF theory and there is more flexibility in fitting data.

E. Molecular mechanism of cooperative binding to haemoglobin[3]

1. Physiological importance of cooperative binding of oxygen

Positive cooperativity is not a device for increasing the affinity of haemoglobin for oxygen; the association constant of free haemoglobin chains for oxygen is far higher than that for the binding of the first mole of oxygen to deoxyhaemoglobin, and is about the same as for binding to oxyhaemoglobin. It is instead a means of changing the affinity over a very narrow range of pressure. It enables haemoglobin to be saturated with oxygen in the lungs and then to unload about 60% of the oxygen to the tissues. This can be done over a relatively narrow range of oxygen pressure because of the steepness of the curve of saturation with oxygen against oxygen pressure in the region of 50% saturation. A simple Michaelis–Menten curve would require a far greater range of pressures. (The Hill constant is a measure of the steepness of the saturation curve at 50% saturation; h is close to 3 for the sigmoid curve of haemoglobin but would be 1 for a hyperbolic curve.)

2. Atomic events in oxygenation of haemoglobin

The two pairs of α and β sub-units are arranged tetrahedrally around a twofold axis of symmetry (Fig. 8.2). There is a cavity at the centre of the

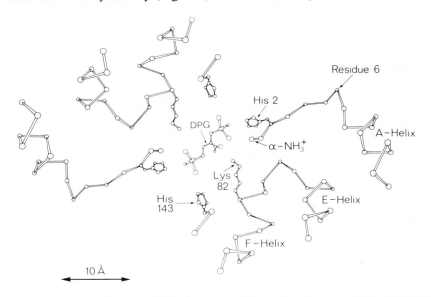

FIG. 8.11. Binding of 2,3-diphosphoglycerate (DPG) between the β chains in the central cavity of human haemoglobin. (A. Arnone and M. F. Perutz, *Nature, Lond.* **249,** 34 (1974))

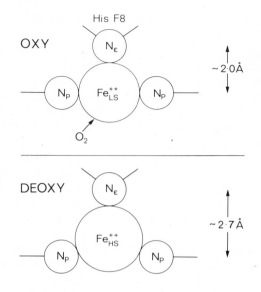

Coordination of iron in oxy (low spin)
and deoxy (high spin) haemoglobin.

Fig. 8.12. Movement of imidazole ring of His F8 on the binding of oxygen to deoxyhaemoglobin. The radius of the Fe^{2+} ion shrinks, allowing it to slip between the pyrrole nitrogens (Np) into the plane of the pyrrole ring

molecule through which the axis passes. Organic phosphates, such as 2,3-diphosphoglycerate, which are allosteric effectors, bind in this cavity in deoxyhaemoglobin in a 1:1 stoichiometry (see Fig. 8.11).[21-23] The negatively charged organic phosphate sits between the two β sub-units, forming four salt linkages with each (with Val-1, His-2, Lys-82, and His-143). Deoxyhaemoglobin is stabilized by four pairs of salt bridges between the interfaces of the two α chains and two salt bridges between the α and β chains.

On oxygenation, it is seen at low resolution that the sub-units rotate relative to each other and the twofold axis by about 10°. The haems change their angles of tilt by a few degrees and the helical regions move 2–3 Å relative to each other. At higher resolution, it is seen that, apart from some conformational changes in the sub-units, the salt bridges between each sub-unit are broken. Also the α-amino groups of Val-1 of each β sub-unit in the central cavity move apart by 4 Å whilst the cavity narrows, expelling any organic phosphate that is bound. There is also a decrease in the hydrophobic area which is in contact at the interfaces of the sub-units.[24] The 'constraints' between the sub-units in the T-state predicted by Monod, Wyman, and Changeux are thus the salt bridges and additional hydrophobic interactions.

This explains why deoxyhaemoglobin has a low affinity for oxygen compared with myoglobin or artificially prepared single chains.[13] On binding of oxygen, favourable interactions in the quaternary structure are disrupted. Salt bridges are broken and hydrophobic surfaces are exposed. The mode of action of the allosteric effectors is also clear. Organic phosphates bind strongly to a well-defined site in the T state and stabilize it relative to the R state.

The breaking of the salt bridges on oxygenation also explains the *Bohr* effect.[25] On oxygenation at physiological pH there is a release of about 0·7 protons per haem. This is due to the amine bases in the salt bridges having higher pK_as because of the carboxylate ions (see Chapter 4). On disruption of the bridges, the pK_a drop to the normal values, releasing protons.

It seems remarkable that the binding of O_2 causes such extensive structural changes. But nature has provided an ingenious trigger mechanism for this. The Fe^{2+} ion in deoxyhaemoglobin is in the high spin state and its radius is too large for it to be in the plane of the porphyrin ring of the haem (Fig. 8.12). On binding O_2 the Fe^{2+} ion becomes low spin and its radius shrinks by 13%. The Fe^{2+} ion can now fit into the energetically favourable position in the porphyrin plane, causing a movement of about 0·6 Å. The histidine which is ligated to the iron is dragged through this distance and sets off the chain of events leading to the conformational changes.[3,26]

References

1   G. S. Adair, *J. biol. Chem.* **63,** 529 (1925).

2   J. Monod, J.-P. Changeux, and F. Jacob, *J. molec. Biol.* **6,** 306 (1963).

3   M. F. Perutz, *Nature, Lond.* **228,** 726 (1970).

4   J. Monod, J. Wyman, and J.-P. Changeux, *J. molec. Biol.* **12,** 88 (1965).

5   D. E. Koshland, Jr., G. Neméthy, and D. Filmer, *Biochemistry* **5,** 365 (1966).

6   M. Eigen, *Nobel Symposium* **5,** 333 (1967).

7   J. Herzfield and H. E. Stanley, *J. molec. Biol.* **82,** 231 (1974).

8   A. Levitzki, W. B. Stallcup, and D. E. Koshland, Jr., *Biochemistry* **10,** 3371 (1971).

9   R. A. MacQuarrie and S. A. Bernhard, *Biochemistry* **10,** 2456 (1971).

10 M. J. Irwin, J. Nyborg, B. R. Reid, and D. M. Blow, *J. molec. Biol.* **105,** 577 (1976).

11 G. Biesecker, J. I. Harris, J. C. Thierry, J. E. Walker and A. J. Wonacott, *Nature, Lond.* **266,** 328 (1977).

12 A. R. Fersht, R. S. Mulvey, and G. L. E. Koch, *Biochemistry* **14,** 13 (1975).

13 H. R. Bosshard, G. L. E. Koch, and B. S. Hartley, *Eur. J. Biochem.* **53,** 493 (1975).

14 A. Conway and D. E. Koshland, Jr., *Biochemistry* **7,** 4011 (1968).
15 J. Schlessinger and A. Levitzki, *J. molec. Biol.* **82,** 547 (1974).
16 A. R. Fersht, unpublished data.
17 R. Hill, *Proc. R. Soc.* **B100,** 419 (1925).
18 M. M. Rubin and J.-P. Changeux, *J. molec. Biol.* **21,** 265 (1966).
19 S. Edelstein, *Nature, Lond.* **230,** 224 (1971).
20 K. Kirschner, E. Gallego, I. Schuster, and D. Goodall, *J. molec. Biol.* **58,** 29 (1971).
21 M. F. Perutz, *Nature, Lond.* **228,** 734 (1970).
22 A. Arnone, *Nature, Lond.* **237,** 146 (1972).
23 A. Arnone and M. F. Perutz, *Nature, Lond.* **249,** 34 (1974).
24 C. Chothia, S. Wodak, and J. Janin, *Proc. natn. Acad. Sci. U.S.A.* **73,** 3793 (1976).
25 C. Bohr, K. A. Hasselbach, and A. Krogh, *Skand. Arch. Physiol.* **16,** 402 (1904).
26 J. L. Hoard, in '*Hemes and hemoproteins*' (eds. B. Chance, R. W. Estabrook, and T. Yonetani). Academic Press, p. 9 (1966).

Chapter 9

Forces between molecules and enzyme–substrate binding energies

An important feature in enzyme catalysis is that the enzyme binds the substrate and the reactions proceed in the confines of the enzyme–substrate complex. In order to gain some insight into the strength and specificity of the binding we shall discuss in a somewhat empirical and phenomenological manner the interactions between non-bonded atoms. In particular, we shall concentrate on the magnitudes of the energies involved. Besides being responsible for binding, the non-covalent interactions are important in further ways. One important consideration discussed in Chapter 10, is that these interactions can be used for lowering the activation energy of a chemical step instead of directly contributing to the binding energy. Also, these forces play a considerable role in maintaining protein structure.

The obvious interactions between non-bonded atoms are those due to electrostatic forces between charged groups. But even non-polar molecules have an attraction for each other. This was pointed out by van der Waals a century ago as one of the reasons for the breakdown of the ideal gas laws. Some of these interactions between neutral gas molecules have been well characterized and we now know that several 'van der Waals'' compounds exist. For example, the noble gas dimers Ne_2, A_2, and Xe_2 have bond energies of 0·2, 0·92, and 2·2 kJ/mol (0·05, 0·22, and 0·53 kcal/mol) respectively.[1]

A. Interactions between non-bonded atoms

1. Electrostatic interactions

All forces between atoms and molecules are electrostatic in origin, even those between non-polar molecules. But we shall reserve the term for those interactions that occur between charged or dipolar atoms and molecules. Electrostatic interactions are the best understood of all. But, in practice, their quantitative importance is difficult to assess because the interaction energies depend crucially on the dielectric constant of the surrounding medium. This was discussed in Chapter 2 where an example was given of how a repulsive energy of 530 kJ (130 kcal) between two juxtaposed positively charged nitrogen atoms *in vacuo* is reduced by a factor of 16 when surrounded by water. The surrounding medium is polarized by the charges to set up a neutralizing field. The effective dielectric constant of the active site of lysozyme has recently been calculated to be about 5.[2]

The following varieties of electrostatic interaction energies exist:

(a) Between ions with net charges; the energy falls off as $1/r$.
(b) Between permanent dipoles; the energy falls off as $1/r^6$.
(c) Between an ion and a dipole induced by it; the energy falls off as $1/r^4$.
(d) Between a permanent dipole and a dipole induced by it; the energy falls off as $1/r^6$.

2. Non-polar interactions (van der Waals' or dispersion forces)

A typical potential energy curve for the interaction of two atoms is illustrated in Fig. 9.1. There is characteristically a very steeply rising repulsive potential at short interatomic distances as the two atoms approach so closely that there is interpenetration of their electron clouds. This potential approximates to an inverse twelfth-power law. Superimposed upon this is an attractive potential due mainly to the London *dispersion forces*. This follows an inverse sixth-power law. The total potential energy is given by

$$U = A/r^{12} - B/r^6 \qquad (9.1)$$

Some semi-empirical values of A and B, estimated from physical chemical data, that have used in calculating interaction energies are listed in Table 9.1.

The distance dependence of $1/r^6$ for the attractive potential is characteristic of the interaction between dipoles. This is because the attractive dispersion forces are due to the mutual induction of electrostatic dipoles. Although a non-polar molecule has no net dipole averaged over a period of time, at any one instant of time there will be dipoles due to the local fluctuations of electron density. Because the energies depend on the induction of a dipole, polarizability is an important factor in the strength of the interactions between any two atoms.

The interatomic distance at the bottom of the potential well, the most favourable distance of separation, is known as the *van der Waals' contact distance*. A particular atom has a characteristic van der Waals' radius. These radii are additive so that the optimal distance of contact between two atoms may be found by the addition of their two van der Waals' radii. Some values, determined by X-ray diffraction, are listed in Table 9.2.

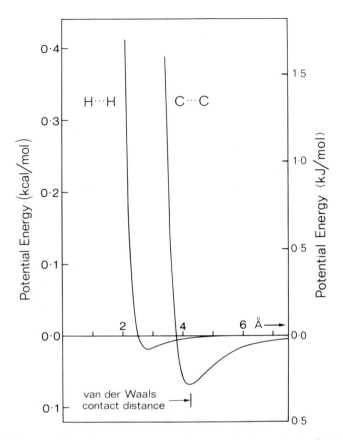

Fig. 9.1. The interaction energies of two hydrogen atoms and two (tetrahedral) carbon atoms in a protein. (Calculated from the data in Table 9.1)

TABLE 9.1. *Values of A and B for the '6–12' equation*

Interaction	A (kcal/mol/Å^{12})	B (kcal/mol/Å^{6})
H...H	$4 \cdot 609 \times 10^{3}$	$18 \cdot 52$
=O...O=	$1 \cdot 537 \times 10^{5}$	$178 \cdot 62$
N...N	$2 \cdot 963 \times 10^{6}$	$1 \cdot 272 \times 10^{3}$
C...C	$2 \cdot 284 \times 10^{6}$	$787 \cdot 7$
C...C[a]	$2 \cdot 399 \times 10^{6}$	$2 \cdot 083 \times 10^{3}$

[a] Carbonyl or carboxyl carbon

Data determined semi-empirically from the observed physicochemical properties of hydrocarbon, amino acid, and amide crystals, A. Warshel and M. Levitt, *J. molec. Biol.* **103**, 227 (1976)

TABLE 9.2. *van der Waals' radii*

Atom	van der Waals' radius (Å)		Minimum radius (Å)[a]
	b	c	
H	1·2	1·4	
O (hydroxyl)	1·5		1·3
O (carbonyl)	1·5	1·7	1·3
O⁻ (carboxyl)	1·5	1·8	
N (amide)	1·55	2	1·3
NH₃⁺ (ammonium)	1·65[d]		
N⁺ (imidazolium)	1·55		
CH (tetrahedral)	1·85[d]	2·1	1·5
C (trigonal)	1·7	1·8	1·45
CH (aromatic)	1·8[d]		
S	1·8		1·5

[a] The minimum contact radius in a protein, M. Levitt, *J. molec. Biol.* **82,** 393 (1974)
[b] A. Bondi, *J. phys. Chem.* **68,** 441 (1965); D. A. Brant and P. J. Flory, *J. Am. chem. Soc.* **87,** 2791 (1965)
[c] A. Warshel and M. Levitt, *J. molec. Biol.* **103,** 227 (1976); these values are close to the potential minimum—the values in the previous column are slightly shorter
[d] Radius includes the bonded hydrogen

Also included are the minimum values, determined from the potential-energy curve and the calculated flexibility of proteins. Some interaction energies for atoms at their optimal separation are given in Table 9.3.

Although the attractive forces are weak and the van der Waals' energies low, they are additive and make significant contributions to binding when summed over a molecule. For example, it is found experimentally from heat of sublimation data that each methylene group in a

TABLE 9.3. *Dispersion (van der Waals') energies for pairs of atoms at optimal separation*

Interaction	van der Waal's energy	
	kJ/mol	kcal/mol
H · · · H	0·0778	0·0186
=O · · · O=	0·217	0·0519
N · · · N	0·572	0·1366
C · · · C	0·284	0·0679
C · · · C[a]	1·891	0·4519

[a] Carbonyl or carboxyl carbon
A. Warshel and M. Levitt, *J. molec. Biol.* **103,** 227 (1976)

crystalline hydrocarbon has 8.4 kJ/mol (2 kcal/mol), and each CH group
in benzene crystals has 6·7 kJ/mol (1·6 kcal/mol) of van der Waals'
energy.[3] It has been calculated that the van der Waals' energy between
the D subsite of lysozyme and the occupying glucopyranose ring is about
−58·5 kJ/mol (−14 kcal/mol).[2]

3. The hydrogen bond

A particularly important bond in biological systems is the hydrogen bond.
This consists of two electronegative atoms, one of which is usually
oxygen, bound to the same proton. The bonds in the systems of interest
are asymmetric, the proton being at its normal covalent bond distance
from the atom to which it is formally bonded, and at a distance from the
other somewhat shorter than the usual van der Waals' contact distance.
The optimal configuration is linear, but bending causes only small energy
losses. In the hydrogen bond between the)NH group and the carbonyl
oxygen in amide crystals, the $O \cdots N$ distance is typically 2·85 to 3·00 Å,
and the $O \cdots H$ distance 1·85 to 2·00 Å.[4,5] The variation of the potential
energy with the $N \cdots O$ distance is similar to that in Fig. 9.1, the
minimum distance of approach being about 2·4 to 2·5 Å. In the
$H_2O \cdots H—OH$ hydrogen bond the distance between the two oxygens is
2·76 Å and the $O \cdots H$ distance is 1·77 Å.

The energies of hydrogen bonds have been variously estimated to be
between 12 and 38 kJ/mol (3 and 9 kcal/mol).[4,6] A reliable value,
estimated from the data in Table 9.4, is 21 kJ/mol (5 kcal/mol) for the

TABLE 9.4. *Dispersion and electrostatic energies in some crystals*

Crystal	Calculated energies Dispersion (van der Waals') kJ/mol	kcal/mol	Electrostatic[a] kJ/mol	kcal/mol	Observed heat of sublimation kJ/mol	kcal/mol
Benzene	−42	−9·9	−5·4	−1·3	−44·8	−10·7
n-Hexane	−54	−12·9	0	0	−50·2	−12·0
n-Pentane	−42	−10·0	0	0	−43·9	−10·5
Perylene	−119	−28·4	−2·5	−0·6	−125·5	−30
Phenanthrene	−85	−20·3	−4·2	−1·0	−91·8	−21·7
Adipamide (4H-bonds)	−86	−20·6	−77	−18·4		—
Formamide (2H-bonds)	−24	−5·7	−43	−10·3	−73·2	−17·5
Malonamide (4H-bonds)	−52	−12·3	−81	−19·4	−120·5	−28·8
Oxamide (4H-bonds)	−45	−10·8	−60	−14·4	−118·0	−28·2
Succinamide (4H-bonds)	−65	−15·5	−76	−18·2	−135·2	−32·3
Urea (3H-bonds)	−28	−6·7	−65	−15·6	−92·8	−22·2

[a] The electrostatic contribution is due to the hydrogen bonds
The energies are calculated from empirical energy functions (see Table 9.3), M. Levitt and
S. Lifson, personal communication

amide/amide NH $\cdots$ O bond.[3] Bonds of this strength are of particular importance since they are stable enough to provide significant binding energy but of sufficiently low strength to allow rapid dissociation. If the activation energy for the breaking of the bond is the whole of the bond strength, then it may be calculated from transition state theory that bonds of energy 12·5, 25·0, and 37.6 kJ/mol (3, 6, and 9 kcal/mol) dissociate with rate constants of 4×10^{10}, 3×10^{8}, and $2 \times 10^{6}\,\mathrm{s}^{-1}$ respectively.

The backbone NH groups of the enzyme are not only used for binding the substrate but may also act as the solvation shell for negative charges developing in the transition state. In particular, it was pointed out in Chapter 1 that there is a binding site for the carbonyl oxygen of the substrates of the serine proteases consisting of two backbone amido NH groups. The oxygen becomes negatively charged during the transition state of the reaction and is stabilized by the dipole moments of the amide groups.

4. The hydrophobic bond[7,8]

The hydrophobic bond is a way of describing the tendency of non-polar compounds, such as hydrocarbons, to transfer from an aqueous solution to an organic phase. This is due not so much to the direct interaction between solvent and solute molecules but to the reorganization of the normal hydrogen bonding network in water by the presence of a hydrophobic compound. Water consists of a dynamic, loose, network of hydrogen bonds. The presence of a non-polar compound causes a local rearrangement in this. In order to preserve the number of hydrogen bonds, each one having an energy of about 25 kJ/mol (6 kcal/mol), the water molecules line up around the non-polar molecule. The hydrophobic solute therefore does not cause large enthalpy changes in the solvent but instead decreases its entropy due to the increase in local order. A hydrophobic molecule is driven into the hydrophobic region of a protein by the regaining of entropy by water.

One convenient way of measuring the hydrophobicity of a molecule is to measure its partition between the organic and aqueous phases when shaken with an immiscible mixture of an organic solvent, often n-octanol, and water. The distribution of the solute between the two phases depends on the competing tendencies of the hydrophobic regions to be squeezed into the organic phase by the hydrophobic bond, and the polar regions, which require solvation to be drawn into the aqueous phase.

a. Hydrophobicity of small groups—the Hansch equation[9,10]
From determinations of the partitioning of several series of substituted compounds between n-octanol and water, Hansch and coworkers found that many of the substituents make a constant, and additive, contribution to the hydrophobicity of the parent compound. If the ratio of the

TABLE 9.5. *Some values of* π

Group[a]	π	Gibbs energy of transfer from n-octanol to water	
		kJ/mol	kcal/mol
CH₃—	0·5	2·85	0·68
CH₃CH₂—	1·0	5·71	1·36
CH₃(CH₂)₂—	1·5	8·56	2·05
CH₃(CH₂)₃—	2·0	11·41	2·73
CH₃(CH₂)₄—	2·5	14·26	3·41
CH₃ ＼ CH— / CH₃	1·3	7·42	1·77
CH₃CH₂ ＼ CH— / CH₃	1·8	2·45	10·27
PhCH₂—	2·63	15·00	3·59
—OH[b]	−1·16	−6·62	−1·58
—NHCOCH₃[b]	−1·21	−6·90	−1·65
—OCOCH₃[b]	−0·27	−1·54	−0·37

[a] Relative to the hydrogen atom
[b] Bound to aliphatic compounds
C. Hansch and E. Coats, *J. Pharm. Sci.* **59**, 731 (1970)

solubility of the parent compound in the organic phase to that in the aqueous phase is P_0, and that of the substituted compound is P, then the hydrophobicity constant π is defined by

$$\pi = \log(P/P_0). \qquad (9.2)$$

Some values of π are listed in Table 9.5. Some points to be noted are:

(i) the values of π for groups that are not strongly electron-donating or -withdrawing are virtually constant and independent of the group to which they are attached. Furthermore, their effects are additive. For example, the methylene group has a π of 0·5, and the addition of each additional methylene adds a further increment of 0·5 to π. 0·5 log units is equal to a change in Gibbs free energy of 2·84 kJ/mol (0·68 kcal/mol). (In this context, the substitution of a methyl group for a hydrogen is the same as the addition of a methylene group since it is equivalent to the interposing of a methylene group between the hydrogen and the rest of the molecule.)

(ii) The behaviour of groups that can conjugate with the benzene ring, such as the nitro and amino groups, is variable and depends on the other groups attached to the ring.

b. Hydrophobicity varies as surface area[11–14]

It has been noted that there is an empirical correlation between the surface area of a hydrophobic side chain of an amino acid and its Gibbs energy of transfer from water to an organic phase. 1 $Å^2$ of surface area gives a hydrophobic energy of 80–100 J/mol (20–25 cal/mol). (The surface of a group is defined as that surface which is the sum of the van der Waals' radii of water and the exterior atoms distant from the group.)

We can rationalize this correlation between surface area and hydrophobicity in terms of a simple model involving the energy of formation of a cavity in water. The surface tension of water is 72 dynes/cm^2, so that to form a free surface of water of 1 $Å^2$ requires 72×10^{-6} ergs or $7 \cdot 2 \times 10^{-22}$ J ($1 \cdot 72 \times 10^{-22}$ cal). Multiplying this by Avogadro's number gives a value of 435 J/$Å^2$/mol (104 cal/$Å^2$/mol). Thus, creating a cavity in water to be occupied by a hydrophobic group costs 435 J/$Å^2$/mol of surface area. This will be offset somewhat by a gain in dispersion energy from the interaction of water with the solute on filling the cavity, and this will also vary with surface area.

The energy of formation of the cavity in water is similar to the total energy change in hydrophobic bond formation and illustrates that this is the driving force of the bond in the partitioning experiments.

B. The binding energies of enzymes and substrates

It is difficult to estimate the contribution of hydrogen bonding, electrostatic linkages, and hydrophobic bonding to the overall binding energy of a substrate and enzyme by extrapolation from the simple physical measurements of the last section. The major cause of the difficulty is that the binding process is an *exchange* reaction; the substrate exhanges its solvation shell of water for the binding site of the enzyme. The net binding energy represents the *differences* between the binding energies of the substrate with water and the substrate with the enzyme.

The evaluation of the net energy of hydrogen bonding is difficult because the substrate is normally hydrogen bonded to water molecules in aqueous solution, as is the enzyme. The formation of hydrogen bonds in the enzyme–substrate complex involves the displacement of hydrogen bonded water molecules. There is thus no net gain in the number of hydrogen bonds.

$$E \cdots H - O \underset{H}{\diagdown} + S \cdots O \overset{H}{\diagup}_{\diagdown H} \rightleftharpoons E \cdots S + (H_2O)_2 \qquad (9.3)$$

But there is an increase in *entropy* on forming *intra*molecular bonds in the complex. The energy of an individual hydrogen bond is composed of a favourable attractive energy term and an unfavourable entropy term because two molecules are linked together to form one (see Chapter 2B4 and this chapter, Section C). But if the substrate is immobilized in the enzyme–substrate anyway, there is no further loss of entropy on forming intramolecular hydrogen bonds. In other words, the loss of entropy has to be 'paid for' only once. Intramolecular hydrogen bonding is favoured by the entropy gain of releasing bound water molecules. A rough estimate is that there is a gain of about 40 J/deg (10 cal/deg) per mole of water released.

The evaluation of the energy of salt linkages is difficult because the ions that are involved are solvated in solution. The solvation energies are very high, a $—CO_2^-$ ion is estimated to be stabilized by about 270 kJ/mol (65 kcal/mol).[2] Also, the energy of a salt bridge depends strongly on the dielectric constant of the surrounding medium. A further factor is that the formation of a buried salt bridge is favoured entropically since water of solvation is released (eqn (9.4)).

$$E—NH_3^+(H_2O)_m + (H_2O)_n\,^-O_2C—S \rightleftharpoons E—NH_3^+ \cdots\, ^-O_2C—S \qquad (9.4)$$
$$+$$
$$(n+m)H_2O$$

The evaluation of hydrophobic bond energies is difficult because there are fundamental differences between hydrophobic bonds measured by partitioning experiments of a solute between an organic solvent and water, and that formed by binding to a protein. The transfer of a solute from the aqueous to the organic phase may be divided up into the notional steps: formation of cavity in organic phase; transfer of solute to the cavity; closing of cavity left in the aqueous phase. The transfer of a hydrophobic substrate to a hydrophobic region in an enzyme involves the occupation of a *preformed* cavity, and probably the transfer of water from this to the aqueous phase.

Rather than use model reactions, it is possible to determine the contributions of the various factors to the binding energy of an enzyme and substrate from direct measurements on enzymic reactions.

1. Estimation of increments in binding energy from kinetics

One way of measuring the contribution of a substituent R in a substrate R–S to binding is to compare the dissociation constants of R–S and H–S from the enzyme. But, as will be seen in the next chapter, this often underestimates the binding energy of the larger substrate since enzymes frequently use binding energy to lower the activation energies of reactions rather than give tighter K_Ms. A much better method is to compare the values of k_{cat}/K_M for the two substrates. This quantity includes both the activation energies and binding energies and avoids the underestimates in using dissociation constants alone. It is shown in eqns (10.2) and (10.10) that

$$\ln(k_{cat}/K_M) = \ln(kT/h) - (\Delta G_0^{\ddagger} + \Delta G_b) \qquad (9.5)$$

(where kT/h is constant at a given temperature T, $\Delta G_0^{\ddagger}$ is the inherent activation energy due to the electronic effects of bond making and breaking, and ΔG_b is the intrinsic binding energy of the substrate to the enzyme). For substrates that are similar, differences in $\Delta G_0^{\ddagger}$ can sometimes be ignored or corrected for by comparison of reactivities in nonenzymic reactions.

2. Estimation of binding energies of small groups using the aminoacyl-tRNA synthetases

The aminoacyl–tRNA synthetases have evolved to discriminate as accurately as possible between their specific amino acids and smaller competitors (Chapter 11). They presumably maximize the binding energies of the groups that are different between the sets of amino acids. The differences in binding energy of the specific substrate and the smaller competitor may be obtained by comparing the values of k_{cat}/K_M for the pyrophosphate exchange reaction (Chapter 7).

a. Alkyl groups

From the data listed in Table 9.6, comparing isoleucine and valine reacting with the isoleucyl-tRNA synthetase, and valine and α-aminobutyric acid with the valyl-tRNA synthetase, it is seen that the additional methylene group on the larger substrates contributes 12–16 kJ/mol (3–3·8 kcal/mol) of binding energy. This is far larger than the value of 2·85 kJ/mol (0·68 kcal/mol) for the transfer of a methylene group from water to n-octanol. Comparing valine with alanine gives a difference in binding energy of 22·6 kJ/mol (5·41 kcal/mol) between the isopropyl and methyl groups, compared with the value of 4·6 (1·1) for the water to n-octanol transfer. The hydrophobic binding energies with the enzyme are five times larger than those in the solution studies.

b. Hydrogen bonding

The tyrosyl-tRNA synthetase binds tyrosine at least 28 000 times more tightly than phenylalanine. There are presumably hydrogen bonding sites in the protein that bind either a water molecule in the absence of substrate or the phenolic hydroxyl of tyrosine. The binding of phenylalanine must require either the displacement of the bound water, leaving two hydrogen bonding sites unbound, or the distortion of the binding pocket. The difference in binding energy of 25·5 kJ/mol (6·1 kcal/mol) reflects both the advantage of forming the hydrogen bonds

TABLE 9.6. *Determination of binding energies from aminoacyl–tRNA synthetases*

Amino acid $(R—CH(NH_3^+)CO_2^-)$ R—	Aminoacyl–tRNA synthetase	k_{cat}^a	K_M (mM)	k_{cat}/K_M relative[a]	$\Delta\Delta G^{\ddagger}$ kJ/mol	(kcal/mol)
CH_3CH_2 $\backslash$ $CH—$ $/$ CH_3	Isoleucyl	1 1	0·005 0·005	1 1		
CH_3 $\backslash$ $CH—$ $/$ CH_3	Isoleucyl	0·77 0·56	0·8 0·39	0·005 0·0072	13·1 12·2	3·14 2·92
CH_3 $\backslash$ $CH—$ $/$ CH_3	Valyl	1	0·14	1		
$CH_3CH_2—$ $CH_3—$	Valyl Valyl	0·36 0·03	30 39	0·0017 $1·1\times10^{-4}$	15·8 22·6	3·78 5·41
$HO—\langle\bigcirc\rangle—CH_2—$	Tyrosyl	1	0·0018	1		
$\langle\bigcirc\rangle—CH_2—$	Tyrosyl	1	50	$3·6\times10^{-5}$	25·3	6·06

[a] Relative to that for the specific substrate in the pyrophosphate exchange reaction
F. H. Bergmann, P. Berg, and M. Dieckmann, *J. biol. Chem.* **236**, 1735 (1961); R. B. Loftfield and E. A. Eigner, *Biochim. biophys. Acta* **130**, 426 (1966); S. L. Owens and F. E. Bell, *J. biol. Chem.* **245**, 5515 (1970); A. R. Fersht, R. S. Mulvey, and G. L. E. Koch, *Biochemistry* **14**, 13 (1975)

between the enzyme and tyrosine, and the disadvantages of binding phenylalanine.

c. Salt linkages

The *binding* of various deaminated amino acids to the aminoacyl-tRNA synthetases has been compared with the corresponding amino acids. The ratios of the binding constants may be used to give a *minimum* estimate of the binding energy of the ammonium ion.[15–17]

It is seen from (9.6), (9.7), and (9.8) that the $—NH_3^+$ group contributes a binding energy of about 17 kJ/mol (4 kcal/mol). This is presumably due to the formation of a salt linkage with an enzyme carboxylate ion.

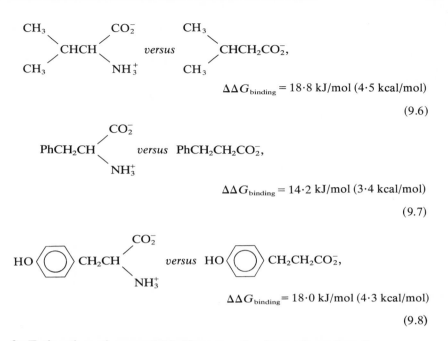

$$\Delta\Delta G_{\text{binding}} = 18\cdot8 \text{ kJ/mol (4·5 kcal/mol)}$$

$$(9.6)$$

$$\Delta\Delta G_{\text{binding}} = 14\cdot2 \text{ kJ/mol (3·4 kcal/mol)}$$

$$(9.7)$$

$$\Delta\Delta G_{\text{binding}} = 18\cdot0 \text{ kJ/mol (4·3 kcal/mol)}$$

$$(9.8)$$

3. Estimation of general binding energies from chymotrypsin

The aminoacyl-tRNA synthetases are unusual in their requirements for extreme specificity. The binding energies obtained from them probably represent the extreme values possible. Some lower values may be obtained from chymotrypsin, an enzyme of broad specificity.

a. Alkyl groups

The values of k_{cat}/K_M for the chymotrypsin-catalysed hydrolysis of a series of esters of the form R-CH(NHAc)CO$_2$Me, where R is an unbranched alkyl chain, increase with increasing hydrophobicity of R.[18] The decrease in the activation energy is 2·2 times greater than the free energy of transfer of the alkyl groups from water to n-octanol (Fig. 9.2).[19] The hydrophobic binding pocket appears to be 2·2 times more hydrophobic than n-octanol.

The inhibition constants of a series of substituted formanilides increase with increasing hydrophobicity. A plot of the logarithms of the constants against π yields a straight line of slope $-1\cdot5$.[20] This again shows that the active site of chymotrypsin is more hydrophobic than n-octanol (Fig. 9.3).

b. Salt linkage

The catalytically active conformation of chymotrypsin is stabilized by a salt bridge between the α-NH$_3^+$ group of Ile-16 and the —CO$_2^-$ of Asp-194. This conformation is in equilibrium with another in which the

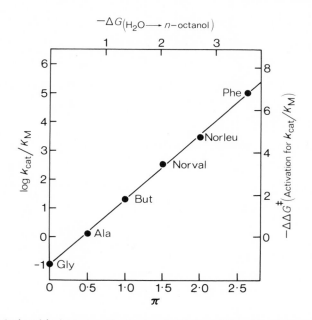

Fig. 9.2. Relationship between the hydrophobicity of the side chain of the amino acid and k_{cat}/K_M for the hydrolysis of N-acetyl-L-amino acid methyl esters by chymotrypsin (data from ref. 16). Energies are in kcal/mol

α-NH_3^+ group is free in solution. By measuring the equilibrium constants between the two forms at high pH where the amino group is unprotonated, and low pH where it is in the —NH_3^+ form, it is found that the salt bridge stabilizes the active conformation by 12·1 kJ/mol (2·9 kcal/mol).[21] This is a buried salt bridge. Those on the surface of haemoglobin have lower stabilization energies.[22]

4. Why are enzymes more hydrophobic than organic solvents?

One reason why chymotrypsin is more hydrophobic than n-octanol is that indicated earlier. The binding pocket of chymotrypsin is present in the absence of substrate, unlike the cavity that has to be made in an organic solvent to accommodate the solute transferred from the aqueous phase. Furthermore it contains some 16 or so molecules of water. Thus hydrophobic bond formation with chymotrypsin is like forming *two* 'normal' hydrophobic bonds, since there are two energetically unfavourable water–hydrophobic interfaces, one for the substrate and one for the enzyme-bound water.

The tight binding of methylene groups by the isoleucyl- and valyl-tRNA synthetases is due to an additional reason. When valine occupies the binding site of the isoleucyl-tRNA synthetase there is a 'hole' in the

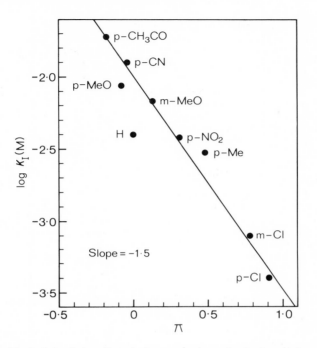

FIG. 9.3. Variation of the dissociation constants of substituted formanilides with π for the partition between n-octanol and water

complex that would be occupied by the additional methylene group in isoleucine if it were bound. It was pointed out earlier that the dispersion energy of a methylene group in a crystalline hydrocarbon is 8.4 kJ/mol (2 kcal/mol). Since proteins are as closely packed as solids,[23] an empty 'hole' adds a further 8.4 kJ/mol to the hydrophobic bond energy of the methylene group.

5. Summary

The dissociation constant of an enzyme and substrate reflects the *relative* stabilities of the substrate bound to the enzyme and free in solution. It depends on the relative strengths of the hydrogen bonding of the enzyme and substrate to each other compared with their separate hydrogen bonding to water molecules, the relative stabilities of the salt linkages between the enzyme and substrate compared with the solvation of the individual ions by water, the relative dispersion energies compared with water, and also the hydrophobic bonding. These differences in energy, summarized in Table 9.7, summed over a whole molecule can be quite considerable. It is of interest that the hydrogen bond and salt linkage with

TABLE 9.7. *Binding energies of groups on substrates*

Group	Enzyme	Contribution to binding energy kJ/mol	kcal/mol
—CH$_2$—	Aminoacyl-tRNA Synthetases[a]	15	3·5
HO— (2 hydrogen bonds)	Aminoacyl-tRNA Synthetases[a]	25	6
—NH$_3^+$ · · · $^-$O$_2$C—	Aminoacyl-tRNA Synthetases[a]	17	4
—CH$_2$—	Chymotrypsin	8	2
—CH$_2$—CH$_2$—	Chymotrypsin	16	3·8
—(CH$_2$)$_3$—	Chymotrypsin	23	5·6
—(CH$_2$)$_4$—	Chymotrypsin	29	6·8
PhCH$_2$—[b]	Chymotrypsin	34	8·2

[a] These represent the maximum values
[b] Relative to the hydrogen atom

the enzyme as well as the hydrophobic bond are favoured entropically by the release of constrained water molecules.

The *absolute* energy of interaction between the enzyme and substrate depends on the dispersion forces and the absolute energies of the hydrogen bonds and salt linkages. Of these, the hydrogen bonds and the salt linkages provide the strongest attractive forces. Since these are also able to stabilize unfavourable charge formation in the transition state, they are especially important in catalysis.

C. Entropy and binding[24]

The pairing of monomeric complementary bases, such as *A* and *U*, is not detectable in aqueous solution due to the competition of hydrogen bonding with water molecules. However, a triplet of bases binds strongly to a complementary anticodon of a tRNA. The triplet *UUC* binds to tRNA^{Phe} with an association constant of 2×10^3 M^{-1}.[25,26] Two tRNA molecules with complementary anticodons associate even more strongly; the association constant between tRNA^{Phe} and tRNA$_2^{\text{Glu}}$ is 2×10^7 M^{-1}.[27] The reason for this increase in the strength of hydrogen bonding is entropy. When a single *A* and *U* associate, they gain the energy of the complementary base pairing but lose the energy of hydrogen bonding with water. They also lose their independent entropies of translation and overall rotation, but in turn there is a gain of entropy as the hydrogen-bonded water molecules are released. When a complementary pair of triplets bind, three times as many molecules of water are freed from hydrogen bonding, but there is still the loss of only one set of overall

TABLE 9.8. *The chelate effect on binding amines to Ni^{2+}*

Ligand	Association constants (M^{-1})					
	K_1	K_2	K_3	K_4	K_5	K_6
NH_3	468	132	41	12	4	0·8
$H_2N(CH_2)_2NH_2$	2×10^7	1×10^6	2×10^4			
$H_2N(CH_2)_2NH(CH_2)_2NH_2$	6×10^{10}	1×10^8				
$H_2N(CH_2)_2NH(CH_2)_2NH(CH_2)_2NH_2$	2×10^{14}					
$(H_2N(CH_2)_2NH(CH_2)_2)_2NH$	3×10^{17}					

The Chemical Society (London) Special Publications 17 and 25 (1964, 1971)

entropies of translation and rotation. The situation is very reminiscent of that of comparing the advantages of an intramolecular reaction over its intermolecular counterpart (Chapter 2B4d). The lesson is that, although single hydrogen bonds are weak in solution, multiple hydrogen bonds may be very stable.

A related phenomenon is that the binding of a dimer X-Y to an enzyme may be far greater than expected from the binding of X and Y separately because the dimer loses only one set of translational and overall rotational entropies.

This phenomenon has been known for many years to inorganic chemists as the *chelate effect*. The magnitude may be illustrated by one of their examples, the replacement by ammonia and polyamines of some or all of the six water molecules that are coordinated to the Ni^{2+} ion. It is seen in Table 9.8 that there are enormous increases in the association constants of the ligands as the number of amino groups increases.

D. Protein–protein interactions

The interfaces of multisub-unit proteins, such as haemoglobin, and of protein–protein complexes, such as the trypsin–trypsin inhibitor complex, are close packed and snug-fitting. All hydrogen bonds and salt linkages are paired. In other words, the two surfaces are complementary to each other. This would appear to be the basis of protein–protein recognition.

Although opinions differ on the overall contribution of hydrogen bonding and salt linkage to the total stabilization energy of protein–protein complexes and it has been suggested that many association constants are entirely accounted for by hydrophobic bonding, there is no doubt that hydrogen bonds and electrostatic interactions are important for specificity.[28-30] Whatever the positive contribution of correctly formed hydrogen bonds and salt bridges in the 'correct' complexes, the presence of *unpaired* hydrogen bond donors/acceptors and ions in 'incorrect' complexes provides considerable driving energy for their dissociation. It was seen from the experiments with the aminoacyl-tRNA synthetases that the lack of the hydrogen bonding hydroxyl on phenylalanine compared with tyrosine causes the binding to the tyrosyl-tRNA synthetase to be weakened by at least 25 kJ/mol (6 kcal/mol). Similarly, the disruption of the salt bridge between the α-ammonium ion of the substrate and the enzyme weakens the binding by about 17 kJ/mol (4 kcal/mol). These results may be extrapolated to the interactions at the interfaces of proteins. For example, the substitution of phenylalanine for the arginine of the soya bean trypsin inhibitor that binds in the specificity pocket of trypsin and forms a salt linkage with the buried aspartate weakens the binding by 13 kJ/mol (3·1 kcal/mol).[31] Further, the pancreatic trypsin inhibitor binds to chymotrypsin, but with 25 kJ/mol (6 kcal/mol) less binding energy than to trypsin.[32,33] This difference reflects mainly the burying of the positively charged lysine side chain in the hydrophobic binding pocket of chymotrypsin.

References

1 G. E. Ewing, *Acct. Chem. Res.* **8,** 185 (1975).
2 A. Warshel and M. Levitt, *J. Molec. Biol.* **103,** 227 (1976).
3 M. Levitt and S. Lifson, personal communication.
4 A. T. Hagler, S. Lifson, and E. Huler in *Peptides, polypeptides, and proteins* (eds. E. R. Blout, F. A. Bovey, M. Goodman, and N. Lotan), John Wiley & Sons, p 35 (1974).
5 A. T. Hagler, E. Huler, and S. Lifson, *J. Am. chem. Soc.* **96,** 5319 (1974).
6 A. Johansson, P. Kollman, S. Rothenberg, and J. McKelvey, *J. Am. chem. Soc.* **96,** 3794 (1974).
7 W. Kauzmann, *Adv. Prot. Chem.* **14,** 1 (1959).
8 M. H. Klapper, *Progr. Bioorg. Chem.* **2,** 55 (1973).
9 T. Fujita, J. Iwasa, and C. Hansch, *J. Am. chem. Soc.* **86,** 5175 (1964).
10 A. Leo, C. Hansch, and D. Elkins, *Chem. Revs* **71,** 525 (1971).
11 R. B. Hermann, *J. phys. Chem.* **76,** 2754 (1972).
12 M. J. Harris, T. Higuchi, and J. H. Rytting, *J. phys. Chem.* **77,** 2694 (1973).
13 J. A. Reynolds, D. B. Gilbert, and C. Tanford, *Proc. natn. Acad. Sci. U.S.A.* **71,** 2925 (1974).
14 C. Chothia, *Nature, Lond.* **248,** 338 (1974).
15 S. L. Owens and F. E. Bell, *J. biol. Chem.* **245,** 5515 (1970).
16 D. V. Santi and V. A. Pena, *J. med. Chem.* **16,** 273 (1973).
17 R. Mulivor and H. P. Rappaport, *J. molec. Biol.* **76,** 123 (1973).
18 J. R. Knowles, *J. theoret. biol.* **9,** 213 (1965).
19 V. N. Dorovskaya, S. D. Varfolomeyev, N. F. Kazanskaya, A. A. Klyosov, and K. Martinek, *FEBS Letts.* **23,** 122 (1972).
20 J. Fastrez and A. R. Fersht, *Biochemistry* **12,** 1067 (1973).
21 A. R. Fersht, *J. molec. Biol.* **64,** 497 (1972).
22 M. F. Perutz, *Nature, Lond.* **228,** 726 (1970).
23 M. I. Page, *Biochem. biophys. Res. Comm.* **72,** 456 (1976).

24 W. P. Jencks, *Adv. Enzymol.* **43,** 219 (1975).
25 J. Eisinger, B. Feuer, and T. Yamane, *Nature New Biology, Lond.* **231,** 126 (1971).
26 O. Pongs, R. Bald, and E. Reinwald, *Eur. J. Biochem.* **32,** 117 (1973).
27 J. Eisinger and N. Gross, *Biochemistry* **14,** 4031 (1975).
28 W. P. Jencks, *Catalysis in chemistry and enzymology.* McGraw-Hill Book Co., pp 351, 399, (1969).
29 C. Chothia and J. Janin, *Nature, Lond.* **256,** 705 (1975).
30 J. Janin and C. Chothia, *J. molec. Biol.* **100,** 197 (1976).
31 M. Laskowski Jr., personal communication.
32 U. Quast, J. Engel, H. Heumann, G. Krause, and E. Steffen, *Biochemistry* **13,** 2512 (1974).
33 J.-P. Vincent, M. Peron-Renner, J. Pudles, and M. Lazdunski, *Biochemistry* **13,** 4205 (1974).

Further reading

W. P. Jencks, *Catalysis in chemistry and enzymology.* McGraw-Hill Book Co., (1969), Chapters 6–9.
W. P. Jencks, Binding energy, specificity, and enzymic catalysis—the Circe effect, *Adv. Enzymol.* **43,** 219–410 (1975).

Chapter 10

Enzyme–substrate complementarity and theories of enzyme catalysis

A. Utilization of enzyme–substrate binding energy in catalysis

It was seen in Chapter 2 that a combination of entropic factors, acid–base catalysis, and electrostatic effects can account for a large fraction of the magnitude of the catalysis by some enzymes. In addition to these catalytic factors, the one outstanding characteristic of enzymes is that they specifically bind their substrates, and the binding energies involved may be very large. Ever since Haldane suggested in 1930 that these energies may be used to distort the substrate to the structure of the products,[1] theoreticians have explored the various ways in which the binding energy of the enzyme and substrate may be used to lower the activation energy of the chemical steps.

Transition-state theory is particularly useful in analysing theories of enzyme catalysis. We shall now apply this approach to the simple Michaelis–Menten mechanism (where $K_M = K_S$, see Chapter 3) to see how the binding energy automatically lowers the activation energy of k_{cat}/K_M and how some of the binding energy may be used to lower the activation energy of k_{cat}.[2]

1. Binding energy lowers the activation energy for k_{cat}/K_M

$$E+S \underset{\Delta G_S}{\overset{K_M}{\rightleftharpoons}} ES \overset{k_{cat}}{\underset{\Delta G^{\ddagger}}{\longrightarrow}} products \tag{10.1}$$

$$E+S \underset{\Delta G_T^{\ddagger}}{\overset{k_{cat}/K_M}{\rightleftharpoons}} ES^{\ddagger} \tag{10.2}$$

It will be recalled from Chapter 3 that the rate constant for the free enzyme reacting with the free substrate to give products is k_{cat}/K_M. Expressed in terms of transition-state theory (Chapter 2), the equilibrium constant between the transition state $ES^{\ddagger}$ and $E+S$ is proportional to the activation energy $\Delta G_T^{\ddagger}$ of k_{cat}/K_M (eqn (10.2)). This activation energy is

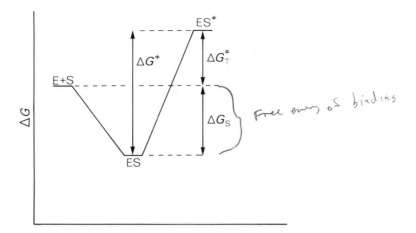

FIG. 10.1. Gibbs energy changes for the scheme

$$E + S \underset{K_S}{\rightleftharpoons} ES \xrightarrow{k_{cat}} products$$

(ΔG_S is algebraically negative and $\Delta G^{\ddagger}$ and $\Delta G^{\ddagger}_T$ positive)

composed of two terms, an energetically unfavourable term $\Delta G^{\ddagger}$, due to the activation energy of the chemical steps of bond-making and -breaking, and a compensating energetically favourable term ΔG_S, due to the realization of the binding energy. That is,

$$\Delta G^{\ddagger}_T = \Delta G^{\ddagger} + \Delta G_S \qquad (10.3)$$

(where ΔG_S is algebraically negative). This is illustrated in Fig. 10.1 for the simple Michaelis–Menten mechanism.

Substituting eqn (10.3) into eqn (2.5) to express k_{cat}/K_M in terms of transition-state theory gives

$$RT \ln(k_{cat}/K_M) = RT \ln(kT/h) - \Delta G^{\ddagger} - \Delta G_S. \qquad (10.4)$$

2. Interconversion of binding and chemical activation energies

The maximum binding energy between an enzyme and a substrate occurs when each binding group on the substrate is matched by a binding site on the enzyme. In this case the enzyme is said to be complementary in structure to the substrate. Since the structure of the substrate changes throughout the reaction, becoming first the transition state and then the products, the structure of the *undistorted* enzyme can be complementary to only one form of the substrate. It may be shown that it is catalytically advantageous for the enzyme to be complementary to the structure of the transition state of the substrate rather than the original structure. If this happens, the increase in binding energy as the structure changes to that of the transition state lowers the activation energy of k_{cat}. Conversely, if the enzyme is complementary to the structure of the unaltered substrate, the decrease in binding energy on the formation of the transition state will increase the activation energy of k_{cat}.

In the next section we shall analyse the consequences of enzyme transition-state complementarity using the procedure of dividing up activation energies into components due to chemical terms and those due to changes in binding energy as the reaction proceeds.

3. Enzyme complementary to transition state implies k_{cat}/K_M is at a maximum

The idea of complementarity in enzyme–substrate interactions was introduced by Fischer with his famous 'lock and key' analogy.[3] In modern terminology this would represent enzyme–substrate complementarity. The currently favoured concept of enzyme–transition state complementarity was introduced by Haldane[1] and elaborated by Pauling.[4]

a. Enzyme complementary to initial substrate
Suppose the maximum amount of intrinsic binding energy available is ΔG_b. In this case it is realized in the initial enzyme–substrate complex so that binding will be good; that is K_M, the dissociation constant of the enzyme–substrate complex, will be low. But the formation of the transition state will lead to a reduction in binding energy as the substrate geometry changes to give a poorer fit, and so will lower k_{cat}. If the adverse energy change caused by the poorer fit is ΔG_R and the free energy of activation due to the chemical bond-making and -breaking involved in k_{cat} is $\Delta G_0^{\ddagger}$, then the observed free energy of activation for k_{cat} is given by

$$\Delta G^{\ddagger} = \Delta G_0^{\ddagger} + \Delta G_R \tag{10.5}$$

and

$$\Delta G_S = \Delta G_b. \tag{10.6}$$

The Gibbs energy of activation for k_{cat}/K_M is given by $\Delta G^{\ddagger} + \Delta G_S$, i.e.

$$\Delta G_T^{\ddagger} = \Delta G_0^{\ddagger} + \Delta G_R + \Delta G_b. \tag{10.7}$$

b. Enzyme complementary to transition state
Here the full binding energy ΔG_b is realized in the transition state. There will be an adverse energy term ΔG_R in the initial enzyme–substrate complex which will increase K_M, but the gain in binding energy as the reaction reaches the transition state will increase k_{cat}. Thus,

$$\Delta G^{\ddagger} = \Delta G_0^{\ddagger} - \Delta G_R, \tag{10.8}$$

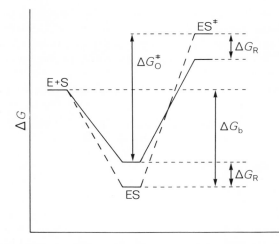

FIG. 10.2. Gibbs energy changes for the scheme in Fig. 10.1. When the enzyme is complementary in structure to (a) the substrate (broken line) and (b) the transition state (solid line). (ΔG_S is algebraically negative and ΔG_0 and ΔG_R positive)

and

$$\Delta G_S = \Delta G_b + \Delta G_R. \qquad (10.9)$$

Again the Gibbs energy of activation for k_{cat}/K_M is given by

$$\Delta G_T^{\ddagger} = \Delta G^{\ddagger} + \Delta G_S,$$

i.e.

$$\Delta G_T^{\ddagger} = \Delta G_0^{\ddagger} + \Delta G_b, \qquad (10.10)$$

and ΔG_R cancels out.

Comparison of eqns (10.7) and (10.10) shows that k_{cat}/K_M is higher for the enzyme being complementary to the transition state rather than to the initial substrate by a factor of $\exp(\Delta G_R/RT)$.

It should be noted that k_{cat}/K_M is independent of the interactions in the initial enzyme-substrate complex as the term ΔG_R drops out of the equations.

4. Experimental evidence for the utilization of binding energy in catalysis and enzyme–transition state complementarity

a. Binding energy
Some of the most instructive evidence comes from kinetic experiments on the serine proteases. It will be recalled from Chapter 1 that these enzymes have a series of sub-sites for binding the amino acid residues of their polypeptide substrates. It is seen in Table 10.1 that as larger groups occupy the leaving group site in chymotrypsin, their binding energy is

used to increase k_{cat}/K_M. Similarly, increasing the length of the polypeptide chain of substrates of elastase increases k_{cat}/K_M. Interestingly, in the examples given, the binding energy of the additional groups does not lower K_M, i.e. is not used for binding the substrate, but is used to increase k_{cat}.

Similar behaviour is observed with pepsin. It is seen in Table 10.1 that

TABLE 10.1. *Interconversion of activation and binding energies*

Enzyme and substrate	k_{cat} (s^{-1})	K_M (mM)	k_{cat}/K_M (s^{-1} M^{-1})
(a) *Chymotrypsin*[a]			
AcTyr-NH$_2$	0·17	32	5
AcTyr-GlyNH$_2$	0·64	23	28
AcTyr-AlaNH$_2$	7·5	17	440
AcProTyr-GlyNH$_2$	4·4	32	140
AcPhe-NH$_2$	0·06	31	2
AcPhe-GlyNH$_2$	0·14	15	10
AcPhe-AlaNH$_2$	2·8	25	114
AcProPhe-GlyNH$_2$	0·76	15	51
(b) *Elastase*[b]			
AcAlaProAla-NH$_2$	0·09	4·2	21
AcProAlaProAla-NH$_2$	8·5	3·9	2200
AcGlyProAla-NH$_2$	0·02	33	0·5
AcProGlyProAla-NH$_2$	2·8	43	64
(c) *Pepsin*[c] (cleavage of Phe-Phe bond in APhePheOP4P)			
PheGly	0·5	0·3	$1·7 \times 10^3$
ZPheGly	25	0·11	$2·2 \times 10^5$
ZAlaGly	145	0·25	$5·8 \times 10^5$
ZAlaAla	282	0·04	7×10^6
ZGlyAla	409	0·11	$3·7 \times 10^6$
ZGlyIle	13	0·07	$1·8 \times 10^5$
ZGlyLeu	134	0·03	$4·2 \times 10^6$
PheGlyGly	6	0·6	1×10^4
ZPheGlyGly	127	0·13	$9·8 \times 10^5$
Mns[d]	0·002	0·1	20
MnsGly[d]	0·13	0·03	$3·7 \times 10^3$
MnsGlyGly[d]	16	0·07	$2·3 \times 10^5$
MnsAlaAla[d]	112	0·06	2×10^6

[a] W. K. Baumann, S. A. Bizzozero, and H. Dutler, *FEBS Letts.* **8**, 257 (1970); *Eur. J. Biochem.* **39**, 381 (1973). 25°, pH 7·9
[b] R. C. Thompson and E. R. Blout, *Biochemistry* **12**, 51 (1973). 37°, pH 9
[c] G. P. Sachdev and J. S. Fruton, *Biochemistry* **9**, 4465 (1970). (0P4P = 3-(4-pyridyl)propyl-1-oxy) 37°, pH 3·5
[d] G. P. Sachdev and J. S. Fruton, *Proc. natn. Acad. Sci. U.S.A.* **72**, 3424 (1975). (Mns = mansyl) 25°, pH 2·4

a wide range of substrates are hydrolysed with K_Ms of about 0.1 mM. The additional binding energy of the groups on the larger substrates is again used to increase k_{cat} rather than decrease K_M. k_{cat}/K_M is accordingly higher for the larger substrates.

A striking series of examples occurs with the chymotrypsin-catalysed hydrolysis of synthetic ester substrates (Table 10.2). As the size of the hydrophobic side chain that fits into the hydrophobic primary binding site of the enzyme is increased, k_{cat}/K_M increases over a range of 10^6. The increase in the binding energy of the larger substrates is distributed between lowering the dissociation constant of the enzyme–substrate complex and increasing the acylation and deacylation rate constants.

It will be seen in Section B how this use of binding energy to increase k_{cat} rather than lower K_M gives higher reaction rates.

TABLE 10.2. *Kinetic parameters for the hydrolysis of N-acetyl amino acid methyl esters by chymotrypsin*[a]

$$(E + RCO_2Me \underset{\rightleftharpoons}{\overset{K_s}{}} E.RCO_2Me \xrightarrow{k_2} RCO\text{-}E \xrightarrow{k_3} E + RCO_2H)$$

Amino acid	k_2 (s^{-1})	k_3 (s^{-1})	K_s (mM)	k_{cat}/K_M ($s^{-1} M^{-1}$)
Gly	0.49	0.14	3380	0.13
But	8.8	1.7	417	21
Norval	35.6	5.93	100	360
Norleu	103	19	34	3×10^3
Phe	796	111	7.6	1×10^5
Tyr[b]	5000	200	17	3×10^5

[a] α-chymotrypsin at 25°, pH 7.8 and 25°. (Data from Table 7.3)
[b] Ethyl ester

b. Transition-state complementarity

Direct evidence for enzyme–transition state complementarity has come from X-ray diffraction experiments on the serine proteases and lysozyme (see Section C4 of this chapter and also Chapter 1), and also from studies on the binding of transition state analogues. This approach was suggested by Pauling in the 1940s but has only recently come into prominence.[4]

5. Transition-state analogues—probes of complementarity[5,6]

The chemist, with his knowledge of organic reaction mechanisms, can guess at the structure of the transition state of an enzymic reaction. Compounds which mimic the transition state may then be synthesized and their binding to the enzyme compared with that of the substrate.

a. Lysozyme and glucosidase

Lysozyme catalyses the hydrolysis of the polysaccharide component of plant cell walls and synthetic polymers of $\beta(1-4)$ linked units of N-acetylglucosamine (NAG)—see Chapter 1. During the hydrolytic reaction it is expected from studies on non-enzymic reactions that a carbonium ion is formed in which the conformation of the glucopyranose ring changes from a full chair to a sofa conformation (see Chapter 1). The transition state analogue (I), in which the lactone ring mimics the carbonium ion-like transition state (II), binds tightly to lysozyme, $K_{diss} = 8\cdot3\times10^{-8}$ M.[7]

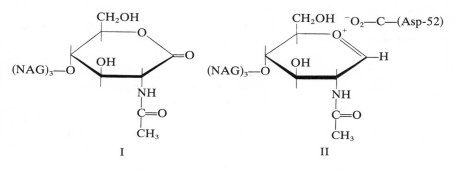

I   II

This may be compared with the dissociation constants of 10^{-5} M and 5×10^{-6} M for NAG$_4$ and NAG-NAM-NAG-NAG binding in the ABCD sub-sites.[7,8] The 100-fold tighter binding of the transition state analogue may be due in part to the electrostatic interaction of the negatively charged Asp-52 with the partial positive charge on the carbonyl carbon of the lactone.

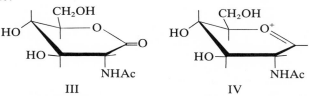

III   IV

Lactone   Carbonium Ion
(sofa conformation)   (sofa conformation)

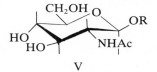

V

Pyranoside
(full chair)

More striking is the binding of the lactone (III) to β-N-acetyl-D-glucosaminidase. The dissociation constant of 5×10^{-7} M is 4000 times smaller than the K_M of 2×10^{-3} M of a pyranoside substrate.[9]

b. Proline racemase
During the racemization of proline (VI) the chiral carbon must at some stage become trigonal. Accordingly, both (VII) and (VIII) bind 160 times more tightly than proline.[10,11]

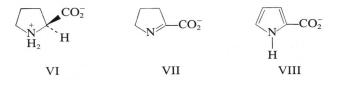

VI	VII	VIII

c. Cytidine deaminase
The dissociation constant of tetrahydrouridine (IX) is about 10 000 times smaller than the combined constants for the reaction products, uridine (X) and ammonia. Tetrahydrouridine presumably resembles the tetrahedral intermediate (XI).[12]

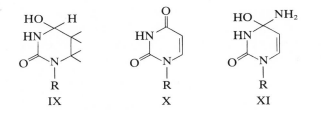

IX	X	XI

d. Assessment of the results of binding transition-state analogues
The transition-state analogues that have been designed so far give a measure of that part of the catalysis that is due to the difference in complementarity of the enzyme for the transition state and substrate. In the four examples given above the transition-state analogues bind between 10^2 and 10^4 times more tightly than the original substrates. This is good evidence that enzymes have evolved to be complementary in structure to the transition state. Furthermore, it shows that k_{cat} may be increased by a factor of 10^2–10^4 at the expense of increasing K_M. Bearing in mind that these synthetic analogues might be extremely inadequate in mimicking the transition-state structure, it is clear that the increase in complementarity between the transition state and the substrate may be worth at least 20 kJ/mol (5 kcal/mol). It will be seen in Chapter 12 that the binding site for the carbonyl oxygen of a substrate of a serine protease is deficient by one hydrogen bond. On forming the transition state for the

reaction the additional hydrogen bond is made. This increase in complementarity must also be worth a factor of about 20–25 kJ/mol (5–6 kcal/mol).

e.   Some possible errors in interpreting transition-state analogue data

The effects of enzyme–transition state complementarity on the binding of transition-state analogues may be masked by extraneous binding artefacts. In Chapter 9 it was seen that small groups can involve large binding energies when a specific binding site is involved. For example, a methylene group may contribute 12 kJ/mol (3 kcal/mol), and a hydroxyl in a hydrogen bond 25 kJ/mol (6 kcal/mol), to the binding energy. Also, where specific binding sites are not involved, fairly large energies may come from general hydrophobic effects; the substitution of a chloro group for an acetyl on a phenyl ring causes it to bind 50 times more tightly to the hydrophobic pocket of chymotrypsin.[13]

Difficulties also arise when dealing with analogues for multisubstrate reactions due to the chelate effect. It was pointed out in the last chapter that multidentate ligands, such as EDTA, bind tightly to metal ions whilst unidentate ligands do not. The difference is due to entropy; the binding of six unidentate ligands leads to the loss of six sets of translational and rotational entropies. The same applies to the binding of a 'multisubstrate' or 'multiproduct' analogue to an enzyme. If A and B bind separately and adjacently to an enzyme with Gibbs energies of association of x and y kJ respectively, then if A–B binds in an identical manner, its free energy of association is given by

$$\Delta G_{ass} = x + y + S \tag{10.11}$$

where S is an energetically favourable term due to only one set of entropies being lost on the binding of A–B compared with two on the binding of A and B separately. The binding of multisubstrate analogues

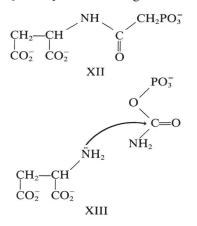

should be very tight without any effects due to enzyme–transition state complementarity. For example, the binding of (XII) to aspartate transcarbamylase is very tight, $K_{diss} = 2\cdot7 \times 10^{-8}$ M. However, this value is only equal to the product of the dissociation constants of succinate and carbamoyl phosphate ($9 \times 10^{-4} \times 2\cdot7 \times 10^{-5}$ M^{-2}), whose reaction it mimics (XIII).[14]

B. Evolution of maximum rate: strong binding of transition state—weak binding of substrate

In the last section it was seen that enzymes have evolved to bind the transition states of substrates more strongly than the substrates themselves. It will now be seen that it is catalytically advantageous to bind substrates *weakly*.

Although enzyme–transition state complementarity maximizes k_{cat}/K_M, this is not a sufficient criterion for the maximization of the overall reaction rate. This is because the maximum reaction rate for a particular concentration of substrate depends on the individual values of k_{cat} and K_M. It is seen in Table 10.3, where some rates are calculated for various values of k_{cat} and K_M subject to k_{cat}/K_M being kept constant, that maximal rates are obtained for K_M greater than [S]. The maximization of

TABLE 10.3. *Illustration of the importance of the evolution of k_{cat} and K_M at constant k_{cat}/K_M and [S]*[a]

($k_{cat}/K_M = 10^6$ M^{-1} s^{-1}, [S] = 10^{-3} M)

K_M (M)	k_{cat} (s^{-1})	Rate[b] (s^{-1})
10^{-6}	1	1
10^{-5}	10	9
10^{-4}	10^2	90
10^{-3}	10^3	500
10^{-2}	10^4	909
10^{-1}	10^5	990
1	10^6	999

[a] The notional processes are: (a) The enzyme has evolved to be complementary to the transition state of the substrate so that k_{cat}/K_M is maximized; (b) whilst maintaining (a), the enzyme evolves to *increase* K_M. The values assigned to k_{cat}/K_M and [S] are arbitrary
[b] Mol of product produced per mol of enzyme per second

rate requires *high* values of K_M. That is, enzymes should have evolved to bind substrates weakly.

1. The principle of maximization of K_M at constant $k_{cat}/K_M{}^2$

This principle contradicts the widely-held belief that strong binding, or low K_M, is an important component of enzymic catalysis. The two additional proofs that follow emphasize the importance of high K_Ms. The graphical illustration indicates the physical reason for this.

a. Graphical illustration of the importance of high K_Ms

Suppose, as on the left-hand side of Fig. 10.3, the substrate is at a higher concentration than the K_M for the reaction. The enzyme–substrate complex is at a lower energy than the free enzyme and substrate, and the activation energy is $\Delta G_T^{\ddagger} + \Delta G$. However, if, as on the right-hand side of the figure, everything is the same except that K_M is now higher than [S], so that the ES complex is at a higher energy than $E + S$, the activation energy is at the lower value of $\Delta G_T^{\ddagger}$.

The low K_M leads to a thermodynamic 'pit' into which the reaction falls and has to climb out. The high K_M leads to the enzyme–substrate complex being on a 'step up the thermodynamic ladder'.

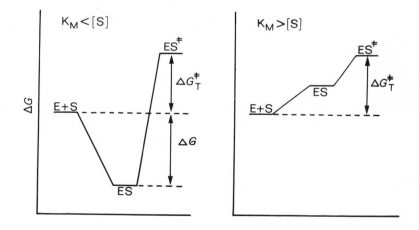

FIG. 10.3. Two cases of enzyme evolution. In both cases the enzymes bind the transition states equally well, but on the left-hand side the substrate is bound strongly and on the right-hand side the enzyme has evolved to bind the substrate weakly ([S] is the same on both sides). The activation energy for the left-hand side is for ES → ES‡, i.e. $\Delta G^{\ddagger} + \Delta G$, whilst on the right it is for $E + S$ → ES‡, i.e. $\Delta G^{\ddagger}$. (The changes in Gibbs energies are for the concentration of substrate used in the experiment and not for standard states of 1 M)

b. Algebraic illustration

In Chapter 3 it was shown that the Michaelis–Menten equation may be cast into the useful form

$$v = [E][S]k_{cat}/K_M \qquad (10.12)$$

to relate the reaction rate to the concentration of free enzyme, [E].

The evolution of an enzyme to give maximum rate may be divided up into two notional steps based on eqn (10.12).

(1) k_{cat}/K_M is maximized by having the enzyme complementary to the transition state of the substrate.
(2) [E] is maximized by having K_M high so that as much of the enzyme as possible is in the unbound form.

There is an evolutionary pressure to increase K_M with a consequent increase in k_{cat}. This evolutionary pressure rapidly decreases as K_M becomes greater than [S]. At $K_M = [S]$, half of the enzyme is unbound, so according to eqn (10.11) the rate is 50% of the maximum possible. At $K_M = 5[S]$, $\frac{5}{6}$ of the enzyme is unbound so the rate is 83% of the maximum. Any further increase in K_M gives only a marginal increase in the rate.

How far a K_M evolves relative to the substrate concentration depends on the change in structure on going from the substrate to the transition state. A limit must eventually be reached when any increase in K_M must be matched by a weakening of transition-state binding. This problem will be most severe for large metabolites present at high concentration.

c. Exceptions to the principle of high K_Ms—control enzymes

Implicit in the above arguments for high K_Ms is the assumption of the maximization of rate. Although this is true for most enzymes, there are cases where rate is subordinate to *control*. Metabolic pathways are characterized by their regulation. This is usually maintained by controlling the activity of certain key enzymes on the pathway. The activities of these control enzymes are often regulated by varying the K_Ms of critical substrates via allosteric effects. The K_Ms for control enzymes have evolved for the purposes of regulation and are not necessarily subject to the rate arguments of the previous section.

A low K_M could sometimes be advantageous for the first enzyme on a metabolic pathway. This would then control the rate of entry to the pathway and prevent it being overloaded and accumulating reactive intermediates. For example, hexokinase, the first enzyme in glycolysis, has a K_M for glucose of 0·1 mM, whilst the concentration of glucose in the human erythrocyte is about 5 mM. A tenfold increase or decrease in the glucose concentration will hardly alter the rate of glycolysis.

TABLE 10.4. *Metabolite concentrations and K_MS for some glycolytic enzymes*[a]

Enzyme	Source	Substrate	Concentration (μM)	K_M (μM)	$K_M/[S]$
Glucose phosphate isomerase	Brain	G6P	130	210	1·6
	Muscle[b]	G6P	450	700	1·6
		F6P	110	120	1·1
Aldolase	Brain	FDP	200	12	0·06
	Muscle[c]	FDP	32	100	3·1
		G3P	3	1000	333
		DHAP	50	2000	40
Triosephosphate isomerase	Erythrocyte[d]	G3P	18	350	19
	Muscle[e]	G3P	3	460	153
		DHAP	50	870	17
Glyceraldehydephosphate dehydrogenase	Brain	G3P	3	44	15
	Muscle[f]	G3P	3	70	23
		NAD	600	46	0·08
		P_i	2000		$>10^g$
Phosphoglycerate kinase	Brain	1,3DPG	<1	9	>9
		ADP	1500	70	0·05
	Erythrocyte[h]	3PG	118	1100	9·3
	Muscle[i]	3PG	60	1200	200
		ADP	600	350	0·6
Phosphoglyceromutase	Brain	3PG	40	240	6
	Muscle[j]	3PG	60	5000	83
Enolase	Brain	2PG	4·5	33	7
	Muscle[k]	2PG	7	70	10
Pyruvate kinase[l]	Erythrocyte[m]	PEP	23	200	9
		ADP	138	600	4·4
Lactate dehydrogenase	Brain	Pyr	116	140	1·2
	Erythrocyte[n]	Pyr	51	59	1·2
		Lac	2900	8400	2·9
		NADH	0·01[o]	10[p]	100
		NAD	33	150	4·6
Glycerophosphate dehydrogenase	Mouse	GlyP	170	37	0·22
	Muscle[q]	GlyP[r]	220	190	0·9
		DHAP	50	190	3·8

Footnotes on facing page.

[a] Abbreviations, G6P = glucose-6-phosphate; F6P = fructose-6-phosphate; FDP = fructose-1,6-diphosphate; G3P = glyceraldehyde-3-phosphate; DHAP = dihydroxyacetone phosphate; P_i = orthophosphate; 1,3DPG = 1,3-diphosphoglycerate; 3PG = 3-phosphoglycerate; 2PG = 2-phosphoglycerate; PEP = phosphoenolpyruvate; Pyr = pyruvate; Lac = lactate; (all D-sugars) GlyP = L-glycerophosphate. Mouse brain enzymes and mouse brain metabolites (O. H. Lowry and J. V. Passonneau, *J. biol. Chem.* **239,** 31 (1964)); human erythrocyte metabolites (S. Minakami, T. Saito, C. Suzuki, and H. Yoshikawa, *Biochem. biophys. Res. Commun.* **17,** 748 (1964)), human erythrocyte enzymes—see below; rat diaphragm metabolites (E. A. Newsholme and P. J. Randle, *Biochem. J.* **80,** 655 (1961); H. J. Hohorst, M. Reim, and H. Bartels, *Biochem. biophys. Res. Commun.* **7,** 137 (1962)); rabbit skeletal muscle enzymes—see below. Metabolite concentrations calculated on an intramolecular water content of 60% for brain and muscle cells, and 70% for erythrocytes. No allowance has been made for compartmentation in the muscle and brain cells, but gross metabolite concentrations are usually close to those in the cytosol (A. L. Greenbaum, K. A. Gumaa, and P. McLean, *Archs Biochem. Biophys.* **143,** 617 (1971)). The values for mouse brain are those immediately on decapitation. The use of peak levels does not cause significant differences

[b] J. Zalitis and I. T. Oliver, *Biochem. J.* **102,** 753 (1967)

[c] W. J. Rutter, *Fed. Proc.* **23,** 1248 (1964); P. D. Spolter, R. C. Adelman, and S. Weinhouse, *J. biol. Chem.* **240,** 1327 (1965)

[d] A. S. Schneider, W. N. Valentine, M. Hattori, and H. L. Heins, *New Engl. J. Med.* **272,** 229 (1965)

[e] P. M. Burton and S. G. Waley, *Biochem. biophys. Acta* **151,** 714 (1968)

[f] M. Oguchi, E. Gerth, B. Fitzgerald, and J. H. Park, *J. biol. Chem.* **248,** 5571 (1973)

[g] The K_M of ~6 mM for P_i refers to high GAP concentrations where the acyl enzyme accumulates. At low concentrations of GAP, the K_M is unmeasurably high (P. J. Harrigan and D. R. Trentham, *Biochem. J.* **143,** 353 (1974)). *Note:* the *unhydrated* forms of G3P and DHAP are most probably the substrates of the reactions. The concentrations tabulated are for both hydrated and unhydrated forms, but both the values of K_M for the unhydrated forms and their concentrations are overestimated in the same ratio (D. R. Trentham, C. H. McMurray, and C. I. Pogson, *Biochem. J.* **114,** 19 (1969); S. J. Reynolds, D. W. Yates, and C. I. Pogson, *Biochem. J.* **122,** 285 (1971))

[h] A. Yoshida and S. Watanabe, *J. biol. Chem.* **247,** 440 (1972)

[i] D. R. Rao and P. Oesper, *Biochem. J.* **81,** 405 (1961)

[j] R. W. Cowgill and L. I. Pizer, *J. biol. Chem.* **223,** 885 (1956); S. Grisolia and W. W. Cleland, *Biochemistry* **7,** 1115 (1968)

[k] F. Wold and R. Barker, *Biochim. biophys. Acta* **85,** 475 (1964)

[l] It is debatable whether or not this is a control enzyme; PEP is certainly well below the K_M in any case. The data quoted are for the presence of 500 μM FDP, when Michaelis–Menten kinetics hold. In the absence of FDP, sigmoid kinetic holds with $K_{0.5}$ 650 μM[m]

[m] S. E. J. Staal, J. F. Koster, H. Kamp, L. van Milligan–Boersma, and C. Veeger, *Biochim. biophys. Acta* **227,** 86 (1971)

[n] J. S. Nisselbaum and O. Bodansky, *J. biol. Chem.* **238,** 969 (1963)

[o] Calculated from the lactate/pyruvate ratio assuming NAD and NADH at equilibrium, using an equilibrium constant of 1.11×10^{-4}, R. L. Veech, L. V. Eggleston, and H. A. Krebs, *Biochem. J.* **115,** 609 (1969)

[p] S. Rapoport, *Essays in Biochemistry* **4,** 69 (1969)

[q] T. P. Fondy, L. Levin, S. J. Sollohub, and C. R. Ross, *J. biol. Chem.* **243,** 3148 (1968)

[r] R. M. Denton, R. E. Yorke, and P. J. Randle, *Biochem. J.* **100,** 407 (1966)

2. Experimental observations on K_Ms

It was seen in Chapter 9 that the binding energy of an enzyme and substrate is potentially very high. However, K_Ms are usually found to be relatively high. An extreme example of this is NAD$^+$. This large substrate has two ribose moieties, one adenine ring, a nicotinamide residue, and a pyrophosphate linkage. If all the potential binding energy of these groups was realized, a dissociation constant of less than 10^{-20} M could be attained. Indeed, the dissociation constant for the binding of the first NAD^+ to the tetrameric glyceraldehyde-3-phosphate dehydrogenase has been found to be immeasurably strong at less than 10^{-11} M.[15] Yet the K_Ms and dissociation constants of NAD^+ with dehydrogenases are often found to be in the range $0 \cdot 1$–1 mM. Even more striking is the dissociation constant of 10^{-13} M for ATP and myosin,[16] which may be compared with the K_Ms of $0 \cdot 1$–10 mM that are often found for ATP.

The comparison of K_Ms with physiological substrate concentrations is difficult in many cases due to a lack of knowledge of the concentrations, but there are some well-characterized examples. One particular case is *carbonic anhydrase* since the concentrations of carbon dioxide and bicarbonate in the blood are easily measured. Under physiological conditions the enzyme is only about 6% saturated with each substrate and the K_M of carbon dioxide is too high to be measured.[17]

a. Substrate concentrations and K_Ms in glycolysis

Good data are available for glycolysis. The glycolytic enzymes are particularly well studied and understood and also metabolite concentrations have been determined for three diverse types of cell (brain, erythrocyte, and muscle). Data for the non-regulatory glycolytic enzymes are listed in Table 10.4 and illustrated in Fig. 10.4. It is seen from the histogram of the figure that the K_Ms tend to be in the range of 1–10 and 10–100 times the substrate concentration. Notable amongst these enzymes is

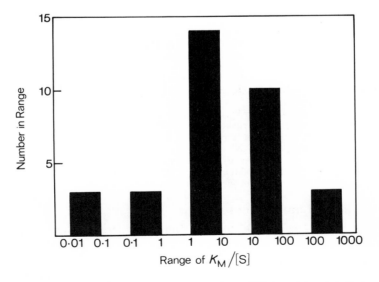

FIG. 10.4. Distribution of the values of $K_M/[S]$ found in glycolysis

triosephosphate isomerase. This well-studied enzyme, which has been described as 'evolutionary perfect'[18] because of its catalytic efficiency, has very high K_Ms for both substrates.

Whether the examples where the K_Ms are below the substrate concentrations are due to an inability to evolve further or to a metabolic reason is not known at present.

3. The perfectly evolved enzyme

We can set up a criterion to judge the state of evolution of an enzyme whose function is to maximize rate by using the two notional steps of Section B1b; i.e. k_{cat}/K_M is maximized, and K_M is greater than [S]. It will be recalled from Chapter 3 that the maximum value of k_{cat}/K_M is the rate constant for the diffusion-controlled encounter of the enzyme and substrate, and from Chapter 4 that this is about 10^8 to $10^9 \, s^{-1} \, M^{-1}$. A perfectly evolved enzyme should have k_{cat}/K_M in the range 10^8 to $10^9 \, s^{-1} \, M^{-1}$ and K_M greater than [S]. Using the data for k_{cat}/K_M listed in Table 10.4 and the substrate concentrations and K_M values mentioned in this chapter, it appears that carbonic anhydrase and triosephosphate isomerase are perfectly evolved for the maximization of rate, agreeing with the conclusions of Albery and Knowles on the isomerase.[18]

There is one further consideration; k_{cat}/K_M cannot be at the diffusion-controlled limit for a reaction that is thermodynamically unfavourable. This point stems from the Haldane equation (Chapter 3H) which states that the equilibrium constant for a reaction in solution is given by the ratio of the values of k_{cat}/K_M for the forward and reverse reactions. Clearly, k_{cat}/K_M for an unfavourable reaction cannot be at the diffusion-controlled limit since k_{cat}/K_M for the favourable, reverse, reaction would have to be greater than the diffusion-controlled limit to balance the Haldane equation. k_{cat}/K_M for an unfavourable reaction is limited by the diffusion-controlled limit multiplied by the unfavourable equilibrium constant for the reaction.

It was pointed out in Chapter 3A3 that when k_{cat}/K_M is at the diffusion-controlled limit, Briggs–Haldane kinetics are obeyed rather than Michaelis–Menten. Thus the more advanced an enzyme is towards the evolution of maximum rate, the more important are Briggs–Haldane kinetics.

C. Molecular mechanisms for the utilization of binding energy

We have discussed in general terms the catalytic advantages of enzyme–transition state complementarity combined with high K_Ms and have seen that this tends to occur in practice. We shall now deal with the specific mechanisms that are used to achieve this.

1. Strain

This is the classical concept of Haldane and Pauling.[1,4] The enzyme has an active site that is complementary in structure to the structure of the transition state of the substrate rather than the substrate itself. On binding the substrate is strained or distorted. In Haldane's words, 'Using Fischer's lock and key simile, the key does not fit the lock perfectly but exercises a certain strain on it' (see Fig. 10.5). Nowadays, a modified

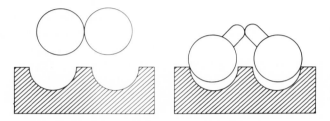

FIG. 10.5. Haldane's picture of strain. The binding site on the enzyme stretches the substrate towards products and compresses the products towards the structure of the substrate

concept, *transition-state stabilization*, is gaining favour. Rather than the substrate being distorted, the transition state makes better contacts with the enzyme than does the substrate so that the full binding energy is not realized until the transition state is reached.

2. Induced fit

In Chapter 8 it was seen that the induced-fit theory nicely described some of the phenomena associated with allosteric enzymes. This theory had been introduced earlier to account for specificity in simple enzymes. It is considered that in the absence of substrate, the enzyme does not have a structure complementary to that of the transition state. However, the enzyme is floppy and the substrate rigid so that on formation of the enzyme–substrate complex, the catalytic groups on the enzyme are aligned in their optimal orientations for catalysis: that is, the structure of the enzyme is complementary to the transition state only after binding has occurred. In the classical strain mechanism, the K_M is increased by binding energy being used to distort the substrate; in induced fit it is increased by binding energy being used to distort the enzyme.

a. Disadvantages of the induced-fit mechanism in non-allosteric enzymes
In the strain mechanism, the value of k_{cat}/K_M is at a maximum since the undistorted enzyme is complementary to the undistorted transition state. Induced fit lowers k_{cat}/K_M because it is only the *distorted* enzyme that is

complementary to the undistorted transition state: k_{cat}/K_M is decreased by the energy required to distort the enzyme.

An alternative way of viewing the induced fit process is to divide it into notional steps. It may be considered that there is an equilibrium between the inactive form of the enzyme (E_{in}) which is the major species ($[E_{in}] \simeq [E_0]$), and a small fraction of active enzyme (E_{act}) in which the catalytic groups are correctly aligned (Scheme 1).

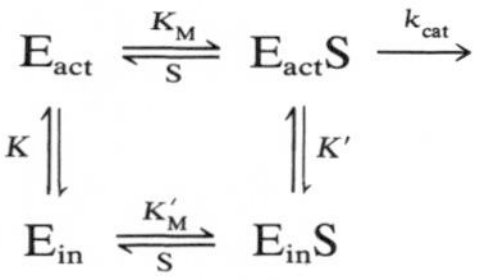

$$E_{act} \underset{S}{\overset{K_M}{\rightleftharpoons}} E_{act}S \overset{k_{cat}}{\longrightarrow}$$

$$K \Big\Updownarrow \qquad \qquad \Big\Updownarrow K'$$

$$E_{in} \underset{S}{\overset{K_M'}{\rightleftharpoons}} E_{in}S$$

Scheme 1

Treating the K_Ms in Scheme 1 as simple dissociation constants and defining $K = [E_{act}]/[E_{in}]$, it may be shown that

$$v = \frac{[E_0][S]k_{cat}K'/(1+K')}{[S]+(K_M K'/K)(1+K)/(1+K')} \tag{10.13}$$

and

$$(k_{cat}/K_M)_{obs} = (k_{cat}/K_M)(K/[1+K]). \tag{10.14}$$

Since $K \ll 1$,

$$(k_{cat}/K_M)_{obs} = K(k_{cat}/K_M). \tag{10.15}$$

The observed value of k_{cat}/K_M is much less than if all the enzyme is in the active conformation.

Eqn (10.13) may be further simplified if $K' \gg 1$, i.e. virtually all of the enzyme is in the active form when bound with substrate. Under these conditions

$$(k_{cat})_{obs} = k_{cat} \tag{10.16}$$

and

$$(K_M)_{obs} = K_M/K. \tag{10.17}$$

k_{cat} is the same as if all the enzyme were in the active conformation, but K_M is far higher.

Induced fit increases K_M without increasing k_{cat}, and decreases k_{cat}/K_M. It mediates against catalysis.

Eqn (10.15) shows that as far as k_{cat}/K_M is concerned, it is as if a small fraction of the enzyme is permanently in the active conformation. This is the same for all substrates since K is substrate independent.

3. Non-productive binding

Although this is not a mechanism for increasing K_M, it is appropriately discussed here since it gives rise to effects that are qualitatively similar to those of strain or induced fit. This theory was originally invoked to account for specificity in the relative reactivities of larger specific substrates compared with smaller non-specific substrates. It is assumed that as well as the productive binding mode at the active site, there are alternative non-productive modes in which the smaller substrates may bind and not react (see Fig. 10.6).

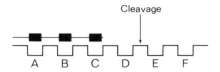

FIG. 10.6. Non-productive binding with substrates and lysozyme. Small substrates may bind at alternative sites along the extended active site of lysozyme, avoiding the cleavage site which has a lower affinity

An example of this is the binding of polysaccharide substrates to lysozyme. In order for reaction to occur, the substrate must bind across sites D and E of the six sub-sites ABCDEF. There is some strain associated with binding in sub-site D and occupying this does not increase the overall binding energy. Trimers and tetramers bind non-productively in ABC or ABCD. However, the favourable binding energy of occupying sites E and F causes hexamers to bind productively in ABCDEF.

As well as this 'gross' non-productive binding, there is an alternative example in which the substrate is bound at the active site, but in the wrong orientation (see Fig. 10.7). This has been proposed to account for the low deacylation rate of non-specific acylenzymes of chymotrypsin.[19]

4. Unimportance of strain, induced fit, and non-productive binding in specificity

Specificity, in the sense of discrimination between competing substrates, is independent of the above three effects. The reasons for this are discussed in detail in Chapter 11. The basic reason is that specificity depends on k_{cat}/K_M, and strain and non-productive binding do not affect this (eqns (10.10) and (3.36)). Eqn (10.15) shows that induced fit just alters k_{cat}/K_M for the active conformation equally for all substrates (i.e. by a factor of K).

5. Experimental evidence concerning the existence and nature of strain and induced-fit processes

a. Steady-state kinetics
Although the results of steady-state kinetic measurements are often interpreted as supporting one of the three mechanisms, the exact one

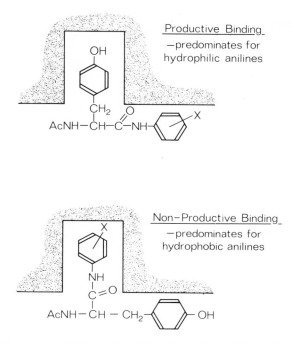

Productive Binding
−predominates for
hydrophilic anilines

Non−Productive Binding
−predominates for
hydrophobic anilines

FIG. 10.7. Wrong-way binding of anilides to chymotrypsin. Synthetic substrates with extraneous hydrophobic residues bind the wrong way round in the hydrophobic binding site of chymotrypsin

depending on the whim of the experimentalist, the evidence is usually ambiguous. The approach generally used is to compare the k_{cat} and K_M values for a series of substrates, as in Tables 10.1 and 10.2, to see if the specificity is manifested in increasing k_{cat} rather than decreasing K_M. If this is found, it is good evidence that one of the processes is occurring, but it does not indicate which one. The problem is that all three mechanisms predict the same result: binding energy is converted into chemical activation energy. The following arguments may be made.

Strain: the additional groups on the larger specific substrates are used to strain the substrate rather than provide binding. (In the transition-state stabilization model this is modified to 'the additional binding energy is not realized until the transition state is reached'.)

Induced fit: the additional groups on the larger specific substrates are used to provide energy for the distortion of the enzyme. The smaller non-specific substrates bind predominantly to the inactive form of the enzyme. k_{cat} is lower for these since less is productively bound, but K_M

is correspondingly lower since the binding energy is not used to convert the inactive to the active conformation.

Non-productive binding: the larger substrate binds in the productive mode only, but the smaller, although binding more weakly to the productive mode, binds in additional non-productive modes lowering the K_M. k_{cat} is correspondingly lower.

Although it is not possible to distinguish between strain and induced fit by a comparison of k_{cat} and K_M values from steady-state kinetics, it is sometimes possible to bring in additional evidence to rule out or favour non-productive binding. One of the clearest examples of this is a study on elastase.[20] The k_{cat} for the hydrolysis of AcProAlaProAla-NH$_2$ is some hundred times higher than that for AcAlaProAla-NH$_2$ (Table 10.1). This is unlikely to be caused by non-productive binding since the presence of the AlaProAla should be sufficient to ensure productive binding. Comparison of the binding of substrates and transition-state analogues suggests that the residues in S_{5-4} destabilize the —CONH$_2$ group that is

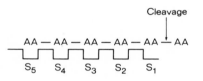

FIG. 10.8. The binding sub-sites in elastase

hydrolysed by about 8 kJ/mol (2 kcal/mol) relative to the transition state. Transition-state analogues which have the —CONH$_2$ replaced by —CHO bind as a tetrahedral adduct to Ser-195, mimicking the tetrahedral intermediate-like transition state of the hydrolytic reaction. Their binding is enhanced by residues in S_{5-4}. It is not possible to say *a priori* whether the mutual destabilization between the S_{5-4} sub-sites and the substrate amide group is due to a conformational change transmitted through the protein or an unfavourable interaction that is relieved on formation of the transition state.

b. Pre-steady-state kinetics and direct binding measurements
There are many well-documented cases of changes of fluorescence and circular dichroism on the addition of substrates and inhibitors to enzymes that are consistent with induced conformational changes.[21] Similarly there are many examples of the measurement of these changes by rapid-reaction techniques. However, the results are not generally interpretable in terms of specific mechanisms.

c. X-Ray diffraction studies
The best information available comes from X-ray diffraction studies on crystalline enzymes.

(i) *Lysozyme: A strain mechanism?* During the hydrolysis of polysaccharide substrates by lysozyme, the sugar ring bound in site D becomes a carbonium ion and its conformation changes from a full-chair to a sofa (Chapter 1 and this Chapter A4b). The original model building studies on the enzyme suggested that the substrate was bound with the residue in the D site forced into the sofa conformation due to unfavourable interactions with the enzyme, i.e. a classical strain mechanism.[22] More recently, Levitt has re-examined the binding of $(NAG)_6$ to lysozyme using accurately refined coordinates of the enzyme and sophisticated calculations of the interactions with the substrate rather than examination of wire models.[8,23] The coordinates of both the enzyme and the substrate

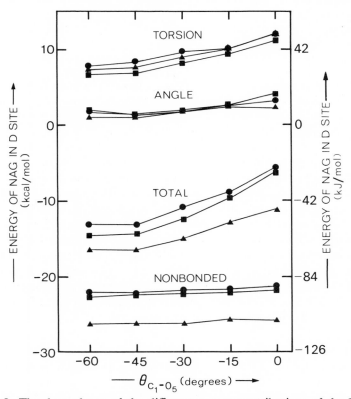

FIG. 10.9. The dependence of the different energy contributions of the D sugar ring on the torsion angle $\theta_{C_1-O_5}$ for various artificially locked $(NAG)_6$ substrates bound to lysozyme (see text). $\theta_{C_1-O_5} = -60°$ for a full-chair and $0°$ for a half-chair (ref. 8)

were optimized by expressing all the bond lengths, bond angles, torsion angles, and non-bonded interatomic distances as empirical energy functions and minimizing the energy by computation. His results for the D sugar ring and D sub-site are given in Fig. 10.9 where the torsion and angle strain in the ring, as well as the non-bonded interactions between the enzyme and substrate, are plotted. Three approaches used are illustrated. First (●), the substrate is assumed to be flexible and it is fitted into a rigid enzyme; second (▲) both the enzyme and substrate are assumed to be flexible, and the pair are convergently energy refined to give a mutual best fit; and third (■) the enzyme is assumed to be partially flexible. It is seen that in all cases the most favourable situation is on the left-hand side of the figure where the substrate is bound as the full chair rather than the sofa on the right-hand side. This suggests then that the substrate binds in its usual full chair conformation in sub-site D and does not take up the sofa conformation until the carbonium ion (or transition state) is formed. The classical strain mechanism of substrate distortion seems unlikely although the transition-state analogue studies suggest that there could be transition state stabilization.

(ii) *Hexokinase: induced fit.* There is good evidence that yeast hexokinase, which phosphorylates glucose with the terminal phosphate of ATP, involves induced fit. X-ray diffraction studies show that the binding of both the sugar and the nucleotide causes extensive changes in tertiary structure. Also the binding of the nucleotide to one form of the enzyme is promoted by sugar binding.[24,25] These results are in accord with earlier solution studies on the ATPase activity of the enzyme in the absence of glucose. Hexokinase phosphorylates glucose at the 6 position with a V_{max} of 800 μmol/min/mg of protein, and a K_M for ATP of 0·1 mM. In the absence of glucose, ATP is hydrolysed with a V_{max} of 0·02 μmol/min/mg and a K_M of 4·0 mM. The addition of lyxose, which cannot be phosphorylated because it lacks the hydroxymethyl group at position 6, increases the V_{max} for the ATPase reaction by a factor of 18 and decreases the K_M of ATP by a factor of 40 to the value for the phosphorylation of glucose.[26,27] Xylose causes changes in the crystal structure of the enzyme similar to those induced by glucose (Fig. 10.10).

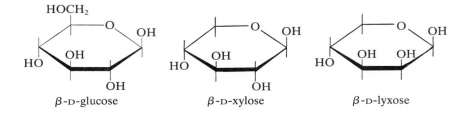

β-D-glucose β-D-xylose β-D-lyxose

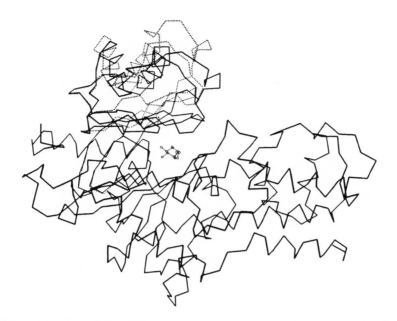

FIG. 10.10. The conformational change in hexokinase induced by glucose binding. The solid lines show the α-carbon backbone of the A isozyme crystallized in
the presence of glucose. The dashed lines show the backbone of that part of the B
isozyme that has a different structure when crystallized in the absence of glucose.
(W. S. Bennett and T. A. Steitz, unpublished observations and *Fedn Proc. Abstr.*,
1977).

(iii) *Serine proteases: transition-state stabilization.* Chymotrypsin and
the serine proteases are among the best-characterized systems because of
high resolution X-ray diffraction studies on the enzymes and also their
complexes with certain naturally occurring polypeptide inhibitors that
mimic substrate binding (Chapter 1).[28-31] The catalytic activity of the
enzyme may be destroyed by chemically converting the nucleophilic
Ser-195 to dehydroalanine by removing the hydroxyl group and a hydrogen atom. The structure of the anhydro-enzyme is otherwise left
unaltered so that it may be used for studies which separate binding from
catalysis.

There is definite evidence for a strain process in peptide hydrolysis.[32]
The crystallographic studies indicate a binding site for the leaving group
of the peptide substrate. The value of k_{cat}/K_M for the hydrolysis of
AcPhe-NH$_2$, AcPhe-GlyNH$_2$, and AcPheAla-NH$_2$ increases considerably
as the size of the leaving group increases to fill the binding site (Table
10.1), but this is due entirely to increases in k_{cat}, with K_M remaining
constant. In the reverse reaction, the attack of the leaving group amine on

AcPhe-chymotrypsin, there is no evidence for $AlaNH_2$ or $GlyNH_2$ binding to the enzyme since saturation kinetics are not observed. Instead they react far more rapidly than ammonia ($NH_3 = 8\ s^{-1}\ M^{-1}$, $GlyNH_2 = 2000\ s^{-1}\ M^{-1}$, and $AlaNH_2 = 6000\ s^{-1}\ M^{-1}$).[32] The binding energy of the Ala and Gly residues is used to lower the chemical activation energy. This is not due to an induced-fit effect since the structure of crystalline trypsin is identical to that of trypsin (and anhydrotrypsin, apart from the hydroxyl of Ser-195) in the crystalline complex with the pancreatic trypsin inhibitor at $1\cdot4$–$1\cdot5$ Å resolution.

There also appears to be no distortion of the substrate on binding. X-ray diffraction studies on the complex of trypsin and the inhibitor are complicated by the inhibitor being distorted in the absence of enzyme,[33] but NMR studies on the binding of small substrates to chymotrypsin indicate that the substrates are unstrained.[34]

The utilization of the binding energy of the leaving group appears to be an example of transition-state stabilization. There is little, if any, distortion of the enzyme or substrate on binding, but the leaving group does not fit snugly into its binding site until the transition state is reached.

(iv) *Specific solvation of the transition state—a backbone contribution.* One feature to emerge from the X-ray diffraction work on the serine proteases is that the amido NH groups of the protein act as the solvation shell for the transition state of the reaction. This leads to a fundamental difference between simple chemical reactions in solution and enzyme-catalysed reactions. In solution the only orientation effects that have to be considered are those between the reagents as the solvent is free to solvate any charges that develop. In the enzyme reaction, on the other hand, there is a precise stereochemical relation between the reacting groups and the effective solvating groups which are part of the enzyme, as illustrated in Fig. 10.11. Hydrogen bonds are particularly suitable for strain and specificity in this sense since their potential varies strongly with distance (Chapter 9). The backbone is also particularly suitable since the rigid positioning of the groups allows the possibility of a strain contribution to catalysis by making weak hydrogen bonds with the substrate but stronger ones with the transition state. It has been proposed that for chymotrypsin and trypsin, the carbonyl oxygen of the reacting amide or ester group of the substrate sits between the backbone NH groups of Ser-195 and Gly-193 but forms a good hydrogen bond with Ser-195 (Chapter 1, and illustrated in Chapter 12B1). The bond with Gly-193 is long and weak, but as the reaction proceeds and the carbonyl oxygen double bond lengthens to become single, the oxygen moves closer to form a strong bond.[32] A similar, but somewhat different, process has been proposed from model building studies on subtilisin.[35] The carbonyl oxygen does not sit between the two NH groups in the enzyme–substrate complex, but swings into position as the tetrahedral intermediate is

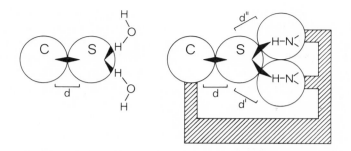

FIG. 10.11. Illustration of the geometric constraints in enzymic catalysis due to the catalyst (C) and the solvating groups (NH) being part of the enzyme structure. The shaded cones represent the permissible angles for orbital overlap or hydrogen bond formation and d the interatomic distance. On the left of the diagram the reaction between the catalyst molecule C and the substrate causes a negative charge to develop which is solvated by water. On the right, the distance between catalyst and hydrogen bond donors is constrained by being part of the enzyme. (Reference 2)

formed. The authors also suggest that another hydrogen bond, that between the N-acylamino group of the substrate and Ser-214, is only made when in the tetrahedral intermediate.

A point worth noting is that specificity in the hydrogen bonding between an enzyme and a substrate is more likely to be caused by the *length* of the bond being critical rather than the *angle* since, as discussed in the previous chapter, the energy of hydrogen bonds varies strongly with distance but only weakly with bending from linearity.[2]

The mechanism of the serine proteases is a good example of what is meant by transition-state stabilization. Further examples of strain and induced fit are discussed in Chapter 12 (*carboxypeptidase*—induced fit; *papain*—strain, induced fit; *glucose-6-phosphate isomerase*—induced fit).

6. Conclusions about the nature of strain. Strain or stress?

Although strain may be manifested in some cases by a genuine distortion of the substrate, it is likely that the strain will generally be distortionless. This could be due to either the substrate and the enzyme having unfavourable interactions which are relieved in the transition state or the transition state having additional binding interactions which are not realized in the enzyme–substrate complex. In both cases there would be forces that *tend* to distort the substrate towards the transition state. As non-bonded interactions have weak force constants (apart from van der Waals repulsion) and enzymes and substrates are flexible, it is difficult to distort the substrate by its interactions with the enzyme. Rotation about single bonds is possible, such as conformational changes in lysozyme substrates, but the stretching of single bonds or the twisting of double bonds appears to be less likely as this requires strong forces. There is ample binding energy available, but for this to be used to distort a substrate the energy must change greatly in magnitude over a short distance, i.e. provide a strong *force*. On the basis of energy calculations, Levitt has suggested the tentative rule 'small distortions of a substrate conformation that cause large increases in strain energy cannot be caused by binding to the enzyme'.[8,36] He suggests also that the largest forces that can be exerted are less than 12 kJ/mol/Å (3 kcal/mol/Å), so that to strain a substrate by about 12 kJ (3 kcal), atoms must be moved by 1 Å. In extreme cases genuine distortion might occur; but, in general, strain will involve the subtle interplay of favourable and unfavourable interactions. It is more likely that the enzyme is distorted than the substrate because the enzyme is less rigid. Indeed, as discussed in this chapter, there are many examples of the distortion of an enzyme on the binding of a substrate. Perhaps some of these distortions are an unavoidable consequence of the flexibility of proteins and the impossibility of constructing an active site precisely complementary to the substrate. Also, a low energy conformational transition where part of the active site closes over the substrate (as with peptides and carboxypeptidase, and NAD^+ and dehydrogenases—Chapter 12) may be a small price to pay for easy access to the active site.

The term 'strain' has a specific meaning in physics and engineering; it implies that an object is physically distorted. Its companion term, 'stress', means that an object is being subjected to forces but is not distorted by them. Using these precise physical terms, it is probably apt to say that in the enzyme–substrate complex the enzyme is often strained whilst the substrate is often stressed.

Strain and stress in enzymes arise from several different causes. We have seen in this chapter and shall see further examples in Chapter 12 that stress and strain may be divided into two processes, substrate destabilization and transition-state stabilization. Substrate destabilization may consist of steric strain, where there are unfavourable interactions between the enzyme and the substrate (proline racemase, lysozyme); desolvation of the enzyme (displacement of two bound water molecules from the carboxylate of Asp-52 of lysozyme); desolvation of the substrate (such as any bound water molecules of a peptide). Transition-state stabilization may consist of the presence of binding modes for the transition state that are not available for the substrate (serine proteases, cytidine deaminase); relief of steric strain; re-establishment of the solvation of the enzyme or substrate (or forming electrostatic bonds) in the transition state.

7. Strain versus induced fit versus non-productive binding

The strain mechanism makes a positive contribution to catalysis by, in addition to providing enzyme–transition state complementarity, increasing k_{cat} and K_M to give increased rates. The induced-fit mechanism mediates against catalysis by increasing K_M without a corresponding increase in k_{cat} compared with all the enzyme being in the active form in the absence of substrate. The non-productive binding mechanism does not affect the catalysis of the specific substrates but just provides additional binding sites for competitive non-specific substrates. This could be deleterious to catalysis if this enabled these to become effective competitive inhibitors of the binding of the specific substrates. Since none of the mechanisms gives any additional specificity, it would seem that in non-regulatory enzymes strain should be the most important of the three processes. Non-productive binding could just be an artefact of systems *in vitro* with no biological importance.[2]

Appendix: the undesirability of accumulating intermediates[2]

It was seen in Chapter 7 that much effort has been put into the detection of chemical intermediates in enzymic reactions. It has been found, though, that these do not accumulate in the reactions of many of the most common hydrolytic enzymes with their natural substrates under physiological conditions (Table 10.5). For this reason, the mechanisms of pepsin and carboxypeptidase are still unresolved. In cases where intermediates do accumulate, it is often through using synthetic highly reactive substrates, such as esters with chymotrypsin, or unusual pH, such as low

TABLE 10.5. *Enzymes and intermediates*

Enzyme (Class)	Substrate	Intermediate	Accumulation[a]
Chymotrypsin (Serine proteases)	Peptides	Acylenzyme	—
Pepsin (Acid proteases)	Peptides	Acylenzyme(?) Aminoenzyme(?)	— —
Carboxypeptidase	Peptides	?	—
Papain (Thiol proteases)	Amides	Acylenzyme	—
Pig liver esterase (Liver esterases)	Aliphatic esters	Acylenzyme	—
Acetylcholine esterase (Choline esterases)	Acetylcholine	Acylenzyme	+
Acid phosphatase	Phosphate monoesters	Phosphoryl-enzyme	+
Lysozyme (Glycosidases)	Polysaccharides	Carbonium ion (ester)	—[b]

[a] Whether or not the physiologically relevant substrates involve the accumulation of an enzyme-bound intermediate even at saturating substrate concentrations. It should be noted that if the physiological concentration of the substrate is below its K_M value, an intermediate does not accumulate even if it would at saturating concentrations.
[b] Intermediate accumulates for artificial leaving group (p-nitrophenol) with glucosidase.

pH with alkaline phosphatase. There is a good theoretical reason for this. It is a corollary of the principle of maximization of rate by the mutual increasing of k_{cat} and K_M that the accumulation of intermediates lowers the reaction rate. Any intermediate which does accumulate lowers the K_M for the reaction, causing saturation at lower substrate concentrations. Another way of looking at it is that if the intermediate accumulates, the rate constant for its decomposition is lower than that for its formation. The reaction rate is therefore lower than it would be if the formation were rate determining.

The accumulation problem is most severe for enzymes that have to cope with high pulses of substrate concentrations, such as the digestive enzymes. If the concentration of the substrate is below the K_M for the reaction under physiological concentrations, no intermediate accumulates *in vivo* in any case, since the enzyme is unbound. But in a test tube experiment where the experimenter can use artificially high concentrations of substrate, an intermediate can sometimes be made to accumulate. An example of this is glyceraldehyde-3-phosphate dehydrogenase. It was seen in Table 10.4 that the aldehyde is below the K_M *in vivo*. But, in the laboratory, the acylenzyme accumulates at saturating substrate concentrations.

References

1   J. B. S. Haldane, *Enzymes*. Longmans, Green and Co. p 182 (1930).
2   A. R. Fersht, *Proc. R. Soc.* **B187,** 397 (1974).
3   E. Fischer, *Ber. dt. chem. Ges.* **27,** 2985 (1894).
4   L. Pauling, *Chem. Engng News* **24,** 1375 (1946), *Am. Scient.* **36,** 51 (1948).
5   R. Wolfenden, *Accts. Chem. Res.* **5,** 10 (1972).
6   G. E. Lienhard, *Science N.Y.* **180,** 149 (1973).
7   I. I. Secemski and G. E. Lienhard, *J. biol. Chem.* **249,** 2932 (1974).
8   M. Levitt, in *Peptides, polypeptides and proteins*, (ed. E. R. Blout, F. A. Bovey, M. Goodman, and N. Lotan). John Wiley and Sons Inc., p 99 (1974).
9   D. H. Leaback, *Biochem. biophys. Res. Comm.* **32,** 1025 (1968).
10   G. J. Cardinale and R. H. Abeles, *Biochemistry* **7,** 3970 (1968).
11   M. V. Keenan and W. L. Alworth, *Biochem. biophys. Res. Comm.* **57,** 500 (1974).
12   R. M. Cohen and R. Wolfenden, *J. biol. Chem.* **246,** 7561 (1971).
13   J. Fastrez and A. R. Fersht, *Biochemistry* **12,** 1067 (1973).
14   K. D. Collins and G. R. Stark, *J. biol. Chem.* **246,** 6599 (1971).
15   J. Schlessinger and A. Levitzki, *J. molec. Biol.* **82,** 547 (1974).
16   H. G. Mannherz, H. Schenck, and R. S. Goody, *Eur. J. Biochem.* **48,** 287 (1974).
17   J. C. Kernohan, W. W. Forrest, and F. J. W. Roughton, *Biochim. biophys. Acta* **67,** 31 (1963).
18   W. J. Albery and J. R. Knowles, *Biochemistry* **15,** 5627, 5631 (1976).
19   R. H. Henderson, *J. molec. Biol.* **54,** 341 (1970).

20 R. C. Thompson, *Biochemistry* **13,** 5495 (1974).
21 N. Citri, *Adv. Enzymol.* **37,** 397 (1973).
22 C. C. F. Blake, L. N. Johnson, G. A. Mair, A. C. T. North, D. C. Phillips, and V. R. Sarma, *Proc. R. Soc.* **B167,** 378 (1967).
23 A. Warshel and M. Levitt, *J. molec. Biol.* **103,** 227 (1976).
24 W. F. Anderson and T. A. Steitz, *J. molec. Biol.* **92,** 279 (1975).
25 S. P. Colowick, *The Enzymes* **9,** 1 (1973).
26 G. DelaFuente, R. Lagunas, and A. Sols, *Eur. J. Biochem.* **16,** 226 (1970).
27 G. DelaFuente and A. Sols, *Eur. J. Biochem.* **16,** 234 (1970).
28 A. Rühlmann, D. Kukla, P. Schwager, K. Bartels, and R. Huber, *J. molec. Biol.* **77,** 417 (1974).
29 R. Huber, D. Kukla, W. Bode, P. Schwager, K. Bartels, J. Deisenhofer, and W. Steigemann, *J. molec. Biol.* **89,** 73 (1974).
30 R. M. Sweet, H. T. Wright, J. Janin, C. H. Chothia, and D. M. Blow, *Biochemistry,* **13,** 4212 (1974).
31 W. Bode, P. Schwager and R. Huber, in *Federation of European Biochemical Societies Tenth Meeting* (ed. P. Desnuelle). North Holland/American Elsevier, **40,** 3 (1975).
32 A. R. Fersht, D. M. Blow, and J. Fastrez, *Biochemistry* **12,** 2035 (1973).
33 J. Deisenhofer and W. Steigemann, *Acta Cryst.* **B31,** 238 (1975).
34 G. Robillard, E. Shaw, and R. G. Shulman, *Proc. natn. Acad. Sci. U.S.A.* **71,** 2623 (1974).
35 J. D. Robertus, J. Kraut, R. A. Alden, and J. J. Birktoft, *Biochemistry* **11,** 4293 (1972).
36 M. Levitt, Ph.D. Thesis, University of Cambridge, p. 270 (1972).

Further Reading

W. P. Jencks, *Adv. Enzymol.* **43,** 219 (1975).

Chapter 11

Specificity and relative reactivity

Specificity is a grossly overworked and often misused word. The most important meaning for the enzymologist refers to the *discrimination* between several substrates competing for the active site of an enzyme; for example, the specificity of a particular aminoacyl-tRNA synthetase for a particular amino acid and a particular tRNA in a mixture of all the amino acids and all the tRNAs. This is the definition of specificity that is relevant to biological systems. It concerns the situation where there is a desired and also an undesired substrate competing for the enzyme, and deals with the problem of how much of the undesired substrate reacts with the enzyme compared with the desired substrate in a mixture of the two. Specificity in this sense is a function of both substrate binding and catalytic rate: if the undesired substrate has a k_{cat} with the enzyme which is a thousand times lower than k_{cat} for the desired substrate, but it binds a thousand times more tightly than the desired substrate, the preferential binding will compensate for the lower rate. For this reason, as discussed below, k_{cat}/K_M is the important kinetic constant in determining specificity since it combines both rate and binding terms.

Another meaning of specificity, which is really a misuse of the term, refers to the activity of an enzyme towards an alternative substrate in the *absence* of a specific substrate, as happens in an experiment *in vitro*. In such a test tube experiment, a substrate is often described as 'poor' because it involves either a high value of K_M or a low value of k_{cat}. In biological systems both k_{cat} *and* K_M are important.

The difference between the two meanings is crucial to the status of strain, induced fit, and non-productive binding in catalysis. As discussed in the last chapter, and amplified below, these do not affect biological specificity since they alter k_{cat} and K_M in a mutually compensating manner without altering k_{cat}/K_M.

A. Limits on specificity

The basic problem in specificity is: how does an enzyme discriminate against a substrate that is smaller than or the same size (isosteric) as the specific substrate? There is no difficulty in discriminating against a substrate that is larger than the specific substrate since the binding cavity at the active site may be constructed to be sufficiently large to accommodate the smaller specific substrate but too small to fit the larger competitor. But a smaller competitor must always be able to bind and cannot be excluded by steric hindrance. There will just be less binding energy available to be used for catalysis. Crude examples of this have been discussed in the sections about the serine proteases. The larger aromatic amino-acid derivatives cannot bind in the small binding pocket of elastase but the smaller amino-acid derivatives can bind to and react with chymotrypsin. But, as discussed at the beginning of the last chapter, the reactions of the smaller substrates involve much lower values of k_{cat} and k_{cat}/K_M. There is also no difficulty in discriminating against substrates of the wrong stereochemistry. As pointed out at the end of Chapter 2, the substitution of an L amino acid by a D amino acid leads to an interchange of two groups around the chiral carbon so that the substrate cannot be bound productively.

Difficult examples of discrimination occur in the reactions of the aminoacyl-tRNA synthetases, enzymes which have to discriminate with high accuracy between the different amino acids. Two examples are the competition between valine and isoleucine for the active site of the isoleucyl-tRNA synthetase, and between threonine and valine for the active site of the valyl-tRNA synthetase (Fig. 11.1).

Valine, being shorter by one methylene group than isoleucine, binds to the isoleucyl-tRNA synthetase, but a hundred times more weakly. Threonine, although isosteric with valine, binds 100 to 200 times more weakly to the valyl-tRNA synthetase because of the burying of the hydroxyl group in the hydrophobic pocket normally occupied by a methyl group of valine.

This situation may be analysed by transition-state theory in the same way as was enzyme–substrate complementarity in the last chapter. The activation energy of the reaction is divided up into contributions from the chemical activation energy, and from the enzyme–substrate binding energy. We shall see that if the smaller or isosteric substrate differs from the specific substrate by lacking an element of structure R, that has a potential binding energy of $\Delta\Delta G_b$, the maximum discrimination possible due to this difference in stereochemistry is $\exp(-\Delta\Delta G_b/RT)$, and that this cannot be amplified by strain, induced fit, a series of conformational changes, an additional series of chemical steps, or two (or more) sites functioning simultaneously. This is first shown for specific examples in Michaelis–Menten kinetics and then generalized for any mechanism. It will be seen that specificity is due to just transition-state binding.

1. Michaelis–Menten kinetics

It was shown in Chapter 3 (Section G2) that specificity for competing substrates is controlled by k_{cat}/K_M. If the rate of reaction of the specific

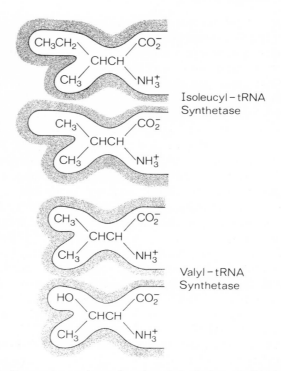

Fig. 11.1. The binding cavity at the active site of the isoleucyl–tRNA synthetase must be able to bind valine as it binds the larger isoleucine. The active site of the valyl–tRNA synthetase cannot reject threonine as it is isosteric with valine

substrate A, is v_A, and that of the competitor B is v_B, then

$$v_A/v_B = [A](k_{cat}/K_M)_A/[B](k_{cat}/K_M)_B. \qquad (11.1)$$

This is translated into terms of binding energy by using eqns (10.3), (10.4), and (10.10), i.e.

$$\ln(k_{cat}/K_M) = \ln(kT/h) - (\Delta G_0^{\ddagger} + \Delta G_b) \qquad (11.2)$$

where $\Delta G_0^{\ddagger}$ is the chemical activation energy and ΔG_b is the binding energy of the enzyme and transition state.

If the additional group R on A is not directly chemically involved in the reaction, $\Delta G_0^{\ddagger}$ will be the same for both A and B, apart from inductive effects. Ignoring these for convenience and setting the difference in binding energy at $\Delta\Delta G_b$ gives from eqn (11.2)

$$(k_{cat}/K_M)_A/(k_{cat}/K_M)_B = \exp(-\Delta\Delta G_b/RT) \qquad (11.3)$$

(where $\Delta\Delta G_b$ is algebraically negative).

Eqn (11.3) gives the *maximum* effect the additional binding energy can make. If the rate of the slow step in the reaction of A is lowered to such an extent in B that another step becomes rate determining, the activation energy has not been lowered by the full amount $\Delta\Delta G_b$. Also, if B binds and reacts in alternative modes, these will be in addition to the mode of A and will add to the overall reaction rate of B.

The following mechanisms have been suggested to cause specificity, but it will be seen that they cannot do so.

a. Strain
It was seen in Chapter 10 that strain does not affect k_{cat}/K_M but just causes compensating changes in k_{cat} and K_M without altering their ratio.

b. Induced fit
It was pointed out in Chapter 10 that k_{cat}/K_M for enzymes involving induced fit is just the value of k_{cat}/K_M for the active conformation scaled down by a constant factor for all substrates (eqns (10.14) and (10.17)). Induced fit does not alter the relative values of k_{cat}/K_M from what they would be if all the enzyme were in the active conformation, and thus does not affect specificity.

c. Non-productive binding
It was shown in Chapter 3E that non-productive binding does not alter k_{cat}/K_M but decreases both k_{cat} and K_M whilst maintaining their ratio. Specificity is unaffected.

d. Series of sequential reactions
Specificity cannot be amplified by there being a series of steps with $\Delta\Delta G_b$ being utilized at each one. A simple way of seeing this is to recall from Chapter 3F, that in eqn (11.4), k_{cat}/K_M is always equal to k_2/K_S irrespective of the number of additional intermediates.

$$E + S \xrightleftharpoons{K_S} ES \xrightarrow{k_2} ES' \xrightarrow{k_3} ES'' \xrightarrow{k_4} ES''' \quad \text{etc.} \quad (11.4)$$

2. The general case
The following formal thermodynamic approach can be used quite generally for analysing binding-energy contributions.

Consider any series of reactions:

$$E + S \rightleftharpoons ES \rightleftharpoons E'S' \rightleftharpoons E''S'' \rightleftharpoons \rightleftharpoons \rightleftharpoons E^{\ddagger}S^{\ddagger} \xrightarrow[\text{determining}]{\text{rate}} \quad (11.5)$$

where E, E', E'' etc. are different states of the enzyme and S, S', S'' different states of the substrates (covalently altered etc). The rate of the reaction may be calculated from transition-state theory ignoring all the intermediate steps and just considering the energetics of the process $E + S \rightarrow E^{\ddagger} S^{\ddagger}$. The Gibbs free energy of activation may be considered to be composed of three terms, a free energy change $\Delta G_E^{\ddagger}$ representing the energy difference between $E^{\ddagger}$ and E, $\Delta G_S^{\ddagger}$ the energy difference between $S^{\ddagger}$ and S, and ΔG_b, the binding energy of $E^{\ddagger}$ and $S^{\ddagger}$.

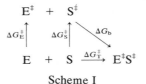

Scheme I

It is seen from the thermodynamic cycle in Scheme I that the activation energy $\Delta G_T^{\ddagger}$ is given by

$$\Delta G_T^{\ddagger} = \Delta G_E^{\ddagger} + \Delta G_S^{\ddagger} + \Delta G_b. \tag{11.6}$$

The rate of reaction is given from transition-state theory:

$$v = (kT/h)[E][S]\exp[-(\Delta G_E^{\ddagger} + \Delta G_S^{\ddagger} + \Delta G_b)/RT]. \tag{11.7}$$

The relative rates of the two substrates A and B reacting are given by substituting the values of the Gibbs energies for A and B into eqn (11.7) and taking the ratio:

$$v_A/v_B = ([A]/[B])\exp[-(\Delta\Delta G_b + \Delta G_A^{\ddagger} - \Delta G_B^{\ddagger})/RT]. \tag{11.8}$$

The difference in the binding energy of A and B, $\Delta\Delta G_b$, comes into the equations only once. $\Delta\Delta G_b$ cannot be used in a cumulative manner at every step to amplify the differences. A reaction might involve several steps, but the specificity due to $\Delta\Delta G_b$ will just be spread out over them.

For similar substrates, $\Delta G_A^{\ddagger}$ will be similar to $\Delta G_B^{\ddagger}$ so that

$$v_A/v_B = ([A]/[B])\exp(-\Delta\Delta G_b/RT). \tag{11.9}$$

This procedure may be extended to include common cosubstrates, e.g. NAD^+, ATP. The Gibbs-energy changes involving these cancel out, as does $\Delta G_E^{\ddagger}$, when the ratios of rates are found from eqn (11.7) or its equivalent.

3. Interacting active sites

Can specificity be increased by more than one molecule of substrate binding to an enzyme with multiple binding sites? This could appear to happen in a test tube experiment with only one substrate present so that absolute rate and not discrimination is measured. But in a biological experiment with the specific and the competitive substrates present, this cannot be so because of the reaction of mixed complexes containing the enzyme and both substrates. This may be proved by the formal thermodynamic approach. In the following example we consider the case of 'half-of-the-sites reactivity' where one molecule of substrate S* binds but does not give products during the turnover of the enzyme, but the second molecule S goes on to react. By comparing the reaction rate of pairs of complexes, e.g. E.A.A* with E.B.A*, it will be seen that the additional binding energy of the larger substrate, $\Delta\Delta G_b$, can be used only 'once'.

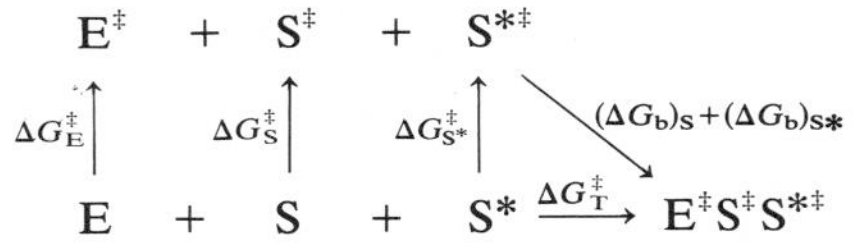

$$E^{\ddagger} \quad + \quad S^{\ddagger} \quad + \quad S^{*\ddagger}$$

$$\Delta G_E^{\ddagger} \Big\uparrow \qquad \Delta G_S^{\ddagger} \Big\uparrow \qquad \Delta G_{S*}^{\ddagger} \Big\uparrow \qquad \searrow (\Delta G_b)_S + (\Delta G_b)_{S*}$$

$$E \quad + \quad S \quad + \quad S^* \xrightarrow{\Delta G_T^{\ddagger}} E^{\ddagger}S^{\ddagger}S^{*\ddagger}$$

Scheme II

According to transition-state theory the rate is given by

$$v = (kT/h)[E][S][S^*]\exp[-(\Delta G_E^{\ddagger} + \Delta G_S^{\ddagger} + \Delta G_{S*}^{\ddagger} + (\Delta G_b)_S + (\Delta G_b)_{S*})/RT].$$

$$(11.10)$$

If A can bind more strongly then B due to the difference in structure contributing a binding energy of $\Delta\Delta G_b$, and if v_{AB*} is the reaction rate when A is bound at the chemically reacting site and B at the other etc., substituting the Gibbs energies into eqn (11.7) and assuming $\Delta G_A^{\ddagger} = \Delta G_B^{\ddagger}$ gives

$$v_{AA*}/v_{BA*} = ([A]/[B])\exp(-\Delta\Delta G_b/RT) \qquad (11.11)$$

and

$$v_{AB*}/v_{BB*} = ([A]/[B])\exp(-\Delta\Delta G_b/RT) \qquad (11.12)$$

so that

$$(v_{AA*} + v_{AB*})/(v_{BA*} + v_{BB*}) = [A]/[B]\exp(-\Delta\Delta G_b/RT). \qquad (11.13)$$

Owing to the mixed complexes of A and B binding to the enzyme, the specificity cannot be enhanced by binding two molecules of substrate simultaneously. Substrate A can enhance activity more than B does by binding at a non-catalytic site, but it enhances activity with B as well as with itself.

4. The stereochemical origin of specificity

Specificity between competing substrates depends on the relative binding of their transition states to the enzyme. Enzyme transition-state complementarity maximizes specificity since it ensures the optimal binding of the desired transition state. This is also the criterion for the optimal value of k_{cat}/K_M, which is not surprising since specificity is determined by k_{cat}/K_M. Maximization of rate parallels maximization of specificity.

The reason why hexokinase phosphorylates glucose in preference to water is that glucose binds well in the transition state whereas there is little binding of water. Whatever the mechanism of the reaction, be it strain or induced fit, the competition between glucose and water is the same. If, for the sake of argument, the V_{max} for the phosphorylation of glucose were the same as that for water, and the binding energy of the glucose used to give a very low K_M, the glucose would be preferentially phosphorylated due its preferential binding to the active site. If, on the other hand, all the binding energy of the glucose were used in lowering the activation energy of V_{max}, this would also lead to its preferential phosphorylation due to its greater reactivity when bound.

There is one situation where strain or induced fit could be useful in a type of specificity. These mechanisms are unimportant where competition between substrates is concerned. If a situation arises where there is *no* specific substrate present, these mechanisms could be of use in providing a low absolute activity of the enzyme towards, say, water. For example, induced fit could prevent hexokinase from being a rampant ATPase in the *absence* of glucose (although its absence is extremely unlikely).

B. Hyperspecificity and editing mechanisms

1. Protein synthesis

The limitations on specificity that are imposed by the differences in binding energy are inadequate for some biological processes. The cell could not tolerate the errors introduced during the replication of the genetic code or during protein synthesis if the fidelity was limited to the constraints outlined in the previous section. The extra methylene group of isoleucine can favour its reaction with the isoleucyl-tRNA synthetase by a factor of only 100 to 200 times over valine. This, combined with a five-fold higher concentration of valine *in vivo*, would give an error rate of one part in 20–40. Yet, it has been found that the overall error rate for the mistaken incorporation of valine into the positions normally occupied by isoleucine in proteins is only one part in 3000.[1] This hyperspecificity is due to the evolution of an *editing mechanism.* In addition to its normal synthetic function the isoleucyl-tRNA synthetase has a hydrolytic capacity. The enzyme will form an enzyme-bound valyl adenylate with k_{cat}/K_M about 150 times smaller than that for the formation of isoleucyl adenylate. But, whereas the addition of tRNA^{Ile} to the correct complex leads to the formation of Ile-tRNA^{Ile}, the addition to the incorrect complex of the valyl adenylate leads to the formation of valine and AMP and no Val-tRNA^{Ile}.[2] In the presence of valine, ATP, and tRNA^{Ile}, the isoleucyl-tRNA synthetase acts as an ATP pyrophosphatase, wastefully, but of necessity, hydrolysing ATP to AMP.

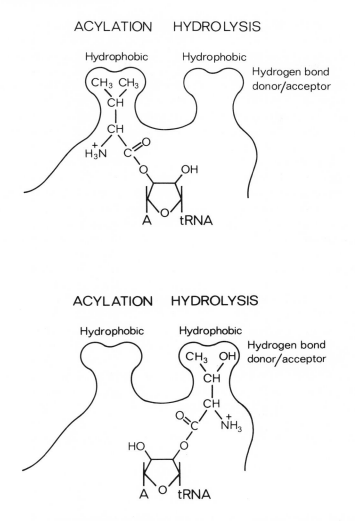

FIG. 11.2. A possible specificity mechanism for the prevention of the misacylation of tRNA^{Val} with threonine. Top, the hydrophobic acylation site discriminates against threonine. Bottom, the hydrolytic site specifically uses the binding energy of the hydroxyl of threonine for a binding or catalytic effect. The translocation may occur as illustrated via a $2' \rightarrow 3'$—OH acyl transfer (ref. 3)

$$E \xrightarrow{\text{Ile, ATP}} E.Ile \sim AMP \xrightarrow{\text{tRNA}^{\text{Ile}}} Ile\text{-}tRNA^{\text{Ile}} + AMP + E \qquad (11.14)$$

$$E \xrightarrow{\text{Val, ATP}} E.Val \sim AMP \xrightarrow{\text{tRNA}^{\text{Ile}}} Val + tRNA^{\text{Ile}} + AMP + E \qquad (11.15)$$

The mechanism for the specificity has been solved for the rejection of threonine by the valyl-tRNA synthetase (Chapter 7D).[3] The enzyme has a separate hydrolytic site for hydrolysing Thr-tRNA^{Val}. The enzyme forms a threonyl adenylate complex with k_{cat}/K_M about 600 times smaller than that for the formation of valyl adenylate. The Thr-tRNA^{Val} is formed but very rapidly hydrolysed with a turnover number of $40 \, \text{s}^{-1}$. Val-tRNA^{Val} is only slowly hydrolysed at $0 \cdot 015 \, \text{s}^{-1}$. The hydrolytic site on the enzyme must clearly have a hydrogen bond donor/acceptor to specifically bind threonine. The hydrophobic valine binds far less well to the hydrolytic site, hydrolysing some 3000 times more slowly. Nature has evolved a specificity mechanism to utilize the difference in structure twice. The difference is used once to discriminate in favour of the correct substrate in the synthetic step, and a second time to discriminate in favour of the incorrect substrate for the destructive step (Fig. 11.2).

The most difficult job of discrimination is between isoleucine and valine by the isoleucyl-tRNA synthetase. The problem is that valine is five times more abundant than isoleucine in *E. coli*.[4] In this case there may be a double check mechanism in which most of the valyl adenylate is destroyed before the transfer to tRNA^{Ile}. The isoleucyl-tRNA synthetase then uses its hydrolytic activity to mop up any Val-tRNA^{Ile} that escapes the first editing step.[5] An analogy may be made between an editing mechanism and a 'double sieve' (Fig. 11.4).

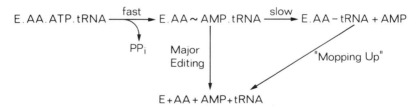

FIG. 11.3. A possible double check mechanism with the major editing step occurring before the transfer of the amino acid to tRNA

2. Editing mechanisms in DNA replication[6]

DNA is replicated with great fidelity. The mutation frequency in *E. coli* is 10^{-6}–10^{-8} or less, and the accuracy of DNA replication must be equal to this or better.[7] This is supported by *in vitro* studies which show the error rates in DNA polymerases from *E. coli* and T4 phage are less than one part in 10^5.[7,8] This accuracy is due to an elaborate series of editing mechanisms.

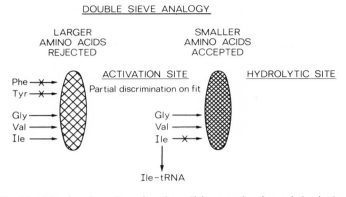

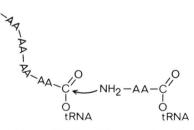

Fɪɢ. 11.4. The 'double sieve' analogy for the editing mechanism of the isoleucyl–tRNA synthetase. The active site for the formation of the aminoacyl adenylate can exclude amino acids larger than isoleucine but not those that are smaller. By the same token, a hydrolytic site that is just large enough to accept valine can exclude isoleucine but accept valine and all the smaller amino acids

There is a fundamental difference between protein synthesis and DNA synthesis that enables errors in the latter to be corrected with relative ease. During protein synthesis the growing end of the polypeptide chain is activated and transferred to the next amino acid in the sequence. There is no means of removing an incorrectly added residue and reactivating the polypeptide. Error correction has to be made before polymerization. But in the synthesis of DNA, the monomeric nucleotide is activated and added to the unactivated growing chain. This has enabled the evolution of a mechanism for the editing of errors after polymerization has occurred.

Fɪɢ. 11.5. Protein synthesis involves the transfer of the activated polypeptide chain to the next amino-acid residue

DNA synthesis proceeds in the 5′–3′ direction with the nucleotides being added to the 3′-hydroxyl of the polynucleotide. At the same time, all prokaryotic DNA polymerases have a 3′–5′ exonuclease activity which works in the opposite direction from synthesis (see Fig. 11.6). There is strong evidence that this is an editing function for the excision of incorrect, mismatched, bases. First, there is the evidence that the exo-

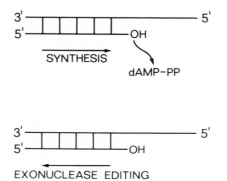

Fig. 11.6. DNA synthesis involves the transfer of the activated deoxynucleotide monophosphate from its triphosphate to the 3′—OH of the growing chain. Editing takes place in the opposite direction

nuclease activity is greatest for mismatched bases or single stranded DNA.[9,10] Second, the mutation frequency in T4 phage correlates with the measured rate of the exonuclease activity catalysed by the DNA polymerase. 'Mutator' mutants have a very high mutation rate and code for a DNA polymerase with a low exonuclease activity (Table 11.1).[7] 'Antimutators' have a high resistance to mutation and a high exonuclease activity. It is seen in Table 11.1 that the antimutators are very wasteful in the high amount of deoxynucleoside triphosphate hydrolysed to incorporate a single base.[11] Third, the 'reverse transcriptase' from avian myeloblastoma (RNA) virus which synthesizes DNA copied from an RNA template does not have a 3′–5′ exonuclease activity and the error rate for the misincorporation of cytidine into polydeoxythymidine has been measured at 1/600.[8] It should be noted that the editing has to be done by the polymerase and not a subsequent repair enzyme since once the double stranded DNA is synthesized, it would not be known which one of the bases in a mismatched pair was incorrect.

The editing mechanism must be somewhat different from that in the aminoacyl-tRNA synthetases: DNA polymerase has to be able to bind all four deoxynucleotides for its synthetic activity so it cannot have a site specifically for the incorrect base as the valyl-tRNA synthetase has for threonine.

As well as errors occurring during synthesis, aberrant base pairing may be caused by damage through external sources such as X-rays, ultraviolet irradiation, oxidation, and chemical modification. Repair is possible in these cases since the DNA molecule is in a dynamic state. A strand may be cut, bases removed and then replaced by the polymerase, and the join sealed by DNA ligase.[6] In these cases the incorrect base is recognized

TABLE 11.1 *Correlation of mutation and exonuclease activity of DNA polymerase of T4 phage*

Strain of phage	Phenotype	dTTP wastefully hydrolysed
		dTMP incorporated
L56	Mutator	0·005
L98	Mutator	0·01
74D	Wild Type	0·04
L42	Antimutator	1·6
L141	Antimutator	13

N. Muryczka, R. L. Poland, and M. J. Bessman, *J. biol. Chem.* **247**, 7116 (1972)

since it is chemically different from the four naturally occurring ones in DNA. There is also a system that can remove deoxyuridine residues that have been mistakenly incorporated instead of thymidine.[12] This situation

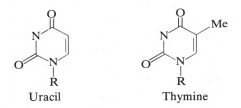

Uracil   Thymine

is similar to that of the isoleucine/valine case in protein synthesis, uracil being smaller than thymine by one methylene group. Uracil is removed from the DNA by a uracil glycosidase which excises the base from the sugar ring. This is similar to the hydrolytic activity of the isoleucyl-tRNA synthetase towards Val-tRNA^{Ile}. In both cases the hydrolytic site is too small by the size of a methylene group to accommodate the substrate which is to be left intact. In DNA synthesis, the editing is performed by a separate enzyme since the editing can be performed after polymerization. As this luxury is not permitted for protein synthesis, the hydrolytic function is on the synthetase so that correction can occur before the misacylated tRNA leaves the enzyme.[13]

3. 'Kinetic proofreading'

The basic requirement of an editing mechanism is that there is an intermediate or product formed in the reaction that is thermodynamically unstable with respect to hydrolysis so that the undesired reagents may be hydrolytically destroyed. Editing introduces a destructive branch point in the reaction pathway through which the undesired products may be channelled. We saw above that this is accomplished in the aminoacylation of tRNA by the aminoacyl-tRNA synthetase having a distinct hydrolytic site for the hydrolysis of mischarged tRNA and possibly also of the precursor aminoacyl adenylate.

Hopfield has proposed a specific chemical mechanism which may be used generally for editing, kinetic proofreading.[14] The distinctive characteristic of kinetic proofreading is that there is no hydrolytic site on the enzyme but the undesired intermediates diffuse into solution where they hydrolyse non-enzymically. The principles of the mechanism may be illustrated by using the aminoacylation of tRNA as a hypothetical example (11.16), although, as seen above, the editing does not occur by kinetic

$$
\begin{array}{ccccc}
 & \text{ATP} \quad \text{PP}_i & & & \\
\text{E.AA} & \xrightarrow{\qquad} & \text{E.AA-AMP} & \xrightarrow{\quad k_3 \quad} & \text{AA-tRNA} \\
k_1 \big\Updownarrow k_{-1} & & k_4 \big\Updownarrow k_{-4} & & \\
\text{E+AA} & & \text{E+AA-AMP} & \xrightarrow{\quad k_h \quad} & \text{AA+AMP}
\end{array}
\qquad (11.16)
$$

proofreading but by the 'double-sieve' mechanism using a distinct hydrolytic site. A high energy intermediate, the aminoacyl adenylate, is generated by the hydrolysis of ATP. The enzyme-bound intermediate may either react with tRNA to give products or diffuse into solution where it may hydrolyse. According to kinetic proofreading, the incorrect intermediate (e.g. valyl adenylate with the isoleucyl-tRNA synthetase) diffuses into solution and is hydrolysed faster than it reacts to give products, that is, $k_3 \ll k_{-4}$. For the desired intermediate, $k_3 > k_{-4}$ so that the products are successfully formed. In this way, the better binding of the specific substrate is used twice over, once in the binding in the Michaelis complex and once again in the binding of the intermediate. Thus, if the discrimination in the initial complex is f, a second factor of f may be attained to give a total of f^2. However, in order for the discrimination at the second step to be at its full value of f, $k_3 < k_{-4}$ for the *correct* substrate so the efficiency of its production is low. These limitations do not apply to the double-sieve mechanisms (Fig. 11.4). This is because by having a separate hydrolytic site on the enzyme the differences between two substrates may be used in two different ways. For example, in the rejection of valine by the isoleucyl-tRNA synthetase, advantage is taken of the greater size of isoleucine to give better binding than valine to the aminoacylation site. But advantage is taken of the smaller size of valine to give it easy access to the destructive site while it is difficult to cram the larger isoleucine into this. Kinetic proofreading uses the differences between the two substrates in the same way twice over. There are as yet no examples where the kinetic proofreading model has been shown unambiguously to occur, but it remains an interesting and plausible possibility.

References

1 R. B. Loftfield and M. A. Vanderjagt, *Biochem. J.* **128,** 1353 (1972).
2 A. N. Baldwin and P. Berg, *J. biol. Chem.* **241,** 831 (1966).
3 A. R. Fersht and M. Kaethner, *Biochemistry* **15,** 3342 (1976).
4 P. Raunio and H. Rosenqvist, *Acta chem. Scand.* **24,** 2737 (1970).
5 A. R. Fersht, *Biochemistry,* **16,** 1025 (1977).
6 A. Kornberg, *DNA synthesis.* W. H. Freeman and Co. (1974).
7 Z. W. Hall and I. R. Lehman, *J. molec. Biol.* **36,** 321 (1968).
8 N. Battula and L. A. Loeb, *J. biol. Chem.* **249,** 408 (1974); **250,** 4405 (1975).
9 D. Brutlag and A. Kornberg, *J. biol. Chem.* **247,** 241 (1972).
10 H. Koessel and R. Roychoudhury, *J. biol. Chem.* **249,** 4094 (1974).
11 N. Muzyczka, R. L. Poland, and M. J. Bessman, *J. biol. Chem.* **247,** 7116 (1972).
12 T. Lindahl, *Proc. natn. Acad. Sci. U.S.A.* **71,** 3649 (1974).
13 J. Ninio, *Biochimie* **57,** 587 (1975).
14 J. J. Hopfield, *Proc. natn. Acad. Sci. U.S.A.* **71,** 4135 (1974).

Chapter 12

Structure and mechanism of selected enzymes

In this chapter we shall discuss the mechanisms of most of the enzymes whose crystal structures have been solved at high resolution. The emphasis is on how kinetic and structural work have been combined to produce satisfactory descriptions of the reaction mechanisms and the general lessons learned about enzyme catalysis. It also provides a non-systematic introduction to some of the experimental approaches used in enzymology. Unfortunately, there are many very interesting enzymes that are not discussed here because their crystal structures have not yet been solved. Any short book must be selective since there are over 1500 known enzymes, and the basis for inclusion in this chapter is the possibility of relating three-dimensional structure and mechanism.

The approximate details of most of the mechanisms are known; for example, the presence and position of the catalytic groups on the enzyme and the overall chemical route of the reaction, although, in certain cases, either or both are obscure. The better understood examples are often those where covalent intermediates occur and have been characterized. Apart from delineating the chemical mechanism, this provides considerable additional structural information: it immediately locates the position of the substrate relative to a catalytic group on the enzyme.

The crucial problem in the structural work is to obtain the structure of the enzyme–substrate complex. Without this, it is not possible to obtain the fine details of the reaction, such as whether there is distortion of the enzyme or substrate, and the precise location of the substrate relative to the catalytic groups. The structure of the enzyme by itself provides only a framework on which to hang hypotheses. Unfortunately, as described in Chapter 1, it is not generally possible to solve directly the structures of productively bound enzyme–substrate complexes, and so substrate analogues and model building must be used as a substitute. This has the disadvantage that strain and distortion effects may be overlooked. Sometimes, as with ribonuclease, the structure of the analogue is very similar to that of the real substrate. In a few rare cases, trypsin and triosephosphate isomerase, it is possible to solve the structures of the productive complexes because of a favourable equilibrium between the substrate and product.

It is important to begin on a note of caution. The nature of science is such that experimentalists push their experimental data to the limits of reliability, and sometimes beyond this point. All experiments, including those from X-ray crystallography, are subject to *interpretation*: one crystallographer may interpret a particular feature in the electron density map as being signficant, another may consider it an artefact of statistical noise. The worst interpreters of all are authors who interpret other peoples interpretations.

The preparation of this chapter was greatly aided by several colleagues allowing my use of their unpublished data. I would like to thank the following (in order of contents) for their help: C.-I. Brändén, H. Dutler, and B. V. Plapp (alcohol dehydrogenase); G. Biesecker and A. Wonacott (glyceraldehyde-3-phosphate dehydrogenase); G. Petsko (elastase); N. Andreeva, V. Antonov, and T. Hofmann (pepsin); D. C. Phillips and J. R. Knowles (triosephosphate isomerase); H. C. Watson (phosphoglycerate mutase).

A. Dehydrogenases

The dehydrogenases we shall discuss in this section catalyse the oxidation of alcohols to carbonyl compounds. They utilize either NAD^+ or $NADP^+$ as coenzymes. The complex of the enzyme and coenzyme is termed the *holo*enzyme, whilst the free enzyme is called the *apo*enzyme. Some dehydrogenases are specific for just one of the coenzymes, whilst a few will use both. The reactions are readily reversible so that carbonyl compounds may be reduced by NADH or NADPH. The rates of reaction in either direction are conveniently measured by the appearance or disappearance of the reduced coenzyme since it has a characteristic ultraviolet absorbance at 340 nm. The reduced coenzymes also fluoresce when excited at 340 nm, providing an even more sensitive means of assay.

The chemistry of the reduction of NAD^+ has been solved most elegantly.[1] Oxidation of the alcohol involves the removal of two hydrogen atoms. One is transferred directly to the 4 position of the nicotinamide ring of the NAD^+, the other is released as a proton (12.1).[2,3] It is generally thought that the hydrogen is transferred as a hydride ion H^-, but a radical intermediate cannot be ruled out. For convenience, we shall assume it is the hydride transfer.

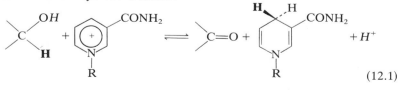

$$(12.1)$$

The transfer is also stereospecific. Using deuterated substrates, it is found that some dehydrogenases will transfer to one side of the ring, and other enzymes the opposite side (12.2). The enzymes are classified as 'A' or 'B' on this basis (Table 12.1)—see Chapter 2H.

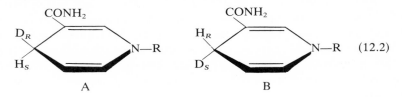

$$ (12.2) $$

The structures of several dehydrogenases have now been solved. These,[4-7] and their physical and kinetic[8] properties have been recently

TABLE 12.1. *Coenzyme specificity of some dehydrogenases*

Dehydrogenase	Coenzyme required	Stereospecificity class
Glutamate	NAD^+ or $NADP^+$	B
Glucose-6-phosphate	$NADP^+$	B
3-Glycerophosphate	NAD^+	B
Glyceraldehyde-3-phosphate	NAD^+	B
Malate (soluble)	NAD^+	A
Alcohol	NAD^+	A
Lactate	NAD^+	A
Isocitrate	$NADP^+$	A

reviewed in depth. Some generalizations may be made. As discussed at the end of Chapter 1, the sub-units may be divided into two domains: a catalytic domain which can be very variable in structure, and a nucleotide binding domain which is formed from a similar overall folding of the polypeptide chain for all the dehydrogenases. The detailed geometry of the nucleotide binding domain varies considerably from one enzyme to another. However, the coenzyme binds in a similar extended, open, conformation in all cases (Fig. 12.1). The most significant variation concerns which side of the nicotinamide ring faces the substrate, the class A enzymes being opposite to the class B.

1. Alcohol dehydrogenases[5]

The alcohol dehydrogenases are zinc metalloenzymes of broad specificity, oxidizing a wide range of aliphatic and aromatic alcohols to their corresponding aldehydes and ketones using NAD^+ as coenzyme (see (12.11)). The two most studied enzymes are those from yeast and horse liver. The

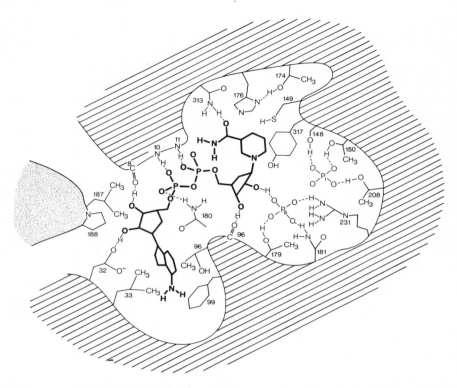

FIG. 12.1. The binding of NAD⁺ to glyceraldehyde-3-phosphate dehydrogenase from *Bacillus stearothermophilus* (Ref. 58).

crystal structures of the *apo* and *holo* horse liver enzymes have been solved at $2 \cdot 4^9$ and $4 \cdot 5$ Å respectively. The molecule is a symmetrical dimer, composed of two identical chains of molecular weight 40 000. Each chain contains one binding site for NAD⁺ but two sites for Zn^{2+}. Only one of the zinc ions is directly concerned with catalysis. The yeast enzyme, on the other hand, is a tetramer of molecular weight 145 000, and each chain binds one NAD⁺ and one Zn^{2+}. Despite these differences, preliminary sequence data indicate a good deal of homology between the two enzymes. It is often assumed that the same overall reaction mechanism holds for both enzymes, although details, such as which step is rate determining, the pH dependence, and Hammett plots, differ between the two.

a. Structure of the active site of liver alcohol dehydrogenase[5]
The Zn^{2+} ion sits at the bottom of a hydrophobic pocket formed at the junction of the catalytic and nucleotide-binding domains. It is ligated by

the sulphur atoms of Cys-46 and Cys-174, and a nitrogen atom of His-67. The fourth ligand is an ionizable water molecule which is hydrogen bonded to the hydroxyl group of Ser-48, which, in turn, is hydrogen bonded to His-51. It is known from the pH dependence of the binding of NAD^{+10} and from the direct determination of the proton release on binding NAD^{+11} that the *apo*enzyme has a functional group of pK_a about 9·6 that is perturbed to a pK_a of about 7·6 in the *holo*enzyme. Crystallographic studies suggest that this is the ionization of the zinc-bound water molecule. The nicotinamide ring of the NAD$^+$ is bound close to the zinc ion at the bottom of the pocket.

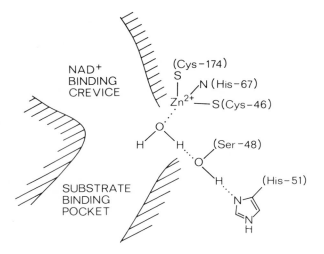

Fig. 12.2. Sketch of the active site of horse liver alcohol dehydrogenase (courtesy of C-I. Bränden)

b. Structure of the enzyme–substrate complex
The structures of the reactive ternary complexes of the enzyme, NAD$^+$, and alcohol, and enzyme, NADH, and aldehyde (or ketone) have not been determined directly but have been obtained by model building. It is suggested that the oxygen atom of the alcohol or aldehyde (ketone) binds directly to the Zn^{2+} ion with the hydrophobic side chain binding in the hydrophobic 'barrel' of the pocket.[5] This hypothesis is supported by new model-building studies that correlate the structure and reactivity of a series of substituted cyclohexanone derivatives,[12] and spectroscopic[13] and binding[14] studies which indicate that the carbonyl group binds directly to a positively charged or acidic centre (the spectrum of a chromophoric aldehyde is perturbed in the appropriate manner, and the binding of substituted benzaldehydes is increased by electron-donating substituents).

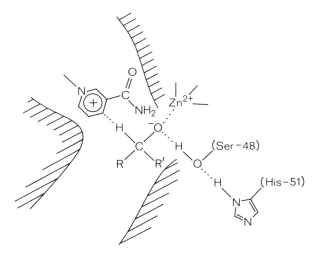

Fɪɢ. 12.3. A proposed model for the productively bound ternary complex of horse liver alcohol dehydrogenase. It is suggested that the ionized alcohol is suggested displaces the zinc-bound water molecule shown in Fig. 12.2 (courtesy of C-I. Brändén)

There is some uncertainty about the coordination number of the Zn^{2+} ion and the ionization state of the alcohol in the ternary complex. In one hypothesis, the alcohol displaces the zinc-bound water molecule or hydroxide ion and binds as the ionized alcoholate ion:[5]

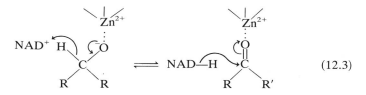

$$\text{(12.3)}$$

An alternative proposal,[15] discussed in the following section, invokes the binding of the undissociated alcohol to the Zn^{2+} ion without the displacement of the water molecule, the coordination number of the zinc increasing to 5.

The model for the ternary complex is constructed around the following interatomic distances: Zn^{2+} to the C-1 carbon of the alcohol = 3·3 Å; Zn^{2+} to the centre of the nicotinamide ring = 4·5 Å; C-1 carbon of the alcohol to the C-4 carbon of the nicotinamide ring = 3·5 Å.[12]

c. The kinetic mechanism
The steady-state and stopped-flow kinetic studies on the horse liver enzyme are now considered as 'classical' experiments. They have shown

that the oxidation of alcohols is an ordered mechanism, the coenzyme binding first, with the dissociation of the enzyme–NADH complex being rate determining.[10,16,17] Both the transient-state and steady-state methods have detected that the initially formed enzyme–NAD$^+$ complex isomerizes to a second complex.[17,18] In the reverse reaction, the reduction of aromatic aldehydes involves rate-determining dissociation of the enzyme–alcohol complex,[17,19] whereas the reduction of acetaldehyde is limited by the chemical step of hydride transfer.

The enzyme–product complexes of the yeast enzyme dissociate rapidly so that the chemical steps are rate determining.[20] This permits the measurement of kinetic isotope effects on the chemical steps of this reaction from the steady-state kinetics. It is found that the oxidation of deuterated alcohols RCD$_2$OH, and the reduction of benzaldehydes by deuterated NADH (i.e. NADD) are significantly slower than the reactions with the normal isotope ($k_H/k_D = 3$–5).[14,20] This shows that hydride (or deuteride) transfer occurs in the rate-determining step of the reaction. The rate constants of the hydride transfer steps for the horse liver enzyme have been measured from pre-steady-state kinetics and found to give the same isotope effects.[21,22]

A major chemical uncertainty in the mechanism concerns the coordination number of the Zn^{2+} ion and the ionization state of its ligands in the ternary complexes. Evidence about this has come from structure–reactivity studies and the pH dependence of catalysis. It has been found that k_{cat} for the oxidation of alcohols by the yeast enzyme and k_{cat} for the reduction of benzaldehydes by the liver enzyme are relatively insensitive to electron donation or withdrawal.[14,15,20,23,24] It is suggested that this is due to general acid–base catalysis in the hydride transfer reaction (i.e. a synchronous movement of a proton to or from the oxygen atom of the substrate neutralizes the charge due to the hydride transfer). However, there is no amino-acid side chain that seems sufficiently close to the substrate to perform this role. It has been suggested that the carbonyl group of the substrate does not bind directly to the zinc ion but binds to the zinc-bound water molecule.[25] Although this would nicely account for the general acid–base catalysis, the zinc-bound water molecule acting as the general acid and the zinc-bound hydroxyl as the general base, the geometry of the complex does not appear to be compatible with the crystal structure.[26] However, there is a neat alternative proposal that

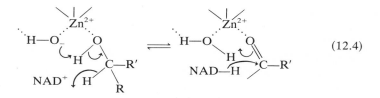

$$(12.4)$$

retains the zinc-bound water molecule as the catalyst and direct coordination of the substrate to the zinc. It is suggested that the binding of the substrate to the zinc does not displace the water molecule but the coordination of the zinc increases to 5 (eqn (12.4)).[15]

The pH dependence of catalysis has been studied to search for the ionization of the catalytic group. k_{cat} for the oxidation of p-methylbenzyl alcohol by the yeast enzyme increases at high pH, depending on the basic form of a group of pK_a 8·25, whilst k_{cat}/K_M for the reduction of acetaldehyde increases at low pH, depending on the concentration of the acid form of a group of pK_a 8·25.[27] This is not inconsistent with mechanism (12.4) since zinc-bound water molecules in model compounds ionize in this region.[28] The pH dependence of k_{cat} for the reactions of the native horse liver enzyme is more difficult to analyse. The rate of reduction of aldehydes varies little between pH 6 and pH 10,[15] and this has been adduced as evidence for mechanism (12.3).[29] If mechanism (12.4) holds, the pK_a of the zinc-bound water molecule in the ternary complex with NADH and aldehyde must be relatively high, and the pK_a of the zinc-bound hydroxide ion in the ternary complex with NAD^+ and aldehyde must be relatively low.[15] Similarly, if mechanism (12.3) holds, the zinc-bound alcoholate ion must have an unusually low pK_a.

A point of controversy concerns whether the horse liver enzyme exhibits half-of-the-sites reactivity. Although the enzyme binds two moles of NAD^+ with equal affinity, it is claimed that only one site of the enzyme reacts during the turnover of the enzyme.[19,30] However, several studies have presented evidence that both sites function simultaneously.[21,31]

2. L-Lactate dehydrogenase[4,6,32]

L-Lactate dehydrogenase catalyses the reversible oxidation of L-lactate to pyruvate using NAD^+ as coenzyme (eqn (12.5)). The enzyme, isolated

$$H-\overset{\overset{\displaystyle CH_3}{|}}{\underset{\underset{\displaystyle CO_2^-}{|}}{C}}-OH + NAD^+ \rightleftharpoons \overset{\overset{\displaystyle CH_3}{|}}{\underset{\underset{\displaystyle CO_2^-}{|}}{C}}=O + NADH + H^+ \qquad (12.5)$$

from many species, is a tetramer of molecular weight 140 000. There are two forms of the enzyme; the H_4, predominating in heart muscle, and the M_4, predominating in skeletal muscle.[33,34] These are *isozymes*, multiple molecular forms of the same enzyme. The amino acid composition of the polypeptide chain that constitutes the M_4 form is significantly different from that of the H_4 form and the two have different kinetic properties. Despite this, the sites for the association of the sub-units must be very similar since the hybrids M_3H, M_2H_2, and MH_3 may be formed in the expected statistical distribution.[35] There are no sub-unit interactions in catalysis so that the kinetic properties of, say, the H_3M form are identical to those expected from a $3:1$ mixture of H_4 to M_4. The crystal structure of the *apo*enzyme from dogfish has been solved at $2\cdot0$ Å resolution and its complex with NAD-pyruvate at $2\cdot8$ Å.[36] The molecule is symmetrical and its sub-units structurally equivalent.

a. Structure of the enzyme–substrate complex
The structures of the ternary complexes have been deduced from crystallographic studies[37] of the binding of NAD-pyruvate, a covalent analogue of pyruvate plus NADH:[38]

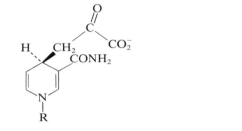

$$(12.6)$$

It seems most likely that the active ternary complexes are of the form (12.7).

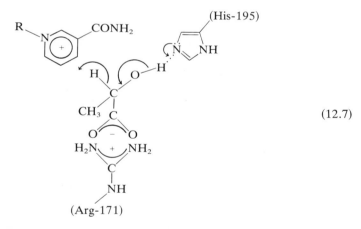

$$(12.7)$$

The carboxylate group of the substrate forms a salt bridge with the side chain of Arg-171. The hydroxyl group of lactate forms a hydrogen bond with the unprotonated imidazole ring of His-195, whilst the carbonyl group of pyruvate forms a hydrogen bond with the protonated imidazole. As well as orientating the substrate, His-195 acts as an acid–base catalyst, stabilizing the negative charge that develops on the oxygen of pyruvate during reduction, and removing the proton from lactate during oxidation.

The NAD-pyruvate is formed by the enzyme from NAD^+ and the *enol* form of pyruvate[39] (the *keto* tautomer is the substrate for the normal reaction).[40] The mechanism for the reaction is presumably catalysed by His-195 in a manner similar to that of the oxidation of lactate (12.8).

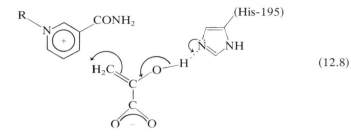

$$(12.8)$$

b. The kinetic mechanism[6,41]

The enzyme binds lactate or pyruvate only in the presence of coenzyme.[42] The mechanism is therefore ordered, coenzyme binding first. Pertinent to this is the observation from the crystallographic studies that coenzyme binding induces a conformational change in which residues 98 to 114 of the chain move through a relatively large distance to close over the active site in the ternary complex.[36,37] Rapid-reaction studies show that the mechanism differs somewhat from that of horse liver alcohol dehydrogenase, although the dissociation of the enzyme–NADH complex is rate determining in both.[41,43] With alcohol dehydrogenase, the dissociation of the aldehyde from the ternary complex E.RCHO.NADH is faster than the rate of hydride transfer, so that the two ternary complexes do not have time to equilibrate. However, the hydride transfer steps are very rapid in the reactions of lactate dehydrogenase and the dissociation of pyruvate is slow so that the two ternary complexes equilibrate. Furthermore, the equilibrium position favours lactate and NAD^+ at neutral pH. On mixing lactate with the *holo*enzyme at pH 7, 20% of the bound NAD^+ is reduced during the first millisecond of reaction as the equilibrium between [E.NAD^+.Lactate] and [E.NADH.Pyruvate] is rapidly attained. As the pyruvate dissociates, the equilibrium is displaced towards products so that all four bound NAD^+ molecules are reduced. There is then the slower dissociation of NADH, the rate-limiting step in the steady state (12.9). This behaviour could easily be mistaken for half-of-the-sites

$$\text{E.NAD}^+.\text{Lactate} \overset{\text{fast}}{\rightleftharpoons} \text{E.NADH.Pyruvate} \overset{\text{slow}}{\longrightarrow} \text{E.NADH} \overset{\text{slowest}}{\longrightarrow} \text{E} \qquad (12.9)$$
$$\qquad\qquad\qquad\qquad\qquad\qquad\qquad\quad \downarrow \qquad\qquad\quad \downarrow$$
$$\qquad\qquad\qquad\qquad\qquad\qquad\quad \text{Pyruvate} \qquad \text{NADH}$$

reactivity, but all four sites appear to be independent.[6,41,43] Coenzymes also bind independently to each site.

The catalytically important His-195 is unusually reactive towards diethyl-pyrocarbonate. This had enabled the pK_a ($= 6.7$) in both the *apo* and *holo*enzymes to be determined directly from the pH dependence of the rate of modification.[44] There is evidence that lactate binds preferentially to the *holo*enzyme containing the un-ionized histidine whilst pyruvate binds preferentially to the enzyme.NADH complex containing protonated histidine.

3. Malate dehydrogenase[7]

Malate dehydrogenase catalyses the reversible oxidation of malate to oxaloacetate using NAD^+ as coenzyme (12.10). The crystal structure of

$$
\begin{array}{c}
CO_2^- \\
| \\
H-C-OH \\
| \\
CH_2CO_2^-
\end{array}
+ NAD^+ \rightleftharpoons
\begin{array}{c}
CO_2^- \\
| \\
C=O \\
| \\
CH_2CO_2^-
\end{array}
+ NADH + H^+ \qquad (12.10)
$$

the soluble, or cytoplasmic, enzyme has been solved at 2·5 Å resolution.[45] The amino-acid sequence has not yet been fitted to the electron density, but, as described in Chapter 1, the overall folding of its polypeptide chain is very similar to that of lactate dehydrogenase, although it is only a dimer of molecular weight 70 000. It presumably has the same catalytic mechanism as the lactate. dehydrogenase as it too has a catalytically essential histidine residue which may be modified with diethylpyrocarbonate.[46] Two moles of NADH or NAD^+ are bound with equal affinity.[47] The apparent negative cooperativity of coenzyme binding found for one enzyme preparation may have been due to the presence of two forms of the enzyme.[48,49]

4. Glyceraldehyde-3-phosphate dehydrogenase[50]

Glyceraldehyde-3-phosphate dehydrogenase, a tetrameric enzyme of molecular weight 150 000 containing four identical chains, catalyses the reversible oxidative phosphorylation of glyceraldehyde-3-phosphate to 1,3-diphosphoglycerate using NAD^+ as coenzyme (12.11).

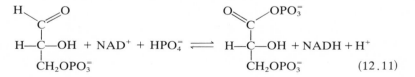

The reaction pathway consists of a series of reactions. The currently accepted mechanism for this (eqns (12.12) to (12.16)), first proposed in 1953,[51] is supported by extensive pre-steady-state[52] and steady-state kinetic studies.[53]

$$
NAD^+.E-SH + RCHO \rightleftharpoons NAD^+.E-S-\underset{\underset{H}{|}}{\overset{\overset{OH}{|}}{C}}-R \qquad (12.12)
$$

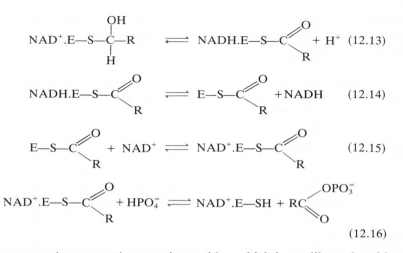

$$\underset{\overset{|}{\underset{H}{}}}{NAD^+.E{-}S{-}\overset{\overset{OH}{|}}{C}{-}R} \rightleftharpoons NADH.E{-}S{-}C\overset{\diagup O}{\diagdown R} + H^+ \quad (12.13)$$

$$NADH.E{-}S{-}C\overset{\diagup O}{\diagdown R} \rightleftharpoons E{-}S{-}C\overset{\diagup O}{\diagdown R} + NADH \quad (12.14)$$

$$E{-}S{-}C\overset{\diagup O}{\diagdown R} + NAD^+ \rightleftharpoons NAD^+.E{-}S{-}C\overset{\diagup O}{\diagdown R} \quad (12.15)$$

$$NAD^+.E{-}S{-}C\overset{\diagup O}{\diagdown R} + HPO_4^= \rightleftharpoons NAD^+.E{-}SH + RC\overset{\diagup OPO_3^=}{\diagdown O}$$
$$(12.16)$$

The enzyme has a reactive cysteine residue which is readily acylated by acyl phosphates to form a thioester (the reverse of reaction (12.16)).[54] The first step in the reaction sequence is the formation of a hemithioacetal between the cysteine and the substrate. This has the effect of converting the carbonyl group, which is not easy to oxidize directly, into an alcohol which is readily dehydrogenated by the usual procedure (12.13). The thioester that is formed in reaction (12.13) reacts with orthophosphate to give the acylphosphate (12.16). However, the acyl transfer is very slow unless NAD^+ is bound to the enzyme.[55,56] The replacement of NADH by NAD^+ in reactions (12.14) and (12.15) is therefore a necessary part of the reaction sequence. It is of interest that the dissociation of the complex of the acylenzyme and NADH (12.14) is the rate-determining step in the sequence at saturating reagent concentrations at high pH.[55] A consequence of this replacement of NADH by NAD^+ before the release of acylphosphate is that the free *apo*enzyme does not take part in the reaction. Also, because acylation of the enzyme by the diphosphate is activated by NAD^+, the *holo*enzyme initiates the reductive dephosphorylation of 1,3-diphosphoglycerate.

The Michaelis complexes of the *holo*enzyme with the aldehyde or the diphosphoglycerate, and the acylenzyme with orthophosphate, are not included in the scheme because their dissociation constants are too high for their accumulation.

a. Structures of the enzyme–substrate complexes
The crystal structures of the *holo*enzymes from lobster[57] and *Bacillus stearothermophilus*[58] have been solved at 2·9 and 2·7 Å respectively whilst the human *holo*enzyme[59] and the bacterial *apo*enzyme[58] have been solved at low resolution. The structures of the enzyme–substrate complexes have been deduced from model building experiments on the

lobster and bacterial enzymes.[57,58] The following description is a compo-
site of these using the specific details of the bacterial enzyme for convenience.

Two binding sites for sulphate ions were identified at the active site of
the enzyme that had been crystallized from ammonium sulphate (Fig.
12.1).[60] A chemically and stereochemically reasonable model for the
course of the reaction may be constructed by assuming that these are the
binding sites for the phosphate residue of the substrate and the nucleophilic phosphate in the deacylation reaction (12.16). The aldehyde
group of the substrate can form a hemithioacetal with Cys-149 when the
3-phosphate is placed to make hydrogen bonds with the hydroxyl of
Thr-179, the positively-charged side chain of Arg-231, and the 2' hydroxyl of the ribose ring that is attached to the nicotinamide of NAD$^+$ (Fig.
12.1). The C-2 hydroxyl of the substrate can then form a hydrogen bond
with Ser-148 whilst the C-1 hydroxyl forms another with a nitrogen of
His-176. These interactions orient the substrate so that the C-1 hydrogen
points towards the C-4 position of the nicotinamide ring, less than 3 Å
away. In this mode of binding. the dehydrogenation reaction may take
place as described earlier for lactate dehydrogenase, with His-176 as the
general base catalyst (12.17).

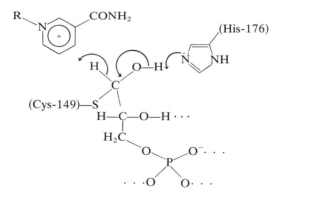

(12.17)

The transition state for the attack of orthophosphate on the thioester
can be stabilized by hydrogen bonds to the attacking phosphate from the
hydroxyls of Ser-148, Thr-150, and the C-2 hydroxyl of the substrate,
and also from the amido nitrogens of Cys-149 and Thr-150. The presence
of this specific binding site for phosphate explains why the thioester is
phosphorolysed rather than hydrolysed. The sulphur atom of the
hemithioacetal is close enough to the C-4 carbon of the nicotinamide ring
of NAD$^+$ to be polarized by its positive charge. This perhaps explains the
activation of the acyl transfer reactions on NAD$^+$ binding.

b. Symmetry of the enzyme and cooperativity of ligand binding

There is considerable controversy in the literature about the symmetry of the dehydrogenase, the cooperativity of ligand binding, and half-of-the-sites versus full-site reactivity.[61-68] The binding of NAD^+ to the enzyme is definitely cooperative; there is strong negative cooperativity in binding to the rabbit muscle and bacterial enzymes,[61,65] although there is positive cooperativity in binding to the yeast enzyme at some temperatures.[69] Glyceraldehyde-3-phosphate, on the other hand, binds independently to all four sub-units.[68] Half-of-the-sites reactivity is found for the reactions of artificial substrates only; 1,3-diphosphoglycerate, for example, acylates all four reactive cysteines with a single rate constant.[55,63,65,68] It has been suggested from some of the kinetic and binding studies that the enzyme exists as a 'dimer of dimers', having two pairs of structurally different sub-units.[67,70] The *interpretation* of the electron density of the high-resolution crystal structure of the lobster *holo*enzyme supports this view,[57] but the more recent study on the bacterial enzyme strongly suggests that all four sub-units are structurally identical and that the enzyme has precise 222 symmetry.[58] The bacterial *apo*enzyme also has this symmetry, although this has been determined only at low resolution.[58] A comparison of the bacterial *apo* and *holo*enzymes shows that the binding of NAD^+ causes a large movement in the coenzyme domain which contracts the volume of the molecule. The structural origins of the negative cooperativity have yet to be elucidated.

5. Some conclusions about dehydrogenases

The structural studies have given a clear and chemically satisfying description of the stereochemical and catalytic requirements of the hydride transfer reaction. In three of the examples there is an acid–base catalyst that forms a hydrogen bond with the carbonyl or alcohol group of the substrate, helps orientate it correctly, and stabilizes the transition state for the reaction (12.18). Liver alcohol dehydrogenase is similar, with either

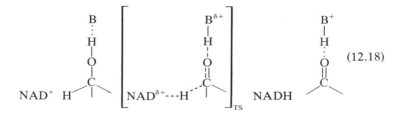

(12.18)

the zinc-bound water molecule or the Zn^{2+} ion taking the place of BH^+ in the stabilization of the transition state and orientation of the substrate.

A consequence of the direction of the hydrogen bonding is that the alcohol binds preferentially to the basic form of the catalyst (B), whilst the aldehyde binds preferentially to the acidic form (BH^+). The pK_a of B is lowered in the $E.NAD^+.RCH_2OH$ ternary complex and raised in the $E.NADH.RCHO$ complex. This means that the proton that is produced during the oxidation does not leave the ternary complex but is taken up by the catalytic group and vice versa. The proton escapes into solution only when there is a change in substrate binding.[18,41,63,71]

The specificities of the enzymes are also nicely explained; the enantiomers of the substrates of L-lactate and D-glyceraldehyde-3-phosphate dehydrogenases cannot be productively bound, whilst the hydrophobic pocket of alcohol dehydrogenase will not bind the charged side chains of lactate etc. However, we do not know if conformational changes occur during catalysis or if there is strain.

A general kinetic feature is that NADH usually binds more tightly than NAD^+. The structural features responsible for this are not clear, although the charged nicotinamide ring is clearly more hydrophilic than the reduced form in NADH. The tight binding leads to the dissociation of the enzyme.NADH complexes being largely rate determining at saturating concentrations of reagents at physiological pH. Furthermore, although the equilibrium constant for the oxidation reaction in solution greatly favours NAD^+ and alcohol, the tighter binding of the NADH causes the equilibrium constant for the enzyme-bound reagents to be less unfavourable: it was seen that the equilibrium constant between the two ternary complexes in the reactions of lactate dehydrogenase is not far from unity.

B. Proteases

The proteases may be conveniently classified according to their activities and functional groups. The serine proteases are endopeptidases that have a reactive serine residue and pH optima around neutrality. The acid proteases are endopeptidases that have catalytically important carboxylates and pH optima at low pH (apart from chymosin, whose activity extends to neutral pH). The thiol proteases are endopeptidases that differ from the serine proteases by having reactive cysteine residues. The carboxypeptidases are zinc-containing exopeptidases which are specific for certain C-termini residues of proteins and function at neutral pH. Apart from leucine aminopeptidase, which has a molecular weight of about 250 000, the proteases are small monomeric enzymes of molecular weight 15 000 to 35 000, readily amenable to kinetic and structural study. Because of this they are amongst the best-understood enzymes. Despite catalysing the same reaction, the different classes utilize different mechanisms. Some are well understood and have chemical models, the others are more obscure.

The notation of Berger and Schechter (eqn (1.6)) is used throughout this section to describe the binding sub-sites. (The scissile bond of the peptide substrate sits across the S_1 and S_1' sub-sites with its C-terminal side occupying the S_1' to S_n' sub-sites and the N-terminal side occupying S_1 to S_n.)

1. Serine proteases

These enzymes have been discussed in various parts of this text. Some major topics are: the enzymes as a family, specificity (Chapter 1C); structures of the active site, enzyme–substrate complex, acylenzyme, and enzyme–product complex (Chapter 1D); proof of the reaction pathway, reaction kinetics (Chapter 7B); pH dependence of catalysis and the state of ionization of the active site (Chapter 5F, G2a); utilization of binding energy to increase k_{cat} (Chapter 10A4); transition state stabilization, specific solvation of the transition state (Chapter 10C5c). The following is a summary of this.

The hydrolysis of ester or amide substrates catalysed by the serine proteases involves an acylenzyme intermediate in which the hydroxyl group of Ser-195 is acylated by the substrate. The formation of the acylenzyme is the slow step in the reaction of saturating concentrations of amide substrates but the acylenzyme often accumulates in the hydrolysis of esters. The attack of Ser-195 on the carboxyl group of the substrate almost certainly forms a high energy tetrahedral intermediate.

$$E{-}OH.R{-}\overset{\displaystyle O}{\overset{\|}{C}}{-}X \longrightarrow E{-}O{-}\overset{\displaystyle O^-}{\underset{\displaystyle R}{\overset{|}{\underset{|}{C}}}}{-}X \underset{HX}{\longrightarrow} E{-}O{-}\overset{\displaystyle O}{\overset{\|}{C}}{-}R \xrightarrow{H_2O}$$

$$E{-}O{-}\overset{\displaystyle O^-}{\underset{\displaystyle OH}{\overset{|}{\underset{|}{C}}}}{-}R \longrightarrow E{-}OH + R{-}CO_2H \qquad (12.19)$$

There is more direct experimental evidence about the mechanism of catalysis and the structures of the intermediates in the reactions of the serine proteases than there is about any other enzyme or families of enzymes. One of the major reasons for the structural knowledge is that it is possible to solve the crystal structures of the co-crystallized complexes of trypsin and some naturally occurring polypeptide inhibitors which mimic substrates (Chapter 1D). We know from these studies that the active site of the enzyme is complementary in structure to the transition state of the reaction, a structure that is very close to the tetrahedral adduct of Ser-195 and the carbonyl carbon of the substrate. Furthermore, the structure of the enzyme is not distorted on binding the substrate. NMR studies on the binding of small peptides show that these are also not distorted on being bound. (The high-resolution study on the crystal structure of the complex between the pancreatic trypsin inhibitor and trypsin shows clearly that the reactive peptide bond is distorted toward its structure in the tetrahedral intermediate. However, this bond is distorted *before* binding to the enzyme, the inhibitor being 'designed' to bind as tightly as possible to the enzyme, i.e. a natural transition-state analogue.)

a. The charge relay system
It has long been thought that the imidazole base of His-57 increases the nucleophilicity of the hydroxyl of Ser-195 by acting as a general base

$$HN \overset{\frown}{} N: \curvearrowright H-O \diagup \qquad\qquad (12.20)$$

catalyst: the activity falls off at low pH according to the ionization of a base about pK_a 7, a characteristic value for a histidine residue; His-57 is modified by the affinity label tosyl-L-phenylalanine chloromethyl-ketone with an irreversible loss of enzymic activity (Chapter 7G).[72] It came as a complete surprise when the crystallographers found that the carboxylate of Asp-102 was also involved at the active site to give a catalytic triad

$$-C \overset{O^{-}}{\underset{O}{\diagup}} \cdots HN \overset{\frown}{} N \cdots HO \diagup \qquad\qquad (12.21)$$

dubbed the 'charge relay system'.[73] Although the carboxyl group is completely buried in the interior of the protein, it is surrounded by polar residues and buried water molecules. It was suggested that the buried carboxyl is ionized in the active protonic state of the enzyme and this was later confirmed (Chapter 5G2f).[74] However, there was a second surprise when NMR[75] and, more recently, infrared[76] procedures indicated that the group ionizing with the pK_a of about 7 is the buried aspartate and not His-57 (Chapter 5G2). If this evidence is correct, and there is no reason to doubt it, the imidazole of His-57 remains unprotonated ın chymotrypsin down to pH 2 since a second ionization at low pH is not seen in the pH-activity profile.[77]

How this affects catalysis is not known. It has been suggested that a proton is transferred between Asp-102 and His-57 during the reaction (12.22).[75]

$$\qquad\qquad (12.22)$$

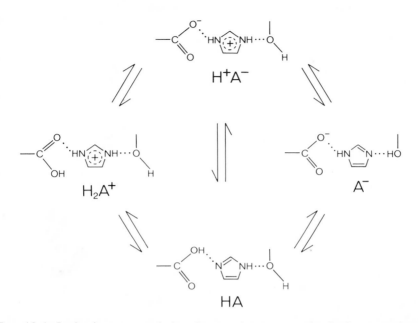

Fig. 12.4. Ionization states of the charge relay system. A^- is the catalytically active state. H_2A^+ is not observed in the reactions of chymotrypsin even as low as pH 2. If the low pH form is H^+A^-, the buried Asp-102 serves just to increase the basic strength of His-57 and to constrain its position. If the low pH form is HA, there is the possibility that during catalysis the proton transfer between Ser-195 and His-57 is accompanied by a simultaneous transfer between His-57 and Asp-102

A most interesting experiment would be to replace somehow Asp-102 by an asparagine and then measure the catalytic activity. The author's guess is that the pK_a of His-57 would be about 5 and the value of k_{cat} would drop by a factor of 10 (based on a lowering of the pK_a of the catalytic base by about 2 pH units combined with a Brönsted β-value of about 0·5 for general base catalysis).[74]

b. Structure and reactivity of the substrate
The structural requirements for a substrate to be reactive have been determined by measuring the values of k_{cat} and K_M for a wide range of ester substrates and the association constants of reversible inhibitors.[78] The inherently high reactivity of esters causes relatively poor ester substrates to be hydrolysed at a measureable rate. Thus esters have been most useful for working out the steric requirements of the acyl portion of the substrate. Amides and peptides are so unreactive that the only ones amenable to study are the derivatives of the specific substrates

TABLE 12.2. *Structural requirements in the deacylation of acylchymotryp-sins (at 25°)*

Acylchymotrypsin (RCO$_2$E) R—	k_{cat} (s^{-1}) (for deacylation)	k_{OH^-} (s^{-1} M^{-1}) (for hydrolysis of RCO$_2$CH$_3$)
CH$_3$—	0·01	0·19
C$_6$H$_5$CH$_2$CH$_2$—	0·178	0·15
CH$_2$(NHCOCH$_3$)—	0·12	2·48
L-C$_6$H$_5$CH$_2$CH(NHCOCH$_3$)—	111	1·94

A. Dupaix, J.-J. Bechet, and C. Roucous, *Biochem. biophys. Res. Commun.* **41,** 464 (1970); I. V. Berezin, N. F. Kazanskaya, and A. A. Klyosov, *FEBS Letts.* **15,** 121 (1971)—see Table 7.3.

phenylalanine, tyrosine, and tryptophan. These studies may now be combined with those from X-ray diffraction.

(i)   *The deacylation step.*   Listed in Table 12.2 are data for the deacylation of various acylenzymes. Further values for amino acids are given in Table 7.3 (p. 184). It is seen that the most reactive derivative is that of acetyl-L-phenylalanine. As discussed in Chapter 1, chymotrypsin has a well-defined binding pocket for the aromatic side chain of the amino acid and a hydrogen bonding site (the C=O of Ser-214) for the NH of the CH$_3$CONH— of the substrate (see Fig. 12.5). On replacing the C$_6$H$_5$CH$_2$— and CH$_3$CONH— groups of acetylphenylalanine by hydrogen atoms to give the simple acetyl group, the deacylation rate drops by

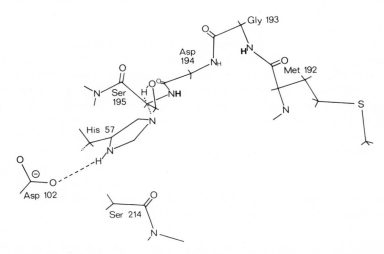

FIG. 12.5. Some of the amino-acid residues of chymotrypsin that are involved in binding and catalysis

a factor of 10^4 (although a factor of 10 of this is caused by the inductive effect of the CH_3CONH- group as seen from the hydroxide-ion-catalysed rate constants listed in the last column of Table 12.2). Interestingly, it is seen in Table 12.2 that *both* the aromatic ring *and* the acylamino group are required for high reactivity. Acetylglycine-chymotrypsin deacylates only 12 times faster than acetyl-chymotrypsin, and the increase is seen from the hydroxide-ion-catalysed rate constants to be caused solely by the inductive effect of the acylamino group rather than by any binding effect. Similarly, β-phenylpropionyl-chymotrypsin deacylates only 17·8 times faster than acetyl-chymotrypsin. The reason why both the aromatic ring and the acylamino group are required for high reactivity has been nicely accounted for by X-ray diffraction studies. As described in Chapter 1 and illustrated below, the carbonyl oxygen of a polypeptide substrate sits between the backbone NH groups of Ser-195 and Gly-193. This mode of binding has been recently found for the specific acylenzyme carbobenzoxy-L-alanine-elastase.[79] However, it was found that the carbonyl oxygen of the non-specific acylenzyme indolylacryloyl-chymotrypsin is not productively bound in this manner.[80] Instead, there is a water molecule forming a hydrogen bonded bridge between the carbonyl oxygen and the catalytic nitrogen atom of His-57 (see Fig. 12.6). For reaction to occur, the carbonyl oxygen must swing into the hydrogen bonding site between Ser-195 and Gly-193 and the bound water molecule must attack the carbonyl carbon. Thus, the acylamino portion and the aromatic ring are together required to anchor the carbonyl group in the productive mode. If either of the anchors is missing, the carbonyl oxygen takes up a non-productive binding mode.

(ii) *The acylation step.* It was pointed out in Chapter 10 that the binding energies of the S_2, S_3, S_4, and S_5 sub-sites often increase k_{cat} for the hydrolysis of polypeptide substrates rather than lower K_M. The reason for this is not known. The binding energy of the S_1' site is also used to increase k_{cat} rather than give tighter binding. It has been suggested that this is a result of unfavourable interactions between the leaving group and the enzyme that are relieved on forming the tetrahedral intermediate (Chapter 10C5c).[81] Possibly better hydrogen bonds are formed between the backbone NH groups and the negatively charged oxygen of the tetrahedral intermediate than with the carbonyl oxygen of the substrate.

c. Description of the reaction mechanism
The kinetic and structural data may be combined to give the following qualitative description of the mechanism of acylation of chymotrypsin by a good polypeptide substrate.[81]

The substrate binds in the specificity pocket of the enzyme with the N-acylamino hydrogen binding to the carbonyl group of Ser-214. Any further residues in the N-acylamino chain bind in the sub-sites that are

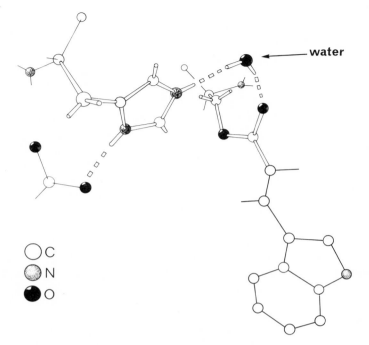

FIG. 12.6. The crystal structure of indolylacryloyl-chymotrypsin (R. Henderson, *J. molec. Biol.* **54,** 341 (1970)). Note that the carbonyl oxygen of this non-specific acylenzyme is not bound between the NH groups of Ser-195 and Gly-193 but is non-productively linked to His-57 by a hydrogen-bonded water molecule. This is the acylenzyme that was found to deacylate at the same rate in solution and in the crystal—Chapter 1—(G. L. Rossi and S. A. Bernhard, *J. molec. Biol.* **49,** 85 (1970))

available. The reactive carbonyl group sits with its oxygen between the backbone NH groups of Ser-195 and Gly-193 (Fig. 12.7). However, it is possible that the hydrogen bond between the oxygen and Gly-193 is long and weak. The leaving group fits into the S'_1 site at the cost of some crowding between it and the side chain of Ser-195. It has been suggested that this causes the reactive hydroxyl group of Ser-195 in chymotrypsin to rotate towards the position it takes up in the transition state. However, the hydroxyl group in trypsin appears to be partly rotated towards this position in the absence of substrate. The first chemical step in the reaction is the attack of the hydroxyl of Ser-195 on the carbonyl carbon of the substrate to form the tetrahedral intermediate (Fig. 12.8). During this, the proton on the hydroxyl is transferred to the imidazole of His-57, and the proton on the other nitrogen of the ring may be simultaneously transferred to the carboxylate of Asp-102. As the bond between Ser-195 and the

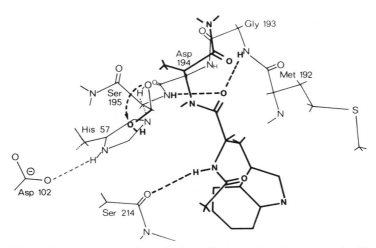

Fɪɢ. 12.7. N-Acetyl-ʟ-tryptophan alaninamide bound to chymotrypsin. Note the hydrogen bonds formed by the carbonyl group of the substrate and the backbone NH groups of Gly-193 and Ser-195. The hydroxyl of Ser-195 rotates from the 'up' position found in the crystals of the enzyme towards the substrate

carbonyl carbon is formed, the C=O bond lengthens to become a single bond. The oxygen, bearing a negative charge, moves closer to the NH of Gly-193, forming a shorter and stronger hydrogen bond. The transition state is stabilized relative to the Michaelis complex because of the relief of strain between the leaving group and the side chain of Ser-195, and by

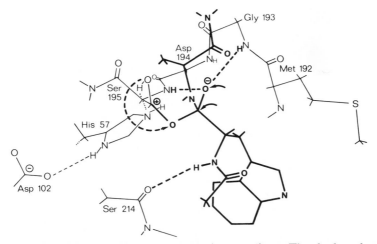

Fɪɢ. 12.8. Formation of the tetrahedral intermediate. The hydroxyl rotates further to form the bond to the substrate. The carbonyl carbon and oxygen atoms of the substrate move during this process, possibly forming better hydrogen bonds with the backbone of the enzyme

the better hydrogen bond with Gly-193. The tetrahedral intermediate collapses to form the acylenzyme and expel the leaving group (Fig. 12.9). The leaving group cannot bind in the S_1' site in the acylenzyme as this would force the amino group to be too close to the carbonyl carbon. Thus, in the reverse reaction, the attack of the leaving group on the acylenzyme, the binding energy to the S_1' site is only realized in the transition state. Deacylation takes place by the charge relay system

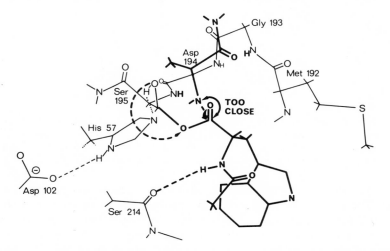

Fig. 12.9. Collapse of the tetrahedral intermediate to expel alaninamide and form the acylenzyme. The alaninamide cannot remain bound in the leaving group site (S_1') because of an unfavourably close contact with the carbonyl group of the acylenzyme

activating the attack of water. Another tetrahedral intermediate is formed which collapses to expel Ser-195 and give the enzyme–product complex.

(*Added in proof.* Direct evidence has now been presented for the occurrence of a tetrahedral intermediate and a concerted proton transfer from Ser-195 to His-57 and from His-57 to Asp-102 during the hydrolysis of an anilide substrate catalysed by a serine protease (M. W. Hunkapiller, M. D. Forgac, and J. H. Richards, *Biochemistry* **15,** 5009 (1976)).)

Despite this detailed knowledge, many important questions still remain unanswered. For example, we do not know how the binding energies of the sub-sites for the N-acylamino chain are sometimes used to increase k_{cat} rather than decrease K_M (Table 10.1). We do not know the contribution to catalysis of the buried Asp-102 in the charge relay system: what would be the activity of chymotrypsin in which the aspartate is converted to asparagine?

d. The zymogens

Some of the serine proteases are stored in the pancreas as inactive precursors which may be activated by proteolysis. Trypsinogen, for example, is converted to trypsin by the removal of the N-terminal hexapeptide on the cleavage of the bond between Lys-6 and Ile-7 by enterokinase. Chymotrypsinogen is activated by the tryptic cleavage of the bond between Arg-15 and Ile-16. (In this case, further proteolysis by the chymotrypsin that is released during the activation leads to the different forms of the enzyme.)

The mechanism of the activation and the reasons for the inactivity of chymotrypsinogen have been nicely explained by comparing the crystal

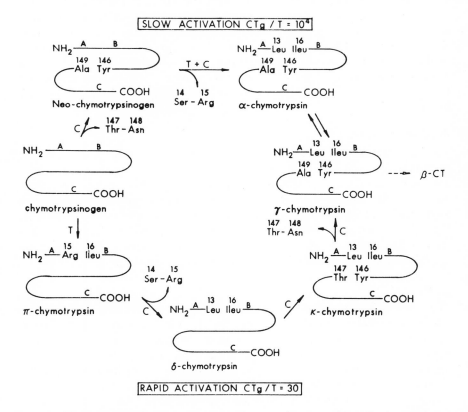

FIG. 12.10. The activation of chymotrypsinogen (D. D. Miller, T. A. Horbett, and D. C. Teller, *Biochemistry* **10**, 4641 (1971)). T = trypsin, C = chymotrypsin, and CTg = chymotrypsinogen. During the 'rapid' activation, there is sufficient trypsin to activate all of the zymogen before the accumulated chymotrypsin autolyses. During the slow activation, the small fraction of trypsin activates the zymogen slowly and the chymotrypsin that is initially produced cleaves the remaining unactivated zymogen to form a neochymotrypsinogen

structures of the enzyme and zymogen.[82–84] The zymogen has the charge relay system and it ionizes in the same manner as in the enzyme.[74,85] However, the activity of the zymogen is extremely low, being devoid of proteolytic activity and only as reactive as a solution of imidazole towards synthetic substrates.[86,87] The reason for this is that the substrate binding pocket is not properly formed in the zymogen, and the important NH group of Gly-193 points in the wrong direction for forming a hydrogen bond with the substrate.[83] This is an important lesson about enzyme catalysis. Enzyme catalysis does not depend on just the presence of an unusually reactive catalytic group on the enzyme but is due to the correct alignment of the substrate and ordinary catalytic groups.

The conformational change that forms the binding pocket and rotates Gly-193 results from a movement of Ile-16 as its α-ammonium group forms a salt bridge with the buried carboxylate of Asp-194. The activation process may be mimicked and studied by the effects of pH on the salt bridge. This deprotonates at high pH and is destabilized so that the enzyme takes up a zymogen-like conformation. The energy difference between the two conformations is small and their equilibrium is delicately balanced.[88]

2. Thiol proteases[89–94]

The thiol proteases are widely distributed in nature. The plant enzymes papain (from papaya), ficin (from figs), and bromelain (from pineapple) are members of a structurally homologous family. They are not homologous with the bacterical thiol protease clostripain (from *Clostridium histolyticum*) and streptococcal proteinase (from haemolytic streptococci). Perhaps the two groups will be found to be related in the same way as the mammalian and bacterial serine proteases. In mammals, the thiol proteases cathepsin B1 and B2 are found packaged in lysozomes.

a. Papain[89–92]

This enzyme is composed of a single polypeptide of 212 amino acids with a molecular weight 23 406.[95] Kinetic studies have shown that the active site can accommodate seven amino acids, four on the acyl side of the cleaved bond (S_4 to S_1) and three on the amino side (S'_1 to S'_3).[96] Unlike the serine proteases that have S_1 as the primary specificity site, papain is specific for hydrophobic amino acids in the S_2 site. There is also a specificity for isoleucine or tryptophan in the S'_1 site.[97] Esters, and presumably peptides, are hydrolysed through an acylenzyme pathway in the same manner as with the serine proteases, except that Cys-25 is acylated.[98–101] A plot of k_{cat}/K_M against pH follows a bell-shaped curve with optimal activity at about pH 6. This is caused by the ionization of His-159 and Cys-25 with pK_a values of 4·2 and 8·2. Denoting the histidine by 'Im' and the cysteine by 'RSH', the ionic form [RSH.HIm$^+$] is inactive at low pH whilst the ionic form [RS⁻.Im] is inactive at high pH. The catalytically active form at neutral pH is one of the tautomers [RSH.Im] or [RS⁻.HIm⁺]: one cannot distinguish between two ionic states bearing the same net charge by examining a pH dependence ('principle of kinetic equivalence', Chapter 2F). The pH dependence of k_{cat} for deacylation follows the ionization of a base of pK_a about 4. This may be attributed to His-159 since the cysteine is blocked in the acylenzyme. The reaction mechanism is of the form:

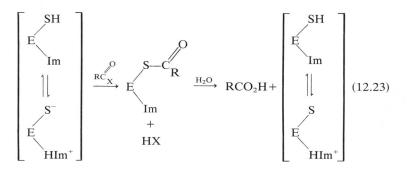

$$(12.23)$$

(i) *Structure of the active site of papain.*[89,102–105] It is seen in the crystal structure that the molecule is formed from two domains with a deep cleft between them. The binding site for the substrate straddles this cleft; although Cys-25 and His-159 are in close contact, they are on opposite sides of the cleft. The fairly deep pocket of the S_2 site for the hydrophobic amino acids is lined with the hydrophobic side chains of Tyr-67, Pro-68, and Trp-69 of one domain, and those of Phe-207, Ala-160, Val-133, and Val-157 of the other. It was once thought that the pK_a of 4·2 that is found in the pH-activity profile is due to the ionization of a carboxylate side chain since they usually ionize in this region. But the nearest carboxyl group to Cys-25, that of Asp-158, is 7.5 Å away.[92] This is too distant to act as an acid–base catalyst, unlike the imidazole ring of His-159 which is correctly placed. The low pK_a of the histidine is presumably due to its being partly buried in a hydrophobic region. There is no equivalent of the charge relay system of the serine proteases; the imidazole ring of His-159 does not interact with a buried carboxylate.

Model-building studies nicely explain the observed stereospecificity of the enzyme.[105] D-amino-acid residues cannot be accommodated in the sub-sites because of steric interference with the bulk of the enzyme. The enzyme is not an exopeptidase since the free carboxylate of the substrate would be only 3–4 Å from the carboxylate of Asp-158 with consequent electrostatic repulsion. These studies also suggest a strain mechanism. The leaving group of the substrate appears to be forced against the

α-CH$_2$ group of His-159 in the enzyme–substrate complex but this interaction is relieved on forming the tetrahedral intermediate. In support of this hypothesis it has been shown that substrate analogues that have a sterically small group in the leaving-group position bind considerably more tightly than those which have bulkier residues.[92,105]

The specificity for large hydrophobic residues in the S$_2$ sub-site is manifested in increased values of k_{cat} rather than tighter binding. Lowe and Yuthavong suggested that the binding of a residue such as phenylalanine in the S$_2$ site forces the cleft to open somewhat and increase the strain at the active site.[105] An outward movement of the walls of the cleft has been subsequently found in the crystal structure of the enzyme inhibited by the chloromethylketone derivative of N-benzyloxycarbonyl-L-phenylalanine-L-alanine.[104] This structure also shows that there is a binding site for the carbonyl oxygen of the scissile peptide bond. This comprises the backbone NH group of Cys-25 in an analogous manner to that in the serine proteases, but the other hydrogen bond is to the side chain —NH$_2$ of Gln-19.

The exact nature of the acid–base catalysis at the active site is not known. Whilst the deacylation of the acylenzyme undoubtedly involves the general-base-catalysed attack of water by the imidazole ring of His-195, the role of the histidine in the acylation reaction is not known for certain.[106,107] It has been found that in simple chemical systems the combination of a base B and a thiol RSH reacts as RS$^-$ and BH$^+$. There is no known example of general-base catalysis of the attack of a thiol. However, the effect of the enforced proximity of a general base on the reactivity of the thiol is not known.[106]

The rate-determining step in the hydrolysis of amides and anilides appears to be the general-acid-catalysed breakdown of the tetrahedral intermediate.[108,109] The evidence for this is that (a) k_{cat} and k_{cat}/K_M are higher for substrates containing the more basic anilines ($\rho = -1\cdot04$),[108] showing that the nitrogen of the substrate becomes protonated during the transition state (the better the base, the easier the protonation), and (b) k_{cat}/K_M for the hydrolysis of benzoyl-L-arginine amide is $2\cdot4\%$ greater

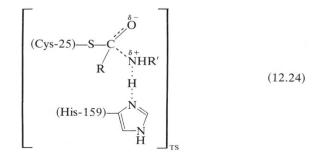

(12.24)

for the substrate containing ^{14}N compared with that containing ^{15}N in the leaving group ($k_{^{14}N}/k_{^{15}N} = 1 \cdot 024$), a value expected for the nearly complete cleavage of a C—N bond.[109] (Note: rate-determining breakdown does not mean that the intermediate accumulates, but merely that it reverts to starting materials faster than it proceeds to products.)

3. Carboxypeptidases[110–112]

Bovine pancreatic carboxypeptidase A is a metalloenzyme, containing one atom of zinc bound to its single polypeptide chain of 307 amino acids and molecular weight 34 472.[113] It is an exopeptidase, removing the C-terminal amino acids from polypeptide substrates, specific for the large hydrophobic amino acids such as phenylalanine. The closely related carboxypeptidase B (see Chapter 1C3) specifically removes lysine and arginine residues. The two enzymes are structurally almost identical apart from the presence of an aspartate residue in the B form which binds the positively charged side chain of the substrate.[114] The following discussion refers to the A form of the enzyme but should hold for both.

a. Structure of the active site of carboxypeptidase A
The crystal structure of the enzyme has been solved at $2 \cdot 0$ Å resolution.[115] The active site consists of a shallow groove on the surface of the enzyme leading to a deep pocket, lined with aliphatic and polar side chains and parts of the polypeptide chain, for binding the C-terminal amino acid. The catalytically important zinc ion is ligated by the basic side chains of Glu-72, His-196, and His-69. In about 20% of the molecules, the phenolic oxygen of Tyr-248 is a fourth ligand.[116] The side chain of this residue is conformationally very mobile, a matter leading to much discussion in the literature and debate concerning the similarities of solution and crystal structure.[116,117] The phenolic side chain may rotate about its C_α—C_β bond, and about 80% of its electron density in the map of the crystal structure is found in an orientation on the surface of the molecule pointing into solution.[116] This residue is also so mobile in carboxypeptidase B that its position cannot be determined in the crystal structure.[114]

A model for the binding of substrates has been extrapolated from the structure of the complex of the enzyme and glycyl-L-tyrosine which had been solved by the difference Fourier method at $2 \cdot 0$ Å resolution (Fig. 12.11).[110,115] The dipeptide is hydrolysed only very slowly and is presumably bound non-productively. It is reasoned that the slow hydrolysis rate is caused by the free amino group of the substrate binding to the carboxylate of Glu-17 via the intervening water molecule. This prevents the carboxylate acting as a general base or nucleophile in the reaction (see below). The remaining features of the complex are used in the construction of the model for productive binding: the aromatic side chain

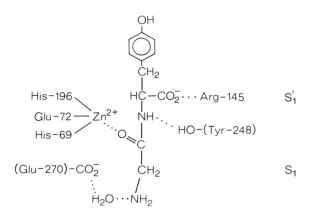

F_{IG}. 12.11. The (partly) non-productively bound complex of a dipeptide and carboxypeptidase A (Courtesy of **W. N. Lipscomb**).

binds in the binding pocket; the carboxylate ion of the C-terminal forms a salt linkage with Arg-145; the carbonyl oxygen of the scissile bond becomes the fourth ligand of the zinc ion; and the phenolic oxygen of Tyr-148 is within about 3 Å of the scissile bond, the side chain having rotated through about 120° from its predominant orientation in the free enzyme. Using this as a basis, the polypeptide chain may be extended to give the structure in Fig. 12.12.

b. The reaction mechanism

This has not been established to the same level of detail as for the serine and thiol proteases. The precise roles of all the catalytic groups are not known unambiguously. There is little doubt that the zinc ion acts as an electrophilic catalyst to polarize the carbonyl group and stabilize the negative charge that develops on the oxygen (Chapter 2B7).[118] The ionized carboxylate of Glu-270 is implicated in catalysis from the pH-rate profile. The activity follows a bell-shaped curve with pH, with an optimum at pH 7·5, depending on the basic form of a group of pK_a 6 and the acid form of a group of pK_a 9·1 in the free enzyme.[119,120] The lower pK_a is that of Glu-270, the higher pK_a has not yet been assigned unambiguously. The hydroxyl of Tyr-248 probably acts as a general acid in the reaction.[110] There is no direct evidence for this, but this residue is certainly essential for the peptidase activity. Modification of the tyrosine side chain by acetylation or diazotization destroys the peptidase activity but *enhances* the esterase activity of the enzyme.[121] Interestingly enough, the esterase activity is retained when the zinc is replaced by mercury, cadmium, or lead, although this destroys the peptidase activity.[122]

A major question is whether the Glu-270 acts as a nucleophilic

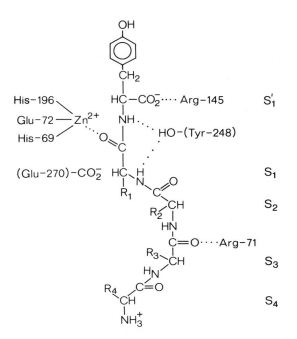

FIG. 12.12. The postulated productively-bound complex of carboxypeptidase A and a polypeptide substrate (Courtesy of W. N. Lipscomb).

catalyst, forming a mixed anhydride with the substrate (eqn (12.25)) or acts as a general base to activate the attack of water on the substrate (eqn (12.26)).[110]

$$\text{E—CO}_2^- + \text{RCONHCH(R')CO}_2^- \longrightarrow \text{E—CO}_2\text{OCR} \xrightarrow{\text{H}_2\text{O}} \text{E—CO}_2^- + \text{RCO}_2^-$$
$$+$$
$$\text{NH}_2\text{CH(R')CO}_2^- \qquad (12.25)$$

$$\text{E—CO}_2^- + \text{H}_2\text{O} + \text{RCONHCH(R')CO}_2^- \longrightarrow \text{E—CO}_2^- + \text{RCO}_2^- + \text{NH}_3^+\text{CH(R')CO}_2^-$$
$$(12.26)$$

The problem is that there is little direct positive evidence that can be used to distinguish between the two possibilities. Trapping and rapid reaction experiments using the physiological substrates have never been able to detect an intermediate acylenzyme. Very recently, two interesting experiments have been described that give conflicting evidence. One experiment, using the ester substrate *O-(trans-p*-chlorocinnamoyl)-L-β-phenyllactate, provides evidence in favour of a two-step mechanism involving an acylenzyme intermediate between the cinnamic acid and a carboxyl group of the enzyme.[123] As intermediates have never been detected at normal temperatures, experiments were performed at temperatures as low as $-60°$ C. The reaction was monitored by the changes in

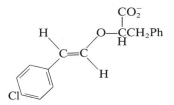

O-(*trans*-*p*-chlorocinnamoyl)-L-β-phenyllactate

the spectrum of the cinnamoyl group. This is a similar spectral probe to the furylacryloyl group discussed in Chapter 7. Its spectrum depends on its precise chemical environment; it is likely that there are spectral shifts on forming the enzyme–substrate complex, acylenzyme, and enzyme–product complex, but the nature of the shifts are not readily predictable. It was found on mixing the substrate with *excess* enzyme at higher temperatures that the substrate is hydrolysed in a single exponential process. However, at low temperatures, about $-40°$ C, the reaction is biphasic. The rate constant for the second phase increases more rapidly than that for the first phase with increasing temperature, so that although the second phase is slower than the first phase at low temperature, it is faster at high temperature. The two phases were interpreted as being the formation and subsequent hydrolysis of an acylenzyme. At normal temperatures the deacylation rate is relatively rapid so the acylenzyme does not accumulate; at low temperatures the deacylation rate is slowed down sufficiently for the acylenzyme to accumulate. However, the spectral changes of the cinnamoyl group cannot be assigned to specific chemical or physical events with the same ease that the increase in, say, the absorbance of *p*-nitrophenol indicates that a nitrophenyl ester is being cleaved. It could be argued that the changes in absorbance are due to substrate-induced conformational changes in the enzyme that perturb the spectrum of the cinnamoyl group. The authors therefore presented further evidence consistent with the formation of a covalent intermediate. At $-58°$ C, where the state corresponding to the acylenzyme is stable, they added a strongly binding competitive inhibitor of the enzyme, L-benzylsuccinate. It does not displace the cinnamoyl compound from the enzyme, indicating that the cinnamoyl group is covalently bound. There is also evidence that the covalent intermediate is an acid anhydride: on partial denaturation of the acylenzyme with urea, deacylation takes place with a rate constant consistent with the hydrolysis of an acid anhydride.

Evidence against the acylenzyme mechanism has been adduced from another recent experiment using isotope exchange measurements.[124] If the hydrolysis occurs by the anhydride route of eqn (12.25), then the synthesis of the peptide in the reverse reaction requires the initial formation of the anhydride from the enzyme and RCO_2^-. The enzyme should thus be able to catalyse the exchange of ^{18}O between the substrate and water (eqn (12.27)). However, exchange does not occur in the

$$RC^{18}O_2^- + E{-}CO_2^- \rightarrow E{-}CO_2{}^{18}OCR \rightarrow RCO^{18}O^- + E{-}CO_2^-$$

$$H_2{}^{18}O \qquad\qquad H_2O$$

$$(12.27)$$

absence of added free amino acid $NH_3^+CH(R')CO_2^-$, and occurs in its presence by the resynthesis of the peptide. The exchange is not stimulated by the analogue $HOCH(R')CO_2^-$. This rules out the anhydride pathway unless the added amino acid is an activator of the exchange reaction (and its hydroxyl analogue is not) or the $H_2{}^{18}O$ released in eqn (12.27) does not exchange with the medium but remains attached to the enzyme. This, and the observation that methanol cannot substitute for water in the solvolytic reaction, has led to the proposed mechanism (12.28).[124]

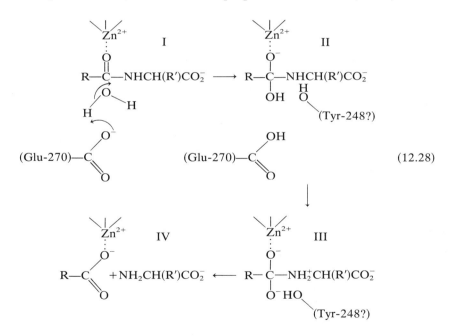

$$(12.28)$$

This mechanism accounts for the inability of methanol to substitute for water since the intermediate III has two negatively charged oxygens. The hydroxyl of Tyr-248 possibly acts as a 'bridge' for the transfer of the proton from the hydroxyl group of the tetrahedral intermediate in II to the nitrogen atom.

Perhaps the two sets of experiments are not in conflict. In view of the observations mentioned earlier that chemical modification of the enzyme or replacement of the metal ion affects the esterase and peptidase activities in different ways, it is possible that the cinnamoyl ester is hydrolysed by the acylenzyme route whilst peptides involve general base catalysis by Glu-270.

c. The zymogen

Procarboxypeptidase A is activated by the removal of a peptide of some 64 residues from the N-terminus by trypsin.[125] This zymogen has significant catalytic activity. As well as hydrolysing small esters and peptides,[126,127] procarboxypeptidase removes the C-terminal leucine from lysozyme only seven times more slowly than does carboxypeptidase. Also, the zymogen hydrolyses BzGly-L-Phe with $k_{cat} = 3$ s^{-1} and $K_M = 2.7$ mM, compared with values of 120 s^{-1} and 1.9 mM for the reaction of the enzyme.[126] Unlike the example of chymotrypsinogen, the binding site clearly pre-exists in procarboxypeptidase and the catalytic apparatus must be nearly complete.

4. Acid proteases[128–130]

The acid proteases are so named because they function at low pH. The best known member of the family is pepsin, which has the distinction of being the first enzyme to be named (in 1825 by Schwann). Other members are chymosin (rennin), cathepsin D, *Rhizopus*-pepsin (from *Rhizopus chinensis*), and penicillopepsin (from *Penicillium janthinellum*). Sequence homologies have been noted between pepsin, chymosin, and penicillopepsin.[131,132] The crystal structures of several of the enzymes have been solved at low resolution.[133,134] The structure of pepsin[135] has been recently refined to 2.7 Å resolution and penicillinopepsin[136] has been solved at 2.8 Å. The reaction mechanism of pepsin and penicillinopepsin is by far the most obscure of all the proteases and there are no simple chemical models for guidance. The following refers to both enzymes; they are similar both structurally and kinetically.

a. Pepsin

Pepsin consists of a single polypeptide chain of molecular weight 34 644 containing 327 amino acid residues.[137,138] Ser-68 is phosphorylated, but this phosphate may be removed without significantly altering the catalytic properties of the enzyme.[139] Like other acid proteases, the active site is an extended area that can accommodate at least four or five, and maybe as many as seven, substrate residues.[140,141] The enzyme has a preference for hydrophobic amino acids on either side of the scissile bond. A statistical survey of the bond cleavages in proteins shows that there is a specificity for leucine, phenylalanine, tryptophan, and glutamate (!) in the

S_1 site, and tryptophan, tyrosine, isoleucine, and phenylalanine in the S_1' site.[141] Pepsin rarely hydrolyses esters, the exceptions being esters of L-β-phenyllactic acid and some sulphite esters.

(i) *The reaction mechanism of pepsin.* There are two catalytically active residues in pepsin, Asp-32, and Asp-215. Their ionizations are seen in the pH-activity profile which has an optimum at pH 2–3 and depends upon the acidic form of a group of pK_a about 4·5 and the basic form of a group of pK_a of about 1·1.[142,143] The pK_a values have been assigned from the reactions of irreversible inhibitors that are designed to react specifically with ionized or un-ionized carboxyl groups. Diazo compounds, such as N-diazoacetyl-L-phenylalanine methyl ester, which react with un-ionized carboxyls, react specifically with Asp-215 up to pH 5 or so (eqn (12.29)).[144–6] Epoxides, which react specifically with ionized

$$(\text{Asp-215})-\text{CO}_2\text{H} + \text{N}_2\text{CHCONHCH(Ph)CO}_2\text{Me} \longrightarrow$$

$$(\text{Asp-215})-\text{CO}_2\text{CH}_2\text{CONHCH(Ph)CO}_2\text{Me} \quad (12.29)$$

$$+ \text{N}_2$$

carboxyls, modify Asp-32 (eqn (12.30)). The pH dependence of the rate

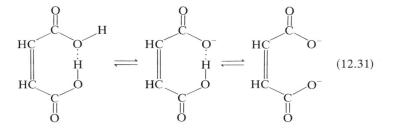

$$(\text{Asp-32})-\text{CO}_2^- + \text{CH}_2-\text{CHR}. \longrightarrow (\text{Asp-32})-\text{CO}_2\text{CH}_2\text{CHR} \quad (12.30)$$

of modification shows that its pK_a is less than 3.[147] It is seen in the high-resolution crystal structure that the carboxyl groups of the two aspartate residues are hydrogen bonded to each other.[145,146] This is similar to the ionization of maleic acid, which has pK_a values of 1·9 and 6·2 (eqn (12.31)).

$$(12.31)$$

Any mechanism proposed for pepsin must account for the observation that both an *amino*enzyme and an *acyl*enzyme are formed during the reaction. The aminoenzyme was first inferred nearly 20 years ago from experiments in which pepsin was found to catalyse transpeptidation as well as hydrolysis.[148] During the hydrolysis of Cbz-Glu-Tyr, such compounds as Cbz-Glu-Tyr-Tyr and Tyr-Tyr are formed. The simplest explanation for this is that an aminoenzyme E—NHR is formed, and that this can transfer the amino group to a free carboxyl group to synthesize a new peptide. Recent partition experiments, in which a common aminoenzyme may be generated from several different substrates, support this hypothesis (eqn (12.32)).[149,150] (In the following, (NH-Tyr) = tyrosine bound by its NH group to another residue, and (Leu-CO) = leucine bound by its carboxyl.)

$$
\begin{array}{c}
\\
\\
E—OH+RCO—(NH\text{-}Tyr) \longrightarrow E—(NH\text{-}Tyr) \\
+ \\
R'CO_2H
\end{array}
\quad
\begin{array}{c}
E—OH+R'CO—(NH\text{-}Tyr) \\
\nearrow^{R'CO_2H} \\
\\
\searrow_{H_2O} \\
E—OH+NH_2\text{-}Tyr
\end{array}
$$

$$(12.32)$$

An acylenzyme intermediate was first postulated in 1962 when it was shown that pepsin catalyses the exchange of ^{18}O from $H_2{}^{18}O$ into the carboxyl groups of Cbz-L-Phe and Cbz-L-Tyr. This was largely ignored until it was recently shown that the hydrolysis of Leu-Tyr-Leu gives the product Leu-Leu, which can be formed from the following acyl transfer:[152,153]

$$
E—OH+\overset{*}{L}eu\text{-}Tyr\text{-}Leu \longrightarrow (\overset{*}{L}eu\text{-}CO)—O—E \xrightarrow{\text{Leu-Tyr-Leu}}
$$

$$
\overset{*}{L}eu\text{-}Leu\text{-}Tyr\text{-}Leu+E—OH
$$

$$
\swarrow \qquad \searrow
$$

$$
\overset{*}{L}eu\text{-}Leu \quad + \quad Tyr\text{-}Leu
$$

$$(12.33)$$

This experiment has been extended by using the double-labelled substrate [^{14}C]Leu-Tyr-[^{3}H]Leu to show that simultaneous amino and acyl transfer takes place. It is found that both [^{3}H]Leu-[^{3}H]Leu and [^{14}C]Leu-[^{14}C]Leu are formed.[154] The ^{14}C-labelled product, which predominates by a factor of three or four, comes from the acyl transfer route, whilst the ^{3}H-labelled product arises from the [^{14}C]Leu-Tyr-[^{3}H]Leu-[^{3}H]Leu produced from the aminoenzyme by mechanism (12.32).

It must be borne in mind that there is no direct evidence for covalent intermediates. The amino group in the aminoenzyme may not be covalently bound, but merely 'activated'. The acyl transfer of eqn (12.33) could occur by the direct attack of Leu-Tyr-Leu on the enzyme-bound Leu-Tyr-Leu. Further, neither of the intermediates has been detected by pre-steady-state kinetics. If they exist, they do not accumulate.

Before attempting to postulate a mechanism for the reaction, there is one further piece of evidence. If an aminoenzyme is formed in which the amino group is linked to the enzyme by an amide bond with one of the carboxyls, then ^{18}O should be incorporated into the enzyme during the hydrolysis in $H_2{}^{18}O$ enriched water (eqn (12.34)). However, it has been

$$E—(Asp\text{-}CO)—NHR \xrightarrow{\ H_2{}^{18}O\ } E—(Asp\text{-}CO^{18}OH) \qquad (12.34)$$

shown by the following experiment from Antonov's laboratory that no ^{18}O is incorporated into the enzyme during the hydrolysis in water

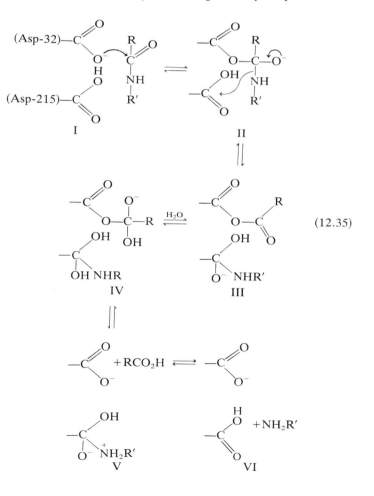

$$(12.35)$$

enriched with 40% $H_2^{18}O$. After catalysing the hydrolysis of AcPhe-TyrOEt, the enzyme was isolated by gel filtration and its two catalytic carboxyls separately blocked by the reactions in eqns (12.29) and (12.30). On hydrolysis of the two esters formed, no ^{18}O is found in the resultant $HO—CH_2CONHCH(Ph)CO_2H$ and $HO—CH_2CH(OH)R$. Either the aminoenzyme contains the amino acid non-covalently bound or possibly the attractive mechanism (12.35) suggested by Antonov holds.[155]

The scheme as written accounts for transpeptidation by amino transfer simply by the reversal of step IV to V. However, in order for acyl transfer to take place, III must also be able to break down to expel $R'NH_2$ and form the acylenzyme $[(Asp-32)—CO_2COR:(Asp-215)—CO_2H]$. Perhaps the possibility of III breaking down to expel either $R'NH_2$ or RCO_2H accounts for an anomaly in partitioning experiments in which a series of substrates generating the common aminoenzyme E-Trp are reacted in the presence of the acceptor AcPhe.[156] The ratio of AcPheTrp:Trp formed on the hydrolysis of a series of peptides XPheTrp in the presence of AcPhe varies with the nature of X (X = AcGly etc.). This is usually taken as evidence against a common intermediate (Chapter 7B2b). However, the relative rates of expulsion of $R'NH_2$ (e.g. Trp) and RCO_2H (e.g. XPhe) will obviously depend on the nature of RCO_2H, and hence the fraction of aminoenzyme formed will be variable. Thus (12.35) need not give constant partition ratios.

The reaction mechanism is far more complicated than those of the other proteases. Nevertheless, all the steps in scheme (12.35) are chemically reasonable, although there are no analogies for the overall mechanism in simple chemical systems.

(ii) *The zymogen.* Pepsin is formed from pepsinogen by the proteolysis of 44 residues from the N-terminus. The zymogen is stable at neutral pH, but below pH 5, it rapidly and spontaneously activates. The activation process takes place by two separate routes, a pepsin-catalysed and an intramolecularly-catalysed process. There is much evidence that pepsinogen may activate itself in a unimolecular process, the active site cleaving the N-terminus of its own polypeptide chain.[157-163] Perhaps the neatest demonstration of this intriguing phenomenon is the auto-activation of pepsinogen that is covalently bound to a sepharose resin.[159] The molecules are immobilized and, in general, not in contact with each other. Yet, on exposure to pH 2, the pepsinogen spontaneously activates. The two routes for activation compete, the bimolecular activation dominating at higher zymogen concentrations and above pH 2·5, the intramolecular activation at low pH. The result of this spontaneous activation of the pure zymogen is that the majority of pepsinogen molecules will be active 10 seconds after mixing with the hydrochloric acid in the stomach.

C. Ribonuclease[164,165]

Bovine pancreatic ribonuclease hydrolyses RNA by a two-step process in which a cyclic phosphate intermediate is formed (eqn (12.36)).

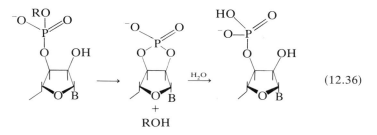

(12.36)

The cyclization step is usually far faster than the subsequent hydrolysis so that the intermediate may be readily isolated. DNA is not hydrolysed as it lacks the 2′ hydroxyl group that is essential for this reaction. There is a strong specificity for the base B on the 3′ side of the substrate to be a pyrimidine: uracil or cytosine.

The enzyme consists of a single polypeptide chain of molecular weight 13 680 containing 124 amino acid residues.[166,167] The bond between Ala-20 and Ser-21 may be cleaved by subtilisin. Interestingly, the peptide remains attached to the rest of the protein, held by non-covalent bonds. The modified protein, called ribonuclease-S, and the native protein, now termed ribonuclease-A, have identical catalytic activities. Because of its small size, availability and ruggedness, ribonuclease is very amenable to physical and chemical study. It was the first enzyme to be sequenced.[166] The crystal structures of both forms of the enzyme have been solved at 2·0 Å resolution.[168,169] Furthermore, because the catalytic activity depends on the ionizations of two histidine residues, the enzyme has been extensively studied by NMR; the imidazole rings of histidines being easily studied by this method (Chapter 5G2a).

The currently accepted chemical mechanism for the reaction was deduced by an inspired piece of chemical intuition before the solution of the crystal structure.[170] It was found that the pH-activity curve is bell shaped with optimal rates around neutrality. The pH dependence of k_{cat}/K_M shows that the rate depends upon the ionization of a base of pK_a 5·22 and acid of pK_a of 6·78 in the free enzyme, whilst the pH dependence of k_{cat} shows that these are perturbed to pK_a values of 6·3 and 8·1 in the enzyme–substrate complex. It was proposed that the reaction is catalysed by concerted general acid–base catalysis by two histidine residues, later identified as His-12 and His-119 (see (12.37) and (12.38)).

In the cyclization step, His-12 acts as a general-base catalyst and His-119 acts as a general acid to protonate the leaving group. Their catalytic roles are reversed in the hydrolysis step; His-119 activates the attack of water by general-base catalysis whilst His-12 is the acid catalyst, protonating the leaving group. This reversal of roles is quite logical.

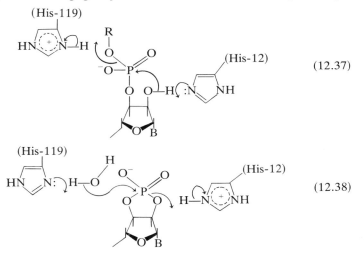

(12.37)

(12.38)

Reaction (12.38) is essentially the reverse of reaction (12.37) except that HOH replaces ROH. It is expected from the principle of microscopic reversibility that a group reacting as a general acid in one direction reacts as a general base in the opposite direction.

1. Structures of the enzyme and enzyme–substrate complex[164,171]

Ribonuclease has a well-defined binding cleft for the substrate. In it are located His-12, His-119, and the side chains of Lys-7, Lys-41, and Lys-66. The structure of the enzyme–substrate complex for the cyclization step has been deduced from the crystal structure of the enzyme and the substrate analogue UpcA (12.39), the phosphonate analogue of UpA.

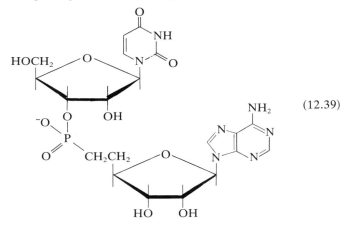

(12.39)

It is a very good analogue, differing from the real substrate only in that a —CH$_2$— group replaces an oxygen atom, so that the structure of its complex with ribonuclease should be close to that of a productively bound enzyme–substrate complex. It is found that His-119 is within hydrogen bonding distance of the leaving group, and His-12 within hydrogen bonding distance of the 2'OH of the pyrimidine ribose. None of the lysine side chains are in contact with the substrate. However, they are thought to be essential to catalysis since (a) activity is lost when they are acetylated, and (b) lysines occur in positions 7, 41, and 66 of all twenty or so ribonucleases of this class that have been sequenced.[172] Perhaps the positive charges on the side chains stabilize the pentacovalent phosphorus intermediate that is formed as the hydroxyl group attacks the phosphate. The pK_a values of His-12 and His-119 have been determined by NMR measurements to be 5·8 and 6·2 respectively at 40°.[173] A considerable fraction of each is in the suitable ionic state at physiological pH for the general acid–base catalysis shown in eqn (12.37) to occur.

A good example of lock-and-key specificity is seen in the pyrimidine binding site.[171] It is possible for both uracil and cytosine rings to make hydrogen bonds with the hydroxyl groups of Ser-123 and Thr-45, and the backbone NH of Thr-45 (12.40). If adenosine derivatives bind in this site, their greater size causes the phosphate to be displaced from the histidines.

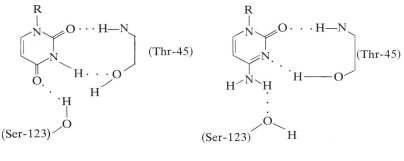

$$(12.40)$$

2. Stereochemistry of the displacement reaction

The nucleophilic displacement on phosphorus in this type of reaction proceeds by an addition–elimination reaction in which a pentacovalent intermediate is formed. This may occur by two routes; the *in-line* mechanism (12.41) in which the attacking nucleophile enters opposite the

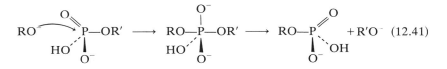

leaving group, or the *adjacent* mechanism (12.42) in which the nucleophile enters on the same side as the leaving group.[174,175] In this case there is an additional step.

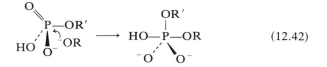

$$(12.42)$$

The trigonal bipyramidal phosphorus must *pseudorotate* so that the leaving group moves to an apical position (12.43). This rearrangement is a

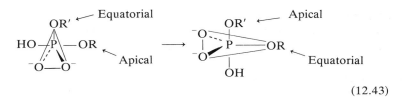

$$(12.43)$$

consequence of microscopic reversibility: the nucleophile enters into an apical position and so the leaving group must leave from an apical position (since in the reverse reaction it must enter apically).

Examination of the enzyme–substrate complex suggests an in-line mechanism. This has been confirmed from some elegant chemical experiments using an optically active substrate, uridine-2'-3'-cyclic-phosphorothioate (12.44).[176–178]

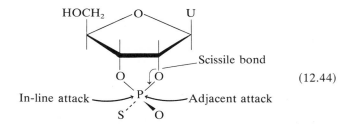

$$(12.44)$$

This compound was crystallized and its structure and absolute stereochemistry determined by X-ray diffraction. On incubating it with ribonuclease in aqueous methanol solution, a methyl ester was formed by the reverse of mechanism (12.37) (eqn (12.45)). The methyl ester was crystallized and its absolute stereochemistry determined by X-ray diffraction to be as in (12.45). This product corresponds to an in-line attack. The methyl ester when incubated with ribonuclease in aqueous solution re-forms the original cyclic-phosphorothioate (12.44). This result is expected from the principle of microscopic reversibility since the forward

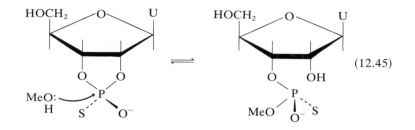

(12.45)

and reverse reactions must go through the same transition state. But it does show directly that the cyclization step involves an in-line attack—an adjacent attack of the ribose hydroxyl in the cyclization of the methyl ester in (12.45) would give the enantiomer of the cyclic-phosphorothioate illustrated.

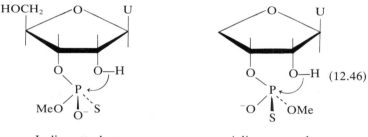

(12.46)

In-line attack   Adjacent attack

D. Staphylococcal nuclease[179,180]

Staphylococcal nuclease is a phosphodiesterase which cleaves DNA and RNA to form 3'-phosphomononucleosides. The enzyme consists of a

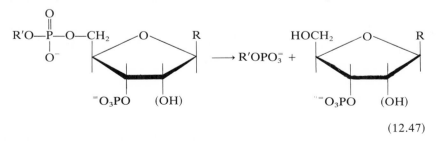

(12.47)

single polypeptide chain of molecular weight 16 900 and contains 149 amino acid residues. The structures of the enzyme and its complex with thymidine-3′,5′-diphosphate have been solved at 2·0 Å resolution.[181,182] Despite this, we do not know the mechanism of the enzyme. After the success of the experiments on ribonuclease, the ignorance about staphylococcal nuclease provides a salutary lesson in that structural knowledge does not automatically solve the problems of mechanism and function. The crystal structure has, however, given the following clues concerning the mechanism.

The 5′-phosphate group of thymidine-3′,5′-diphosphate in the enzyme–inhibitor complex presumably occupies the binding site for the phosphate of a diester substrate. It is seen to be hydrogen bonded to the positively charged side chains of Arg-35 and Arg-87 which will activate it to nucleophilic attack. However, there are none of the usual acid–base side chains to be found in the vicinity of the scissile bond. But there is a calcium ion some 5 Å from the phosphate, bound by the carboxylate groups of Asp-21, Asp-40, and Glu-43. Calcium ions are known to be essential for activity. One possibility is that they provide a metal-bound nucleophilic hydroxide ion (12.48).[183,184] Some support for this idea

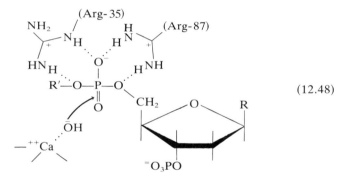

(12.48)

comes from the observation that there is some electron density between the calcium ion and the phosphate group in the map of the crystal structure, and also the pH optimum of the enzyme activity is rather high (the pH-k_{cat}/K_M profile is bell shaped and depends on pK_a values of 8·4 and 9·2).[185] This is very tentative and there is a further cautionary note: two moles of calcium ions bind to the enzyme–inhibitor complex[186] but only one is seen in the crystal structure.

E. Lysozyme[187–189]

Hen egg-white lysozyme is a small protein of molecular weight 14 500 and 129 amino acid residues. It was introduced at the end of Chapter 1 where it was pointed out that the crystal structure of the enzyme stimulated and pre-dated most of the solution studies. A mechanism was proposed for the enzymic reaction based on the structure of the active site and ideas from physical organic chemistry.[190,191] This consists of the following points: there are six sub-sites for binding the glucopyranose rings of the substrate, labelled ABCDEF; the scissile bond lies between sites D and E close to the carboxyl groups of Glu-35 and Asp-52; it is suggested that the reaction proceeds via a carbonium ion intermediate (more strictly, a carboxonium ion) which is stabilized by the ionized carboxylate of Asp-52; the expulsion of the alcohol is general-acid catalysed by the un-ionized carboxyl of Glu-35; furthermore, as discussed in Chapter 10A5b and 10C5c, the sugar ring in site D is distorted to the sofa conformation expected for a carbonium ion; small polysaccharides avoid the strain in the D sub-site by binding in the ABC sites. We shall now see how all of these points, apart from possibly the role of distortion in site D, have been experimentally verified. (The conformation in D was originally called 'half-chair', but 'sofa' is more appropriate.)[192]

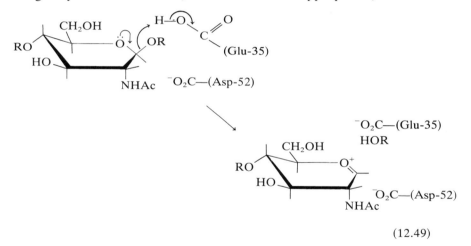

(12.49)

1. The carbonium ion

Alternatives to the carbonium ion mechanism are the direct attack of water on the substrate or the nucleophilic attack of Asp-52 on the C-1 carbon to give an ester intermediate. The single displacement reaction has been ruled out by showing that the reaction proceeds with retention of

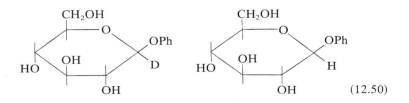

(12.50)

configuration (see Chapter '7C3).[193-195] The carbonium ion or S_N1 mechanism has been substantiated by secondary isotope effects using substrates containing either deuterium or tritium attached to the C-1 carbon rather than hydrogen.[196,197] For example, k_H/k_D for (12.50) is 1·11 compared with values of 1·14 found for a carbonium ion reaction and 1·03 for a bimolecular displacement (S_N2) in simple chemical models.

2. Electrostatic and general-acid catalysis

The pH dependence of k_{cat}/K_M shows that the reaction rate is dependent on an acid of pK_a about 6 and a base of pK_a about 4 in the free enzyme.[198] Asp-52 may be specifically blocked by esterification with triethoxonium fluoroborate (a reaction which requires an ionized carboxylate). The difference titration between the modified and native proteins shows that the pK_a of Asp-52 is 4·5 whilst that of Glu-35 is 5·9.[199] The reaction rate therefore depends on the ionization of the two residues in the manner predicted by the mechanism.

An observation consistent with general-acid catalysis by Glu-35 comes from a study of the reverse reaction. It is found that the rate of reaction of alcohols with the carbonium ion intermediate is virtually independent of their pK_a. This is consistent with the general-base-catalysed attack of the alcohol on the ion, and hence by the principle of microscopic reversibility, the expulsion of the alcoholate ion from the glycoside is general-acid catalysed.[200]

It was emphasized by Vernon in the initial formulation of the lysozyme mechanism that the electrostatic stabilization of the carboxonium ion by Asp-52 is the most important catalytic factor.[191] A recent theoretical study, the first to use the dielectric constant of the protein in such calculations, suggests that the activation energy for k_{cat} is lowered by 37·6 kJ/mol (9 kcal/mol) by the electrostatic stabilization.[201] A high value for this is indicated from experiments showing that the chemical conversion of the carboxyl group of Asp-52 to —CH_2OH abolishes the enzymic activity.[202]

3. Binding energies of the sub-sites

It was originally proposed that non-bonded interactions between the enzyme and the sugar ring in site D distort it to the sofa conformation of the carbonium ion. Various workers have searched for weak binding in this site by estimating the binding energies of the individual sites from binding and kinetic measurements.[187,189] Some estimates for these are given in Table 12.3. Although these values are not precise, it is clear that there is a repulsive energy against the binding of NAM in site D. Also, as

TABLE 12.3. *Binding energies of sub-sites in hen egg-white lysozyme*

Site	Residue binding[a]	Binding energy kJ/mol	kcal/mol
A	NAG	−8	−2
B	NAG	−12	−3
	NAM	−16	−4
C	NAG	−20	−5
D	NAM	+12	+3
	NAG	0	0
E	NAG	−16	−4
F	NAG	−8	−2

[a] NAG = N-acetylglucosamine, NAM = N-acetylmuramic acid

$(NAG)_4$ is found to bind about equally in sites ABC and in ABCD, there is no net binding energy for NAG in site D.[203] The position of binding of small substrates has been located from their interactions with probes, such as the dye Biebrich Scarlet[204] and the lanthanide ion,[203] bound in the cleft, and they have been shown to be predominantly non-productively bound. This would appear to provide very strong evidence for the strain mechanism. However, the evidence has been reviewed by Levitt[205] who, as discussed in Chapter 10C4c, finds by calculation that it is very unlikely that the sugar ring in site D is distorted.[201,205] Also, transition-state analogues that have the sugar ring in site D chemically modified to resemble the carbonium ion bind no more than a hundred times more tightly than substrates containing the unmodified ring (Chapter 10A5).[203,206] Furthermore, it is suggested that the poor binding of the ring in site D is due to the displacement by the ring of two water molecules that are bound to the carboxylate of Asp-52. The strain on binding to the D sub-site is thus 'electrostatic' rather than mechanical. (In any case, as emphasized in Chapter 10, better binding of the transition state does not imply that the bound substrate is distorted.)

The deductions from this remarkable example of X-ray crystallography have not only stood the test of time but have been neatly confirmed by solution studies, except that the emphasis of the strain mechanism is now on the electrostatic component rather than on that due to distortion.

To end on a note of caution, it has been found that the binding of small substrates is a two-step process with a step involving the isomerization of the enzyme–substrate complex.[207,208] It is not clear what is involved in this so-called conformational change and how this fits into the above scheme, but it could be related to the small structural changes that are observed on binding $(NAG)_3$ to the crystalline enzyme.

F. Carbonic anhydrase[209,210]

The carbonic anhydrases catalyse the hydration of carbon dioxide and the dehydration of bicarbonate (12.51). The crystal structures of the human B

$$CO_2 + H_2O \rightleftharpoons HCO_3^- + H^+ \qquad (12.51)$$

and C isozymes have been solved at high resolution.[211,212] They are both metalloenzymes containing one mole of tightly bound zinc per single polypeptide chain of molecular weight about 29 000. The C enzyme, which is the more active, has 259 amino-acid residues in the chain, one less than the B form.[213,214] The sequences of the two are about 60% homologous and their tertiary structures are quite similar. The Zn^{2+} ion is ligated by the imidazole rings of three histidine residues at the bottom of the active site cavity, some 12 Å from the surface. The fourth ligand is probably a water molecule or hydroxide ion, depending on the pH.[211,212]

The C enzyme is an extremely efficient catalyst. In the hydration reaction k_{cat} is 10^6 s^{-1} and K_M for CO_2 is 8·3 mM whilst in the dehydration reaction k_{cat} is $6 \times 10^5 \text{ s}^{-1}$ and K_M for HCO_3^- is 32 mM.[215] The catalytic activity depends on the ionization of a group of pK_a 7 in the free enzyme.[215] The hydration reaction depends upon this group being in the basic form, and dehydration in the acid form. The turnover numbers of the reactions are far higher than the rate constants for the transfer of protons between water and a group of pK_a 7 (about $2 \times 10^3 \text{ s}^{-1}$—Table 4.2). The enzymes will also catalyse the hydration of aldehydes and hydrolyse esters in non-physiological reactions.

There are three questions in particular that have occupied recent studies: (a) what is the ionizing group of pK_a 7 responsible for activity ?; (b) what is the bound substrate in the dehydration reaction (HCO_3^- or H_2CO_3) ?; (c) are the high turnover numbers compatible with the rates of proton transfer?

The classical mechanism invokes a zinc-bound water molecule which ionizes with a pK_a of 7 to give the nucleophilic zinc-bound hydroxide ion at high pH.[210,216] The zinc-bound hydroxide is known to be nucleophilic and to ionize in this region (Chapter 2B7b). Many other proposals have been made[217-219] but the only other groups at the active site that can ionize with this pK_a are the histidines. However, their ionizations have been studied by NMR and found to be inconsistent with the activity-linked ionization.[220] Furthermore, measurements on nuclear quadrupole interactions in the cadmium-substituted enzyme are consistent with the ionization of a metal-bound water molecule with the same pK_a as found in the pH–activity profile.[221]

A chemically reasonable mechanism is the following:

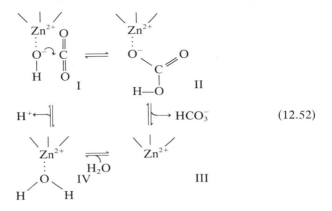

$$(12.52)$$

Any mechanism involving HCO_3^- as substrate requires that at some stage in the reaction there is proton release from the enzyme to the medium. In (12.52) this occurs at the step $IV \rightarrow I$. In an unbuffered solution the rate of this proton transfer step would be too slow to regenerate the free enzyme. For example, from the rate constants in Table 4.2 it is seen that the rate constant for the transfer of a proton from a base of pK_a 7 to water is $2 \cdot 5 \times 10^3 \, s^{-1}$, and to $10^{-7} \, M$ hydroxide ions at pH 7 is $2 \cdot 3 \times 10^3 \, s^{-1}$, values well below the turnover number of $5 \times 10^5 \, s^{-1}$ at this pH. This cannot be circumvented by invoking the combination of the zinc ion with OH^- at step $III \rightarrow IV$ instead of water for similar reasons. The maximum value of a second-order rate constant is about 10^{10}–$10^{11} \, s^{-1} \, M^{-1}$ (Chapter 4, part 2), which combined with the hydroxide concentration of $10^{-7} \, M$ gives a limit of 10^3–$10^4 \, s^{-1}$ for this step at pH 7. A similar calculation indicates that it is unlikely that the undissociated H_2CO_3 is the substrate in the dehydration reaction since its concentration is so low compared with HCO_3^- at pH 7.[216,222] (If this were the substrate, there would be no proton uptake or release in the reaction.) However, a simple calculation shows that low concentrations of buffers in solution account for the transfer rate constants (see Table 4.2).[216,223] This has recently been verified from experiments in unbuffered solutions where it is found that the rate of the dehydration reaction is anomalously slow. The addition of 10 mM buffer restores the maximum reaction rate.[224,225]

Added in proof. The crystal structure of the complex of human carbonic anhydrase B with its competitive inhibitor imidazole has now been solved (K. K. Kannan, M. Petef, K. Fridborg, H. Cid-Dresdener, and S. Lövgren, *FEBS Letts* **73**, 115 (1977)). The imidazole is bound in a hydrophobic pocket with a nitrogen atom directly coordinated to the zinc ion. Most interestingly, it appears to bind as fifth ligand without displacing the zinc-bound H_2O or OH^-. By analogy with this, it is suggested that an oxygen atom of carbon dioxide also binds directly to the zinc and does not displace the bound water. The zinc ion thus both orients the CO_2 and polarizes it as well as providing a zinc-bound nucleophilic water molecule (12.52a).

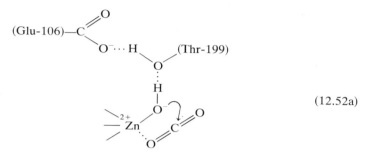

(12.52a)

The expansion of the coordination number of the zinc to 5 on substrate binding is similar to that proposed for horse liver alcohol dehydrogenase (12.4).[15] It is noted by the authors that the bound water molecule is hydrogen bonded to the hydroxyl of Thr-199 which is in turn hydrogen bonded with the carboxylate of Glu-106. This adds a further twist to the problem of the identification of the group ionizing with the pK_a of 7. It could perhaps be the carboxylate of Glu-106, perturbed to a high value by its environment, rather than the zinc-bound water. If this is so, then the nucleophile is not the hydroxide ion as in (12.52a) but the un-ionized water molecule activated by general-base catalysis from Glu-106 transmitted by the intervening hydroxyl of Thr-199.

G. Glycolytic enzymes

Most of the glycolytic enzymes have been crystallized and their crystal structures solved. The current status of the X-ray studies has been recently reviewed by Blake.[226] Unfortunately for our purposes, only a few of the enzymes have been sequenced. Two of these, lactate and glyceraldehyde-3-phosphate dehydrogenases, were discussed earlier in Section A. In this section we shall just deal with some interesting points of a few of the enzymes.

1. Triosephosphate isomerase[227-229]

Triosephosphate isomerase very efficiently catalyses the interconversion of D-glyceraldehyde-3-phosphate and dihydroxyacetone phosphate (12.53). The crystal structure of the enzyme from chicken muscle has been solved at 2·5 Å resolution.[230-231] It is a symmetrical

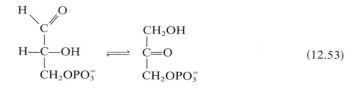

$$(12.53)$$

dimer of molecular weight 53 000 with 247 amino-acid residues in each of the identical chains. There is no evidence for any cooperativity between the sub-units during catalysis.

a. Structure of the enzyme–substrate complex

This enzyme affords a rare opportunity for directly studying an enzyme substrate complex by X-ray diffraction methods. The reaction catalysed is a simple equilibrium involving one substrate and one product. Furthermore, the equilibrium constant $(\sim 20)^{\dagger}$ greatly favours the dihydroxyacetone phosphate. This has enabled the solution of the crystal structure of the complex of the enzyme and the acetone phosphate to be carried out at 6 Å resolution by diffusing the substrate into the crystals and using the difference Fourier method.[230,232] Apart from locating some conformational changes in the enzyme, it was possible to place the substrate relative to two acid–base catalysts in the enzyme and clarify the reaction mechanism. It is known from solution studies (see below) that the reaction occurs via a *cis*enediol intermediate[233,234] with the proton being transferred by a single base, probably the carboxylate of Glu-165.[235–237] The crystal structure shows that the glutamic carboxylate is equidistant from the C-3 and C-2 carbons of the substrate so that it can shuttle a proton between the two. Furthermore, it is also found that the imidazole ring of His-95 is equidistant from the carbonyl oxygen and the hydroxyl oxygen so that it can shuttle a proton between them.

2. Deuterium and tritium tracer experiments and the mechanism of aldose–ketose isomerases

One of the most important techniques for studying the mechanism of these enzymes which transfer a hydrogen between two carbon atoms is

---

†The aldehyde and ketone substrates are partly hydrated in solution. At 25°, about 55% of dihydroxyacetone phosphate and 3·3% of glyceraldehyde-3-phosphate are in the unhydrated *keto* forms which are the substrates for enzymic reactions. The equilibrium constant

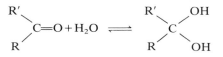

between the unhydrated *keto* forms is therefore about 330:1 in favour of the dihydroxyacetone phosphate (D. R. Trentham, C. H. McMurray, and C. I. Pogson, *Biochem. J.* **114,** 19 (1969); S. J. Reynolds, D. W. Yates, and C. I. Pogson, *Biochem. J.* **122,** 285 (1971)).

the use of isotopically labelled hydrogen. The experiments have increased in complexity and sophistication over the years and are best understood by a historical approach.

a. Exchange of protons with the medium—the enediol intermediate[233,238–240]

The first experiments performed in aqueous solutions enriched with deuterium or tritium showed that up to one mole of isotope is stereo-specifically incorporated into the products of the reactions of mannose-6-phosphate, glucose-6-phosphate, ribose-5-phosphate, and triose-phosphate isomerases (12.54). This type of experiment is always inter-

$$
\begin{array}{c}
\underset{H}{\overset{O}{\diagdown\!\!\diagup}}C \\
| \\
H-C-OH \\
| \\
R
\end{array}
\quad\xrightarrow{\text{E, D}_2\text{O}}\quad
\begin{array}{c}
D \\
| \\
H-C-OH \\
| \\
C=O \\
| \\
R
\end{array}
\qquad (12.54)
$$

preted as ruling out a direct hydride transfer and instead showing that protons are formed which can exchange with the medium. It was suggested that an enediol intermediate, $RC(OH)=C(OH)H$, is formed.[238]

b. Detection of intramolecular proton transfer – cisenediol and a single base[234]

The next development was the discovery that there is also an intramolecular transfer of hydrogen (tritium) between the two carbons (12.55). The simplest interpretation of this is that the tritium from the

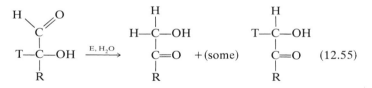

C-2 of the substrate is transferred to a catalysing base, some of the tritium exchanges with protons from the water before it is transferred to the C-1 carbon whilst the rest is incorporated into the C-1 position before it has time to escape into solution. Two important conclusions were drawn from this.

(1) The transfer is mediated by a single base (if a base removed the proton from C-2 and a separate acid residue transferred its proton to C-1, there would be no transfer).

(2) As the proton is transferred by a single base, it must return to the *same* face of the enediol from which it is removed. The stereochemistry

of the products shows that the enediol is *cis* (Chapter 2H). On the basis of this, Rose proposed the following mechanism:[228,248]

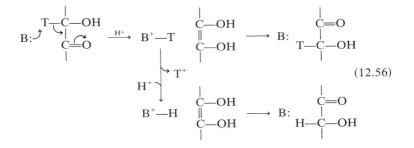

$$(12.56)$$

The base B: in the triosephosphate isomerase mechanism is the carboxylate of Glu-165. The precise role of His-95 is not yet clear: its ionization is not seen in the pH dependence of k_{cat}. (The plot of k_{cat}/K_M against pH for the chicken muscle enzyme is bell shaped with pK_a values of 6 and 9, whilst k_{cat} follows a sigmoid curve showing only the lower pK_a.[241] Also, k_{cat}/K_M for the yeast enzyme is a sigmoid curve showing only the lower pK_a.[242] It was once thought that the pK_a of 6 was due to the ionization of Glu-165, but this is, in fact, the pK_a of the phosphate group of the substrate.[242] The pH dependence of the rate of the chemical modification of Glu-165 shows that its pK_a is $3 \cdot 9$[242,243]—see Chapter 5B2c and Chapter 7G.)

c. Construction of the Gibbs energy profile for the reaction[244]
The most recent development in the use of isotopes in these reactions is the complete steady-state analysis of the exchange of tritium from tritiated water into the product and remaining substrate, the exchange of tritium from tritiated substrate into the solvent and products, and the primary kinetic isotope effects on the reaction rate. From this, it has been possible to determine all the rate constants for mechanism (12.56) and construct the Gibbs energy profile for the reaction (Fig. 12.13).

3. Glucose-6-phosphate isomerase[228,229]

The other isomerase in glycolysis, D-glucose-6-phosphate isomerase, catalyses the interconversion of glucose-6-phosphate and fructose-6-phosphate. The crystal structure of the enzyme from pig muscle has been solved at $3 \cdot 5$ Å and is found to be a symmetrical dimer of molecular weight 120 000.[245]

The reaction sequence is more complicated than that of triosephosphate isomerase. The sugars exist in solution as the cyclic hemiacetals which are equilibrium mixtures of the α and β anomers (the glucose-6-phosphate is 38% α and 62% β, the fructose-6-phosphate is 20% α and

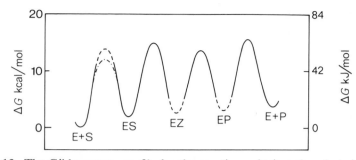

FIG. 12.13. The Gibbs energy profile for the reactions of triosephosphate isomerase (W. J. Albery and J. R. Knowles, ref. 244). S = dihydroxyacetone phosphate,
Z = the enediol, P = glyceraldehyde-3-phosphate. The line $-\cdot-\cdot-\cdot-$ represents the
diffusion-controlled encounter rate. The energy barriers for the bimolecular
reactions are calculated for a concentration of 40 μM, the approximate concentration of dihydroxyacetone phosphate in the cell. Note how all the energy barriers
are about the same and little larger than that for the association of enzyme and
substrate. Further evolution to increase the rate is not possible. This enzyme nicely
satisfies the criterion of Chapter 10B2 for a perfectly evolved enzyme, k_{cat}/K_M is
close to the diffusion controlled limit and K_M is far higher than [S]

80% β).[228] The actual substrate for the isomerization reaction is the
acyclic form in which the ring has been opened.[246–248] Very little of this
exists in solution and so the first step of the reaction is the enzyme-
catalysed opening of the ring. The α anomer reacts more rapidly than the
β, but both are utilized and produced in the reaction.

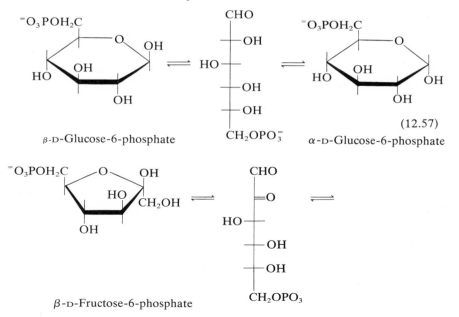

$$(12.57)$$

β-D-Glucose-6-phosphate   α-D-Glucose-6-phosphate

β-D-Fructose-6-phosphate

$$^-O_3POH_2C \quad CH_2OH \quad (12.58)$$

α-D-Fructose-6-phosphate

The reaction pathway is of the following form:[228,234]

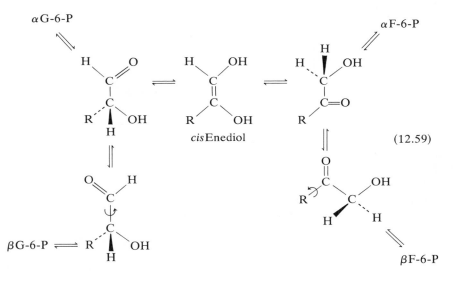

(12.59)

Mannose-6-phosphate is not converted to fructose-6-phosphate by the enzyme, but the conversion of the α-anomer to the β-anomer by ring opening is catalysed by the enzyme at a rate approaching that of the isomerization of glucose-6-phosphate.[249]

Mannose derivatives differ from those of glucose by the H and OH groups about the C-2 carbon being stereochemically interchanged. This presumably displaces the hydrogen atom from the base that catalyses the formation of the cisenediol in (12.59).

The rate of the isomerization reaction depends on the presence of a basic group of pK_a 6·8 and an acidic group of pK_a 9·3 in the free enzyme.[250] It has been tentatively suggested on the basis of heats of ionization that these are due to the imidazole and ammonium side chains of a histidine and a lysine residue respectively.[250] A glutamic residue is modified by the epoxide 1,2-anhydro-D-mannitol-6-phosphate (Chapter 7G).[251] These observations have been synthesized into a mechanism by the following crystallographic study. An interesting point about this is

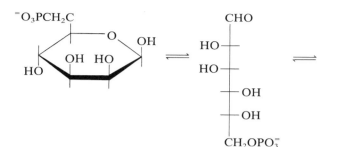

β-D-Mannose-6-phosphate

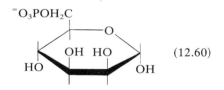

(12.60)

α-D-Mannose-6-phosphate

that the enzyme has not yet been sequenced and so the amino acid side chains cannot be positively identified.

The crystal structure of the complex between the enzyme and the powerful inhibitor arabinonate-5-phosphate has been solved at 3·5 Å resolution.[252] This compound resembles the *cis*enediol intermediate (12.61) and binds a thousand times more tightly than the substrate.[253]

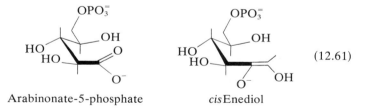

Arabinonate-5-phosphate   *cis*Enediol   (12.61)

The binding site is located between the two sub-units. The phosphate moiety, the ring oxygen, and the C-1 and C-2 hydroxyls bind to one sub-unit whilst there is a movement in the other sub-unit that causes two large side chains to bind to the hydroxyls on the C-2 and C-3 carbons.

It is suggested that the glutamate that is modified by the epoxide is *not* the base that catalyses the proton transfer between the two carbon atoms in the isomerization reaction. It instead catalyses the ring opening reaction and the proton transfer between oxygens of the carbonyl and hydroxyl groups in the steps involving the *cis*enediol (see Fig. 12.14).[252]

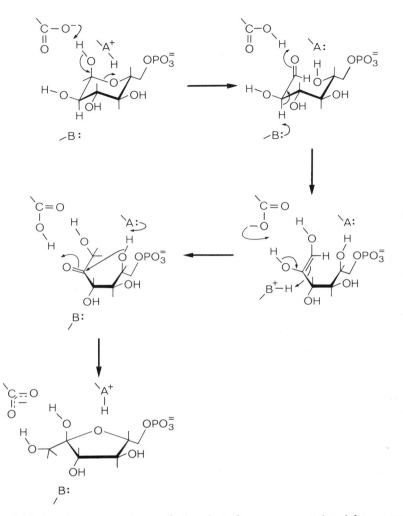

FIG. 12.14. Mechanism of glucose-6-phosphate isomerase postulated from crystallographic studies (ref. 252)

It will be interesting to see, when the enzyme has been sequenced, whether or not the crystallographers guessed correctly about the amino acid side chains.

4. Phosphoglycerate mutase[254]

D-Phosphoglycerate mutase catalyses the interconversion of 2- and 3-D-phosphoglycerate (12.62). The enzyme from muscle is a dimer of molecular weight 54 000. The crystal structure of the yeast enzyme has been

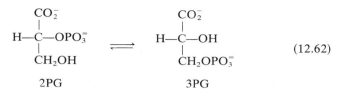

$$(12.62)$$

solved at 3·5 Å and found to be a symmetrical tetramer of molecular weight 110 000.[255]

The reaction sequence involves 2,3-diphosphoglycerate (2,3DPG) as an activator or primer. Its function appears to be to phosphorylate a histidine side chain and produce a catalytically active phosphorylenzyme. These have been isolated for both the yeast and muscle enzymes.[256] The reaction kinetics (ping-pong) are consistent with the scheme (12.63).[257]

$$E + 2,3DPG \longrightarrow E{-}P \xrightarrow{\text{2PG or 3PG}} (E.2,3DPG) \Big\langle \begin{array}{c} 2PG + E{-}P \\ 3PG + E{-}P \end{array} \qquad (12.63)$$

The formation of 2PG from 3PG, for example, involves the transfer of the phosphate from the phosphorylenzyme to the 2 position of the substrate whilst the phosphate on the 3 position is transferred to the enzyme to regenerate the phosphorylenzyme. A tightly bound complex of the enzyme and 2,3DPG is possibly formed during the catalytic cycle.

The mechanism of the reaction comes out beautifully from the crystal structure. The sequence around the active site of the enzyme has been determined and fitted to the electron density. It is seen clearly that there is a binding site for the substrate in which a lysine side chain binds the carboxylate of the substrate. The most interesting feature is that there are two imidazole rings of histidine side chains which are parallel, some 4 Å apart, and span the 2 and 3 positions of the substrate.[258] These provide an obvious mechanism for the transfer and acceptance of the phosphates from the substrate.

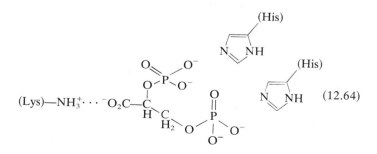

$$(12.64)$$

A phosphate from the diphosphoglycerate can be transferred to one of the histidine residues. The monophosphoglycerate then leaves the enzyme. The phosphoryl group will then rapidly transfer back and forth between the two histines so that either 2PG or 3PG may bind to the phosphorylenzyme. For example,

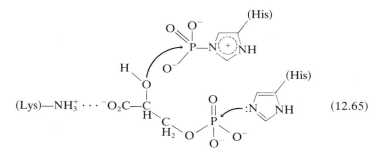

(12.65)

H. Epilogue

A traditional way of viewing enzyme catalysis, formulated from chemical studies long before the determination of the crystal structure of any enzyme, is to attribute it to contributions from the following factors: general acid–base catalysis; metal ion catalysis; nucleophilic catalysis; electrostatic catalysis; propinquity effects (a combination of an intramolecular reaction and correct orientation); strain (distortion of substrate); induced fit (distortion of enzyme). All of these factors have now been found to some extent.

General-acid–base catalysis is the most ubiquitous, being found in the reactions of the dehydrogenases, the serine proteases (and most probably the thiol proteases and carboxypeptidase), ribonuclease, lysozyme, and the aldose–ketose isomerases. Metal ion catalysis, in its classical form of stabilizing an anion, is found in the mechanisms of carboxypeptidase and possibly alcohol dehydrogenase. In the form of a metal-bound hydroxide ion, it is thought to provide a nucleophile in the reactions of carbonic anhydrase and possibly staphylococcal nuclease. Nucleophilic catalysis certainly occurs during the hydrolyses catalysed by the thiol and serine proteases and also in the reactions of many enzymes where Schiff bases are formed between the carbonyl group of the substrate and the side chain of a lysine residue. Transfer of a phosphoryl group is found between the substrates and histidine side chains in phosphoglycerate mutase. Electrostatic catalysis is important in the stabilization of the carbonium ion intermediate in the reactions of lysozyme. Propinquity effects in enzymic reactions are not easily studied and the best information has come from simple model systems and calculations as discussed in Chapter 2. There is no direct evidence for the distortion of a substrate on being bound to an enzyme, but, as discussed in Chapter 10, there are many examples of strain in the form of 'transition-state stabilization'. There are also several examples of the distortion of the enzyme on binding a substrate.

It is still not possible to take a particular reaction and say that the rate enhancement has a factor of x from general-base catalysis, y from the entropic advantage, z from transition-state stabilization etc. The description of an enzymic reaction at this level will no doubt come in the future from the theoretical chemists, the first, tentative, steps having been made in this direction. The ground has been laid for such a treatment by the chemists and kineticists who have elucidated the reaction sequences, and by the spectroscopists and the crystallographers who have provided the structures of the enzymes and complexes. At the present time, we have chemically satisfying descriptions of the mechanisms of many enzymic reactions. Some of these are better understood than the chemical models in solution chemistry because we know the precise disposition of the catalytic groups in the enzymic reactions.

The study of sub-unit interactions in oligomeric enzymes is a product of the crystallographic era. We know the structures of deoxyhaemoglobin and its fully ligated form, and have a good idea of the causes of the cooperativity of ligand binding. But we do not know the kinetic route between the two states and the structures of the intermediates. Phenomena such as half-of-the-sites reactivity and negative cooperativity of ligand binding are well-documented. Their importance is yet to be explained.

References

1   G. Popják, *The Enzymes* **2,** 115 (1970).
2   H. F. Fisher, E. E. Conn, B. Vennesland, and F. H. Westheimer, *J. biol. Chem.* **202,** 687 (1953).
3   M. E. Pullman, A. San Pietro, and S. P. Colowick, *J. biol. Chem.* **206,** 129 (1954).
4   M. G. Rossman, A. Liljas, C.-I. Brändén, and L. J. Banasjak, *The Enzymes* **11,** 61 (1975).
5   C.-I. Brändén, H. Jornvall, H. Eklund, and B. Furugren, *The Enzymes* **11,** 104 (1975).
6   J. J. Holbrook, A. Liljas, S. J. Steindel, and M. G. Rossman, *The Enzymes* **11,** 191 (1975).
7   L. J. Banaszak and R. A. Bradshaw, *The Enzymes* **11,** 369 (1975).
8   K. Dalziel, *The Enzymes* **11,** 2 (1975).
9   H. Eklund, B. Nordstrom, E. Zeppezauer, G. Söderlund, I. Ohlsson, T. Boiwe, B.-O. Söderberg, O. Tapia, C.-I. Brändén, and Å. Åkeson, *J. molec. Biol.* **102,** 27 (1976).
10 K. Dalziel, *J. biol. Chem.* **238,** 2850 (1963).
11 J. D. Shore, H. Gutfreund, R. L. Brooks, D. Santiago, and P. Santiago, *Biochemistry* **13,** 4185 (1974).

12 H. Dutler and C.-I. Brändén (to be published).

13 M. F. Dunn, J. F. Biellman, and G. Bruylant, *Biochemistry* **14,** 3176 (1975).

14 J. P. Klinman, *Biochemistry* **15,** 2018 (1976).

15 R. T. Dworschack and B. V. Plapp, *Biochemistry* (to be published).

16 H. Theorell and B. Chance, *Acta. chem. Scand.* **5,** 1127 (1951).

17 C. C. Wratten and W. W. Cleland, *Biochemistry* **2,** 935 (1963); **4,** 2442 (1965).

18 J. D. Shore, H. Gutfreund, and D. Yates, *J. biol. Chem.* **250,** 5276 (1975).

19 S. A. Bernhard, M. F. Dunn, P. L. Luisi, and P. Schack, *Biochemistry* **9,** 185 (1970).

20 J. P. Klinman, *J. biol. Chem.* **247,** 7977 (1972).

21 J. D. Shore and H. Gutfreund, *Biochemistry* **9,** 4655 (1970).

22 R. L. Brooks and J. D. Shore, *Biochemistry* **10,** 3855 (1971).

23 J. W. Jacobs, J. T. McFarland, I. Wainer, D. Jeanmaier, C. Ham, K. Hamm, M. Wnuk, and M. Lam, *Biochemistry* **13,** 60 (1974).

24 L. F. Blackwell and M. J. Hardmann, *Eur. J. Biochem.* **55,** 611 (1975).

25 D. L. Sloan, J. M. Young, and A. S. Mildvan, *Biochemistry* **14,** 1998 (1975).

26 C.-I. Brändén in *Alcohol and aldehyde metabolizing systems,* Vol. 2 (ed. R. Thurman). Academic Press (1977).

27 J. P. Klinman, *J. biol. Chem.* **250,** 2569 (1975).

28 P. Woolley, *Nature, Lond.* **258,** 677 (1975).

29 J. T. McFarland and Y. H. Chu, *Biochemistry* **14,** 7140 (1975).

30 P. L. Luisi and E. Bignetti, *J. molec. Biol.* **88,** 653 (1974).

31 M. Hadorn, V. A. John, F. K. Meier, and H. Dutler, *Eur. J. Biochem.* **54,** 65 (1975).

32 J. Everse and N. O. Kaplan, *Adv. Enzymol.* **37,** 61 (1973).

33 C. L. Markert and F. Moller, *Proc. natn. Acad. Sci. U.S.A.* **45,** 753 (1959).

34 I. Fine, N. O. Kaplan, and D. Kuftinec, *Biochemistry* **2,** 116 (1963).

35 O. P. Chilson, L. A. Costello, and N. O. Kaplan, *Biochemistry* **4,** 271 (1965).

36 J. L. White, M. L. Hackert, M. Buehner, M. J. Adams, G. C. Ford, D. L. Lentz Jr., I. E. Smiley, S. J. Steindel, and M. G. Rossmann, *J. molec. Biol.* **102,** 759 (1976).

37 M. J. Adams, M. Buehner, K. Chandrasekhar, G. C. Ford, M. L. Hackert, A. Liljas, M. G. Rossmann, I. E. Smiley, W. S. Allison, J. Everse, N. O. Kaplan, and S. S. Taylor, *Proc. natn. Acad. Sci. U.S.A.* **70,** 1968 (1973).

38 J. Everse, R. E. Barnett, C. J. R. Thorne, and N. O. Kaplan, *Archs biochem. Biophys.* **143,** 444 (1971).

39 C. J. Coulson and B. R. Rabin, *FEBS Letts* **3,** 333 (1969).

40 J. H. Griffin and R. S. Criddle, *Biochemistry* **9,** 1195 (1970).

41 J. J. Holbrook and H. Gutfreund, *FEBS Letts.* **31,** 157 (1973).

42 W. B. Novoa and G. W. Schwert, *J. biol. Chem.* **236,** 2150 (1961).

43 J. R. Whitaker, D. W. Yates, N. G. Bennett, J. J. Holbrook, and H. Gutfreund, *Biochem. J.* **139,** 677 (1974).

44 J. J. Holbrook and V. A. Ingram, *Biochem. J.* **131,** 729 (1973).

45 L. E. Webb, E. Hill, and L. J. Banaszak, *Biochemistry* **12,** 5101 (1973).

46 J. J. Holbrook, A. Lodola, and N. P. Illsley, *Biochem. J.* **139,** 797 (1974).

47 J. J. Holbrook and R. G. Wolfe, *Biochemistry* **11,** 2499 (1972).
48 M. Cassman and R. C. King, *Biochemistry* **11,** 4937 (1972).
49 M. Cassman and D. Vetterlein, *Biochemistry* **13,** 684 (1974).
50 J. I. Harris and M. Waters, *The Enzymes* **13,** 1 (1976).
51 H. L. Segal and P. D. Boyer, *J. biol. Chem.* **204,** 265 (1953).
52 P. J. Harrigan and D. R. Trentham, *Biochem. J.* **143,** 353 (1974).
53 R. G. Duggleby and D. T. Dennis, *J. biol. Chem.* **249,** 167 (1974).
54 I. Krimsky and E. Racker, *Science, N.Y.* **122,** 319 (1955).
55 D. R. Trentham, *Biochem. J.* **122,** 59, 71 (1971).
56 L. D. Byers and D. E. Koshland, *Biochemistry* **14,** 3661 (1975).
57 D. Moras, K. W. Olsen, M. N. Sabesan, M. Buehner, G. C. Ford, and M. G. Rossman, *J. biol. Chem.* **250,** 9137 (1975).
58 G. Biesecker, J. I. Harris, J. C. Thierry, J. E. Walker, and A. J. Wonacott, *Nature, Lond.* **266,** 328 (1977).
59 H. C. Watson, E. Duée, and W. D. Mercer, *Nature New Biology, Lond.* **240,** 130 (1972).
60 M. Buehner, G. C. Ford, D. Moras, K. W. Olsen, and M. G. Rossmann, *J. molec. Biol.* **90,** 25 (1974).
61 A. Conway and D. E. Koshland, Jr., *Biochemistry* **7,** 4011 (1968).
62 B. D. Peczon and H. O. Spivey, *Biochemistry* **11,** 2209 (1972).
63 P. J. Harrigan and D. R. Trentham, *Biochem. J.* **135,** 695 (1973).
64 F. Seydoux, S. A. Bernhard, O. Pfenninger, M. Payne, and O. P. Malhotra, *Biochemistry* **12,** 4290 (1973).
65 J. Schlessinger and A. Levitzki, *J. molec. Biol.* **82,** 547 (1974).
66 A. Levitzki, *J. molec. Biol.* **90,** 451 (1974).
67 F. Seydoux and S. A. Bernhard, *Bioorg. Chem.* **1,** 161 (1974).
68 N. Kelemen, N. Kellershohn, and F. Seydoux, *Eur. J. Biochem.* **57,** 69 (1975).
69 L. S. Gennis, *Proc. natn. Acad. Sci. U.S.A.* **73,** 3928 (1976).
70 J. Bode, M. Blumenstein, and M. A. Raftery, *Biochemistry* **14,** 1146 (1975).
71 M. Dunn, *Biochemistry* **13,** 1146 (1974).
72 G. Shoellman and E. Shaw, *Biochemistry* **2,** 252 (1963).
73 D. M. Blow, J. J. Birktoft, and B. S. Hartley, *Nature, Lond.* **221,** 337 (1970).
74 A. R. Fersht and J. Sperling, *J. molec. Biol.* **74,** 137 (1973).
75 M. W. Hunkapiller, S. H. Smallcombe, D. R. Whitaker, and J. H. Richards, *Biochemistry* **12,** 4732 (1973).
76 R. E. Koeppe II and R. M. Stroud, *Biochemistry* **15,** 3450 (1976).
77 A. R. Fersht and M. Renard, *Biochemistry* **13,** 1416 (1974).
78 G. E. Hein and C. Niemann, *J. Am. chem. Soc.* **84,** 4495 (1962).
79 G. Petsko, personal communication.
80 R. Henderson, *J. molec. Biol.* **54,** 341 (1970).
81 A. R. Fersht, D. M. Blow, and J. Fastrez, *Biochemistry* **12,** 2035 (1973).
82 S. T. Freer, J. Kraut, J. D. Robertus, H. T. Wright, and Ng. H. Xuong, *Biochemistry* **9,** 1997 (1970).
83 H. T. Wright, *J. molec. Biol.* **79,** 1, 13 (1973).
84 J. J. Birktoft, J. Kraut, and S. T. Freer, *Biochemistry* **15,** 4481 (1976).

85   G. Robillard and R. G. Shulman, *J. molec. Biol.* **86,** 519 (1974).
86   A. R. Fersht, *FEBS Letts* **29,** 283 (1973).
87   A. Gertler, K. A. Walsh, and H. Neurath, *Biochemistry* **13,** 1302 (1974).
88   A. R. Fersht, *J. molec. Biol.* **64,** 497 (1972).
89   J. Drenth, J. N. Jansonius, R. Koekoek, and B. G. Wolthers, *The Enzymes* **3,** 485 (1971).
90   A. N. Glazer and E. L. Smith, *The Enzymes* **3,** 501 (1971).
91   J. Drenth, J. N. Jansonius, R. Koekoek, and B. G. Wolthers, *Adv. Prot. Chem.* **25,** 79 (1971).
92   G. Lowe, *Tetrahedron* **32,** 291 (1976).
93   T.-Y. Liu and S. D. Elliott, *The Enzymes* **3,** 609 (1971).
94   W. M. Mitchell and W. F. Harrington, *The Enzymes* **3,** 699 (1971).
95   R. E. J. Mitchell, I. M. Chaiken, and E. L. Smith, *J. biol. Chem.* **245,** 3485 (1970).
96   A. Berger and I. Schechter, *Phil. Trans. R. Soc.* **B257,** 249 (1970).
97   M. R. Alecio, M. L. Dann, and G. Lowe, *Biochem. J.* **141,** 495 (1974).
98   A. Stockell and E. L. Smith, *J. biol. Chem.* **227,** 1 (1957).
99   G. Lowe and A. Williams, *Biochem. J.* **96,** 189, 199 (1965).
100 P. M. Hinkle and J. F. Kirsch, *Biochemistry* **10,** 2717 (1971).
101 L. J. Brubacher and M. L. Bender, *J. Am. chem. Soc.* **88,** 5871 (1966).
102 J. Drenth, J. N. Jansonius, and B. G. Wolthers, *J. molec. Biol.* **24,** 449 (1967).
103 J. Drenth, J. N. Jansonius, R. Koekoek, H. M. Swen, and B. G. Wolthers, *Nature, Lond.* **218,** 929 (1968).
104 J. Drenth, K. H. Kalk, and H. M. Swen, *Biochemistry* **15,** 3731 (1976).
105 G. Lowe and Y. Yuthavong, *Biochem. J.* **124,** 107 (1971).
106 A. R. Fersht, *J. Am. chem. Soc.* **93,** 3504 (1971).
107 L. Polgar, *FEBS Letts* **47,** 15 (1974).
108 G. Lowe and Y. Yuthavong, *Biochem. J.* **124,** 117 (1971).
109 M. H. O'Leary, M. Urberg, and A. P. Young, *Biochemistry* **13,** 2077 (1974).
110 J. A. Hartsuck and W. N. Lipscomb, *The Enzymes* **3,** 1 (1971).
111 F. A. Quiocho and W. N. Lipscomb, *Adv. Prot. Chem.* **25,** 1 (1971).
112 W. N. Lipscomb, *Tetrahedron* **30,** 1725 (1974).
113 R. A. Bradshaw, L. H. Ericsson, K. A. Walsh, and H. Neurath, *Proc. natn. Acad. Sci. U.S.A.* **63,** 1389 (1969).
114 M. F. Schmid and J. R. Herriott, *J. molec. Biol.* **103,** 175 (1976).
115 G. N. Reeke, J. A. Hartsuck, M. L. Ludwig, F. A. Quiocho, T. A. Steitz, and W. N. Lipscomb, *Proc. natn. Acad. Sci. U.S.A.* **58,** 2220 (1967).
116 W. N. Lipscomb, *Proc. natn. Acad. Sci. U.S.A.* **70,** 3797 (1973).
117 J. T. Johansen and B. L. Vallee, *Biochemistry* **14,** 649 (1975).
118 B. L. Vallee, J. F. Riordan, and J. E. Coleman, *Proc. natn. Acad. Sci. U.S.A.* **49,** 109 (1963).
119 D. S. Auld and B. L. Vallee, *Biochemistry* **10,** 2892 (1971).
120 J. W. Bunting and S. S.-T. Chu, *Biochemistry* **15,** 3237 (1976).
121 J. F. Riordan, M. Sokolovsky, and B. L. Vallee, *Biochemistry* **6,** 358, 3609 (1967).
122 J. E. Coleman and B. L. Vallee, *J. biol. Chem.* **236,** 2244 (1961).

123   M. W. Makinen, K. Yamamura, and E. T. Kaiser, *Proc. natn. Acad. Sci. U.S.A.* **73,** 3882 (1976).

124   R. E. Breslow and D. Wernick, *J. Am. chem. Soc.* **98,** 259 (1976).

125   J. H. Freisheim, K. A. Walsh, and H. Neurath, *Biochemistry* **6,** 3010, 3020 (1967).

126   J. R. Uren and H. Neurath, *Biochemistry* **13,** 3512 (1974).

127   T. J. Bazzone and B. L. Vallee, *Biochemistry* **15,** 818 (1976).

128   J. S. Fruton, *The Enzymes* **3,** 119 (1971).

129   G. E. Clement, *Prog. Bioorg. Chem.* **2,** 177 (1973).

130   J. S. Fruton, *Adv. Enzymol.* **44,** 1 (1976).

131   J. Sodek and T. Hofmann, *Can. J. Biochem.* **48,** 1014 (1970).

132   V. B. Pedersen and B. Foltmann, *FEBS Letts.* **35,** 250 (1973).

133   N. S. Andreeva, V. V. Borisov, V. R. Melik-Adamyan, V. S. Raiz, L. N. Trofimova, and N. E. Shutskever, *Mol. Biol.* **5,** 908 (1971).

134   E. Subramanian, I. D. A. Swan, and D. R. Davies, *Biochem. biophys. Res. Commun.* **68,** 875 (1976).

135   N. S. Andreeva, A. A. Fedorov, A. A. Gushchina, N. E. Shutskever, R. R. Riskulov, and T. V. Volnova, *Dokl. Akad. Nauk. S.S.S.R.* **228,** 480 (1976).

136   I. N. Hsu, L. T. Delbaere, M. N. G. James, and T. Hofmann, *Nature, Lond.* **266,** 140 (1977).

137   P. Sepulveda, J. Marciniszyn Jr., D. Liu, and J. Tang, *J. biol. Chem.* **250,** 5082 (1975).

138   L. Moravek and V. Kostka, *FEBS Letts* **43,** 207 (1974).

139   G. E. Clement, J. Rooney, D. Zakheim, and J. Eastman, *J. Am. chem. Soc.* **92,** 186 (1970).

140   P. S. Sampath-Kumar and J. S. Fruton, *Proc. natn. Acad. Sci. U.S.A.* **71,** 1070 (1974).

141   A. A. Zinchenko, L. D. Rumsh, and V. K. Antonov, *Bioorg. Chem. (USSR)* **2,** 803 (1976).

142   J. L. Denburg, R. Nelson, and M. S. Silver, *J. Am. chem. Soc.* **90,** 479 (1968).

143   A. J. Cornish-Bowden and J. R. Knowles, *Biochem. J.* **113,** 353 (1969).

144   G. R. Delpierre and J. S. Fruton, *Proc. natn. Acad. Sci. U.S.A.* **54,** 1161 (1965); **56,** 1817 (1966).

145   R. L. Lundblad and W. H. Stein, *J. biol. Chem.* **244,** 154 (1969).

146   R. S. Bayliss, J. R. Knowles, and G. B. Wybrandt, *Biochem. J.* **113,** 377 (1969).

147   J. A. Hartsuck and J. Tang, *J. biol. Chem.* **247,** 2575 (1972).

148   H. Neuman, Y. Levin, A. Berger, and E. Katchalski, *Biochem. J.* **73,** 33 (1959).

149   V. K. Antonov, L. D. Rumsh, and A. G. Tikhodeeva, *FEBS Letts* **46,** 29 (1974).

150   A. G. Tikhodeeva, L. D. Rumsh, and V. K. Antonov, *Bioorg. Chem. (USSR)* **1,** 993 (1975).

151   N. Sharon, V. Grisaro, and H. Neumann, *Archs biochem. Biophys.* **97,** 219 (1962).

152   M. Takahashi and T. Hofmann, *Biochem. J.* **127,** 35P (1972); **147,** 549 (1975).

153 M. Takahashi, T. T. Wang, and T. Hofmann, *Biochem. biophys. Res. Commun.* **57,** 39 (1974). T. T. Wang and T. Hofmann, *Biochem. J.* **153,** 691 (1976).

154 A. K. Newmark and J. R. Knowles, *J. Am. chem. Soc.* **97,** 3557 (1975).

155 V. K. Antonov, *3rd All-Union symposium on structure and function of active centres of enzymes,* Pushchino, 1976.

156 M. S. Silver, M. Stoddard, and M. H. Kelleher, *J. Am. chem. Soc.* **98,** 6684 (1976).

157 M. Bustin, M. C. Lin, W. H. Stein, and S. Moore, *J. biol. Chem.* **245,** 846 (1970).

158 J. Tang, *Biochem. biophys. Res. Commun.* **41,** 697 (1970).

159 M. Bustin and A. Conway-Jacobs, *J. biol. Chem.* **246,** 615 (1971).

160 J. Al-Janabi, J. A. Hartsuck, and J. Tang, *J. biol. Chem.* **247,** 4628 (1972).

161 P. McPhie, *J. biol. Chem.* **247,** 4277 (1972).

162 C. G. Sanny, J. A. Hartsuck, and J. Tang, *J. biol. Chem.* **250,** 2635 (1975).

163 C. W. Dykes and J. Kay, *Biochem. J.* **153,** 141 (1976).

164 F. M. Richards and H. W. Wyckoff, *The Enzymes* **4,** 647 (1971).

165 F. M. Richards and H. W. Wyckoff, *Atlas of molecular structures in biology.* Clarendon Press, Oxford (1973).

166 C. H. W. Hirs, S. Moore, and W. H. Stein, *J. biol. Chem.* **235,** 633 (1960).

167 D. G. Smyth, W. H. Stein, and S. Moore, *J. biol. Chem.* **238,** 227 (1963).

168 G. Kartha, J. Bello, and D. Harker, *Nature, Lond.* **213,** 862 (1967).

169 H. W. Wyckoff, D. Tsernoglou, A. W. Hanson, J. R. Knox, B. Lee, and F. M. Richards, *J. biol. Chem.* **245,** 305 (1970).

170 D. Findlay, D. G. Herries, A. P. Mathias, B. R. Rabin, and C. A. Ross, *Nature, Lond.* **190,** 781 (1961).

171 F. M. Richards, H. W. Wyckoff, W. D. Carlson, N. M. Allewell, B. Lee, and Y. Mitsui, *Cold Spring Harb. symp. Quant. Biol.* **36,** 35 (1971).

172 B. Walter and F. Wold, *Biochemistry* **15,** 304 (1976).

173 J. L. Markley, *Biochemistry* **14,** 3546 (1975).

174 F. H. Westheimer, *Acc. Chem. Res.* **1,** 70 (1968).

175 S. J. Benkovic and K. J. Schray, *The Enzymes* **8,** 201 (1973).

176 D. A. Usher, D. I. Richardson, and F. Eckstein, *Nature, Lond.* **228,** 663 (1970).

177 D. A. Usher, E. S. Erenrich, and F. Eckstein, *Proc. natn. Acad. Sci. U.S.A.* **69,** 115 (1972).

178 F. Eckstein, W. Saenger, and D. Suck, *Biochem. biophys. Res. Commun.* **46,** 964 (1972).

179 F. A. Cotton and E. E. Hazen, Jr. *The Enzymes* **4,** 153 (1971).

180 C. B. Anfinsen, P. Cuatrecasas, and H. Taniuchi, *The Enzymes* **4,** 177 (1971).

181 A. Arnone, C. J. Bier, F. A. Cotton, V. W. Day, E. E. Hazen, Jr., D. C. Richardson, J. S. Richardson, and A. Yonath, *J. biol. Chem.* **246,** 2302 (1971).

182 F. A. Cotton, C. J. Bier, V. W. Day, E. E. Hazen, Jr., and S. W. Larsen, *Cold Spring Harb. symp. Quant. Biol.* **36,** 243 (1971).

183 A. S. Mildvan, *The Enzymes* **2,** 446 (1970).

184 A. S. Mildvan, *Ann. Rev. Biochem.* **43,** 357 (1974).

185   B. M. Dunn, C. Di Bello, and C. B. Anfinsen, *J. biol. Chem.* **248,** 4769 (1973).

186   P. Cuatrecasas, S. Fuchs, and C. B. Anfinsen, *J. biol. Chem.* **242,** 3063 (1967).

187   T. Imoto, L. N. Johnson, A. C. T. North, D. C. Phillips, and J. A. Rupley, *The Enzymes* **7,** 665 (1972).

188   B. Dunn and T. C. Bruice, *Adv. Enzymol.* **37,** 1 (1973).

189   D. M. Chipman and N. Sharon, *Science, N. Y.* **165,** 454 (1969).

190   C. C. F. Blake, L. N. Johnson, G. A. Mair, A. C. T. North, D. C. Phillips, and V. R. Sarma, *Proc. R. Soc.* **B167,** 378 (1967).

191   C. A. Vernon, *Proc. R. Soc.* **B167,** 389 (1967).

192   L. O. Ford, L. N. Johnson, P. A. Machin, D. C. Phillips, and R. Tjian, *J. molec. Biol.* **88,** 349 (1974).

193   J. A. Rupley and V. Gates, *Proc. natn. Acad. Sci. U.S.A.* **57,** 496 (1967).

194   M. A. Raftery and T. Rand-Meir, *Biochemistry* **7,** 3281 (1968).

195   U. Zehavi, J. J. Pollock, V. I. Teichberg, and N. Sharon, *Nature, Lond.* **219,** 1152 (1968).

196   F. W. Dahlquist, T. Rand-Meir, and M. A. Raftery, *Proc. natn. Acad. Sci. U.S.A.* **61,** 119 (1968).

197   L. E. H. Smith, L. H. Mohr, and M. A. Raftery, *J. Am. chem. Soc.* **95,** 7497 (1973).

198   J. J. Pollock, D. M. Chipman, and N. Sharon, *Archs biochem. Biophys.* **120,** 235 (1967).

199   S. M. Parsons and M. A. Raftery, *Biochemistry* **11,** 1623 (1972).

200   J. A. Rupley, V. Gates, and R. Bilbrey, *J. Am. chem. Soc.* **90,** 5633 (1968).

201   A. Warshel and M. Levitt, *J. molec. Biol.* **103,**˙227 (1976).

202   Y. Eshdat, A. Dunn, and N. Sharon, *Proc. natn. Acad. U.S.A.* **71,** 1658 (1974).

203   I. I. Secemski and G. E. Lienhard, *J. biol. Chem.* **249,** 2932 (1974).

204   E. Holler, J. A. Rupley, and G. P. Hess, *Biochemistry* **14,** 1088, 2377 (1975).

205   M. F. Levitt in *Peptides, polypeptides, and proteins* (eds. E. R. Blout, F. A. Bovey, M. Goodman, and N. Lotan). John Wiley & Sons, p. 99 (1974).

206   M. Schindler and N. Sharon, *J. biol. Chem.* **251,** 4330 (1976).

207   E. Holler, J. A. Rupley, and G. P. Hess, *Biochem. biophys. Res. Commun.* **37,** 423 (1969).

208   J. H. Baldo, S. E. Halford, S. L. Patt, and B. D. Sykes, *Biochemistry* **14,** 1893 (1975).

209   S. Lindskog, L. E. Henderson, K. K. Kannan, A. Liljas, P. O. Nyman, and B. Strandberg, *The Enzymes* **5,** 587 (1971).

210   J. E. Coleman, *Progress Bioorg. Chem.* **1,** 159 (1971).

211   A. Liljas, K. K. Kannan, P.-C. Bergsten, I. Waara, K. Fridborg, B. Strandberg, U. Carlbom, L. Järup, S. Lövgren, and M. Petef, *Nature New Biology, Lond.* **235,** 131 (1972).

212   K. K. Kannan, B. Notstrand, K. Fridborg, S. Lövgren, A. Ohlsson, and M. Petef, *Proc. natn. Acad. Sci. U.S.A.* **72,** 51 (1975).

213   B. Andersson, P. O. Nyman, and L. Strid, *Biochem. biophys. Res. Commun.* **48,** 670 (1972).

214 L. E. Henderson, D. Henriksson, and P. O. Nyman, *J. biol. Chem.* **251,** 5457 (1976).

215 H. Steiner, B. H. Jonsson, and S. Lindskog, *Eur. J. Biochem.* **59,** 253 (1975).

216 S. Lindskog and J. E. Coleman, *Proc. natn. Acad. Sci. U.S.A.* **70,** 2505 (1973).

217 S. H. Koenig and R. D. Brown III, *Proc. natn. Acad. Sci. U.S.A.* **69,** 2422 (1972).

218 D. W. Appleton and B. Sarkar, *Proc. natn. Acad. Sci. U.S.A.* **71,** 1686 (1974).

219 J. M. Pesando, *Biochemistry* **14,** 675, 681 (1975).

220 I. D. Campbell, S. Lindskog, and A. I. White, *J. molec. Biol.* **90,** 469 (1974); **98,** 597 (1975).

221 R. Bauer, P. Limkilde, and J. T. Johansen, *Biochemistry* **15,** 334 (1976).

222 R. H. Prince and P. Woolley, *Bioorganic Chemistry* **2,** 337 (1973).

223 R. G. Khalifah, *Proc. natn. Acad. Sci. U.S.A.* **70,** 1986 (1973).

224 D. N. Silvermann and C. K. Tu, *J. Am. chem. Soc.* **97,** 2263 (1975).

225 C. K. Tu and D. N. Silverman, *J. Am. chem. Soc.* **97,** 5935 (1975).

226 C. C. F. Blake, in *Essays in Biochemistry* (eds. P. N. Campbell and W. N. Aldridge). Academic Press, Vol. 11, p. 37 (1975).

227 I. A. Rose, *The Enzymes* **2,** 281 (1970).

228 I. A. Rose, *Adv. Enzymol.* **43,** 491 (1975).

229 E. A. Noltmann, *The Enzymes* **6,** 271 (1972).

230 D. W. Banner, A. C. Bloomer, G. A. Petsko, D. C. Phillips, and C. I. Pogson, *Cold Spring Harb. symp. Quant. Biol.* **36,** 151 (1971).

231 D. W. Banner, A. C. Bloomer, G. A. Petsko, D. C. Phillips, C. I. Pogson, I. A. Wilson, P. H. Corran, A. J. Furth, J. D. Milman, R. E. Offord, J. D. Priddle, and S. G. Waley, *Nature, Lond.* **255,** 609 (1975).

232 D. C. Phillips (personal communication).

233 S. V. Rieder and I. A. Rose, *J. biol. Chem.* **234,** 1007 (1958).

234 I. A. Rose and E. L. O'Connell, *J. biol. Chem.* **236,** 3086 (1961).

235 F. C. Hartmann, *Biochem. biophys. Res. Commun.* **33,** 888 (1968); **39,** 384 (1970).

236 S. G. Waley, J. C. Miller, I. A. Rose, and E. L. O'Connell, *Nature, Lond.* **227,** 181 (1970).

237 S. De La Mare, A. F. W. Coulson, J. R. Knowles, J. D. Priddle, and R. E. Offord, *Biochem. J.* **129,** 321 (1972).

238 Y. J. Topper, *J. biol. Chem.* **225,** 419 (1957).

239 B. Bloom and Y. J. Topper, *Nature, Lond.* **181,** 1128 (1958).

240 I. A. Rose and E. L. O'Connell, *Biochim. biophys. Acta* **42,** 159 (1960).

241 B. Plaut and J. R. Knowles, *Biochem. J.* **129,** 311 (1972).

242 F. C. Hartman, G. M. LaMuraglia, Y. Tomozawa, and R. Wolfenden, *Biochemistry* **14,** 5274 (1975).

243 K. J. Schray, E. L. O'Connell, and I. A. Rose, *J. biol. Chem.* **248,** 2214 (1973).

244 W. J. Albery and J. R. Knowles, *Biochemistry* **15,** 5588, 5627 (1976).

245 H. Muirhead and P. J. Shaw, *J. molec. Biol.* **89,** 195 (1974).

246 M. Salas, E. Vinuela, and A. Sols, *J. biol. Chem.* **240,** 561 (1965).

247 B. Wurster and B. Hess, *Hoppe-Seyler's Z. physiol. Chem.* **351,** 1537 (1970).

248 K. J. Schray, S. J. Benkovic, P. A. Benkovic, and I. A. Rose, *J. biol. Chem.* **248,** 2219 (1973).

249 I. A. Rose, E. L. O'Connell, and K. J. Shray, *J. biol. Chem.* **248,** 2232 (1973).

250 J. E. D. Dyson and E. A. Noltmann, *J. biol. Chem.* **243,** 1401 (1968).

251 E. L. O'Connell and I. A. Rose, *J. biol. Chem.* **248,** 2225 (1973).

252 P. J. Shaw and H. Muirhead, *FEBS Letts* **65,** 50 (1976).

253 J. M. Chirgwin and E. A. Noltmann, *J. biol. Chem.* **250,** 7272 (1975).

254 W. J. Ray, Jr. and E. J. Peck, Jr. *The Enzymes* **6,** 4 (1972).

255 J. W. Campbell, H. C. Watson, and G. I. Hodgson, *Nature, Lond.* **250,** 301 (1974).

256 Z. B. Rose, *Archs biochem. Biophys.* **140,** 508 (1970); **146,** 359 (1971).

257 S. Grisolia and W. W. Cleland, *Biochemistry* **7,** 1115 (1968).

258 H. C. Watson, L. A. Fothergill, and S. I. Winn (personal communication).

Author Index

Subject Index